Distributed Power Amplifiers
for RF and Microwave Communications

For a listing of recent titles in the
Artech House Microwave Library,
turn to the back of this book.

Distributed Power Amplifiers
for RF and Microwave Communications

Narendra Kumar
Andrei Grebennikov

ARTECH
HOUSE

BOSTON | LONDON
artechhouse.com

Library of Congress Cataloging-in-Publication Data
A catalog record for this book is available from the U.S. Library of Congress

British Library Cataloguing in Publication Data
A catalog record for this book is available from the British Library.

ISBN-13: 978-1-60807-831-8

Cover design by John Gomes

© 2015 Artech House
685 Canton St.
Norwood, MA 02062

10 9 8 7 6 5 4 3 2 1

Contents

Preface

The main objective of this book is to present all relevant information and comprehensive design methodologies for wideband amplifiers—specifically, distributed amplifiers in general and their main components in particular—in different RF and microwave applications including well-known historical and recent architectures, theoretical approaches, circuit simulation, and practical implementation techniques. This comprehensive book will be useful for lecturing to promote a systematic way of thinking with analytical calculations and practical verification of wideband amplifiers providing the link between theory and practice of RF and microwave distributed amplifiers. Therefore, this book is recommended to academicians, researchers, and professors, and provides good coverage to practicing designers and engineers because it contains numerous well-known and novel practical circuits, architectures, and theoretical approaches with a detailed description of their operational principles.

Chapter 1 introduces basic two-port networks by describing the behavior of two-port parameters including impedance, admittance and ABD matrices, scattering parameters, conversion between two-port networks, and practical two-port circuits. Lumped elements, particularly inductors and capacitors, are also discussed. Monolithic implementation of lumped inductors and capacitors is usually required at microwave frequencies and for portable devices. Transmission-line theory is introduced and followed by design formulas; curves are given for several types of transmission lines including striplines, microstrip lines, slotlines, and coplanar waveguides. Noise phenomena such as noise figure, additive white noise, low-frequency fluctuations, and flicker noise are discussed at the end of the chapter.

In Chapter 2, the design fundamentals of power amplifiers are presented. Design is generally a complicated procedure where it is necessary to provide simultaneously accurate active device modeling, effective impedance matching depending on the technical requirements and operating conditions, stability during operation, and simplicity in practical implementation. The main characteristics, principles, and impedance matching techniques are described. The quality of power amplifier designs is evaluated by determining the realized maximum power gain under stable operating conditions with minimum amplifier stages, and the requirement of linearity or high efficiency can be considered where it is needed. For stable operation, it is necessary to evaluate the operating frequency domains where the active device may be potentially unstable. To avoid parasitic oscillations, the stabilization circuit technique for different frequency domains (from low frequencies up to high frequencies close to the device transition frequencies) is discussed. The device bias conditions, which are generally different for linearity or efficiency improvement, depend on the power amplifier operating class and the type of active device. The basic classes,

Classes A, AB, B, and C, of power amplifier operations are introduced, analyzed, and illustrated. All necessary steps to provide an accurate device modeling procedure, starting with the determination of a device's small-signal equivalent circuit parameters, are described. A variety of nonlinear models for MOSFET, MESFET, HEMT, and BJT devices including HBTs, which are very attractive for modern monolithic microwave integrated circuits, are described. The procedure for designing for dc biasing is discussed and, finally, an overview of impedance transformers and power combiners and directional couplers is given.

An overview of broadband power amplifiers is given in Chapter 3. The chapter begins with the Bode-Fano criterion, explaining about the bandwidth analysis of a broadband power amplifier. A matching circuit is crucial to providing maximum power transfer from one point to another point, in which transformation using lumped elements, mixed-lumped, and distributed elements will be discussed to give choices to designers trying to meet technical requirements. In addition, transformation with transmission lines and power amplifiers with lossy compensation networks are discussed in this chapter. Push-pull and balanced power amplifier topologies are discussed to give us an understanding of circuit principles and design implementation. At the end of the chapter, several practical broadband RF and microwave power amplifier topologies are introduced.

Chapter 4 introduces the concept of distributed amplification by means of gain and the bandwidth product of an amplifier stage. The concept explains how the gain stages are connected such that their capacitances are isolated, yet the output currents still combine in an additive fashion. The resulting topology forms an artificial transmission line and is extended to an image impedance method. A theoretical analysis of the distributed amplifier is presented with several approaches (i.e., two-port, admittance, and wave theories). The approach via two-port theory considers only a unilateral small-signal transistor model. The admittance method is more general because there is no simplifying assumption regarding the transistor model. Finally, the wave theory method, which uses the normalized transmission matrix approach, has the advantage of displaying the traveling wave nature of a distributed amplifier. The gain/power-bandwidth trade-off is discussed to give an overview of the influence a simple bandpass amplifier circuit has over the bandwidth response. The design methodology for a practical distributed amplifier is presented, which provides guidelines for designers who desire to realize a distributed amplifier in a timely manner, without any tedious optimization at the board level. Layout design guidelines (PCB selection, full-wave simulation, layout optimization, via-hole simulation, and so forth) should be taken into consideration during the design stage.

Chapter 5 introduces the limitations of the conventional distributed amplifier and an analytical approach for achieving high-efficiency performance in distributed amplifiers, where the multicurrent sources must be combined to a single load by presenting optimum virtual impedance to each current source. The systematic generalized design equations are given and a summary of the equations are presented in table form. Obviously, to keep the output impedance of the distributed amplifier closer to 50Ω (and to avoid additional impedance transformation), the magnitude and phase properties of the current source (or transistor) must be adjusted. The adjustment can be made according to the designer's need, the complexity of the design circuit, and so forth. A few design examples of high-efficiency distributed amplifiers are discussed. A parallel coupled-line approach is adopted as a test vehicle for broadband impedance transformation purposes. As an important point, note that the design concept presented provides appropriate guidelines for maximizing the efficiency of a distributed amplifier.

The basic principle and motivation for using stability analyses (with K-factor, feedback and NDF factor, and pole-zero identification methods) with the intent of understanding the strategies are discussed in Chapter 6. The pole-zero identification technique is applied to a distributed amplifier to provide an understanding of the origin of oscillation due to the multiple-loop nature. The analysis considers the distributed amplifier to be a basic feedback oscillator circuit; that is, a Hartley oscillator using a simplified transistor model. The origin of the distributed amplifier oscillation can be traced to the transconductance nature or multiple-loop nature of the feedback network. An explanation of odd-mode oscillation in a distributed amplifier topology is discussed, which is useful for practical applications. Large-signal stability analysis based on pole-zero identification is applied to analyze the parametric oscillations in high-efficiency distributed amplifiers. The parametric oscillation is correlated to a gain expansion phenomenon that directly affects the critical poles of the circuit. Large-signal stability analysis is then used to stabilize a high-efficiency distributed amplifier with a minimum impact on circuit performance.

Chapter 7 introduces the design implementation of a distributed amplifier. A distributed amplifier overcomes the difficulty of conventional amplifiers by offering a broadband frequency response. The concept of a vacuum-tube distributed amplifier based on combining interelectrode capacitances with series wire inductors is discussed using an analytical approach. A distributed amplifier with a microwave GaAs FET, including configuration with microstrip lines, lumped elements, and capacitive coupling, is discussed. A tapered distributed amplifier offers high-efficiency performance and eliminates dummy termination. Other implementations of distributed amplifiers such as power combining, bandpass, and parallel and series feedback configurations are discussed. A cascade distributed amplifier topology minimizes the degeneration at higher frequencies, improves the isolation between inputs and outputs, and reduces the drain-line loading effect. Also, an extended-resonance power-combining technique can be used to form a resonant power combining/dividing structure, which can benefit wideband performance. High gain over a wide frequency range can be achieved by cascading several stages of a single-stage amplifier, a so-called *cascaded distributed amplifier*. Lastly, the operating principles for matrix and CMOS distributed amplifiers are described.

Chapter 8 discusses high-power applications of distributed amplifiers for which power devices are used instead of the small-signal devices, so-called *distributed power amplifiers*. The need to achieve high output power over a wide frequency range can be addressed in both the device technology and circuit design areas. Realization of distributed power amplifiers has posed a challenge due to electrical and thermal limitations, typically with a GaAs or HBT transistor, and AlGaN/GaN technology has established itself as a strong contender for such applications. A dual-feed distributed power amplifier incorporates a termination adjustment approach that allows for efficient power combining at the load termination. The approach is cost effective and results in efficiency-bandwidth improvement for high-power applications. Another suitable topology for distributed power amplifiers is introduced that employs tapering the interstage termination with a cascaded connection. This topology permits various device technologies to be cascaded for the low-power stage to the high-power stage to offer high-gain and high-output power performance simultaneously. Another distributed power amplifier that features vectorially combined current with a load pull impedance determination technique is a feasible solution for high output power applications. A design example for a broadband matching network with minimum reasonable loss and a low-cost implementation over the entire bandwidth response via a real-frequency technique is explained. Lastly, loading compensation with a cascade distributed power amplifier for GaN HEMT is discussed.

CHAPTER 1

Two-Port Network Parameters

Two-port equivalent circuits are widely used in radio-frequency (RF) and microwave circuit design to describe the electrical behavior of both active devices and passive networks [1, 2]. The two-port network impedance Z-parameters and admittance Y-parameters are very important in the characterization of the nonlinear properties of any type of the bipolar or field-effect transistors used as an active device of the power amplifier. The transmission $ABCD$-parameters of a two-port network are very convenient for designing the distributed circuits of, for example, transmission lines or cascaded active or passive elements. The scattering S-parameters are useful for characterizing linear circuits and required to simplify the measurement procedure. Transmission lines are widely used in the matching networks of high-power or low-noise amplifiers, directional couplers, power combiners, and power dividers. Monolithic implementation of lumped inductors and capacitors is usually required at microwave frequencies and for portable devices.

1.1 Impedance, Admittance, and ABCD Matrices

The basic diagram of a two-port transmission system can be represented by the equivalent circuit shown in Figure 1.1, where V_S is the voltage source, Z_S is the source impedance, Z_L is the load impedance, and the linear network is a time-invariant two-port network without independent source. The two independent phasor currents I_1 and I_2 (flowing across input and output terminals) and phasor voltages V_1 and V_2 characterize such a two-port network. For autonomous oscillator systems, in order to provide an appropriate analysis in the frequency domain of the two-port network in the negative one-port representation, it is sufficient to set the source impedance to infinity. For a power amplifier design, the elements of the matching circuits, which are assumed to be linear or appropriately linearized, can be found among the linear network elements, or additional two-port linear networks can be used to describe their frequency domain behavior.

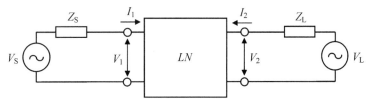

Figure 1.1 Basic diagram of two-port nonautonomous transmission system.

For a two-port network, the following equations can be considered as boundary conditions:

$$V_1 + Z_S I_1 = V_S \tag{1.1}$$

$$V_2 + Z_L I_2 = V_L \tag{1.2}$$

Suppose that it is possible to obtain a unique solution for the linear time-invariant circuit shown in Figure 1.1. Then, the two linearly independent equations, which describe the general two-port network in terms of circuit variables V_1, V_2, I_1, and I_2, can be expressed in matrix form as

$$[M][V] + [N][I] = 0 \tag{1.3}$$

or

$$\begin{aligned} m_{11}V_1 + m_{12}V_2 + n_{11}I_1 + n_{12}I_2 &= 0 \\ m_{21}V_1 + m_{22}V_2 + n_{21}I_1 + n_{22}I_2 &= 0 \end{aligned} \tag{1.4}$$

The complex 2×2 matrices $[M]$ and $[N]$ in (1.3) are independent of the source and load impedances Z_S and Z_L, respectively, and voltages V_S and V_L, respectively. They depend only on the circuit elements inside the linear network.

If matrix $[M]$ in (1.3) is nonsingular with $|M| \neq 0$, this matrix equation can be rewritten in terms of $[I]$ as

$$[V] = -[M]^{-1}[N][I] = [Z][I] \tag{1.5}$$

where $[Z]$ is the open-circuit impedance two-port network matrix. In a scalar form, the matrix of (1.5) is given by

$$V_1 = Z_{11} I_1 + Z_{12} I_2 \tag{1.6}$$

$$V_2 = Z_{21} I_1 + Z_{22} I_2 \tag{1.7}$$

where Z_{11} and Z_{22} are the open-circuit driving-point impedances, and Z_{12} and Z_{21} are the open-circuit transfer impedances of the two-port network. The voltage components V_1 and V_2 due to the input current I_1 can be found by setting $I_2 = 0$ in (1.6) and (1.7), resulting in an open-circuited output terminal. Similarly, the same voltage components V_1 and V_2 are determined by setting $I_1 = 0$ when the input terminal becomes open circuited. The resulting driving-point impedances can be written as

$$Z_{11} = \left. \frac{V_1}{I_1} \right|_{I_2=0} \qquad Z_{22} = \left. \frac{V_2}{I_2} \right|_{I_1=0} \tag{1.8}$$

whereas the two transfer impedances are

$$Z_{12} = \left. \frac{V_1}{I_2} \right|_{I_1=0} \qquad Z_{21} = \left. \frac{V_2}{I_1} \right|_{I_2=0} \tag{1.9}$$

A dual analysis can be used to derive the short-circuit admittance matrix when the current components I_1 and I_2 are considered as outputs caused by V_1 and V_2. If matrix $[N]$ in (1.3) is nonsingular with $|N| \neq 0$, this matrix equation can be rewritten in terms of $[V]$ as

$$[I] = -[N]^{-1}[M][V] = [Y][V] \tag{1.10}$$

where $[Y]$ is the short-circuit admittance two-port network matrix. In a scalar form, the matrix of (1.10) is written as

$$I_1 = Y_{11}V_1 + Y_{12}V_2 \tag{1.11}$$

$$I_2 = Y_{21}V_1 + Y_{22}V_2 \tag{1.12}$$

where Y_{11} and Y_{22} are the short-circuit driving-point admittances, and Y_{12} and Y_{21} are the short-circuit transfer admittances of the two-port network. In this case, the current components I_1 and I_2 due to the input voltage source V_1 are determined by setting $V_2 = 0$ in (1.11) and (1.12), thus creating a short-circuited output terminal. Similarly, the same current components I_1 and I_2 are determined by setting $V_1 = 0$ when the input terminal becomes short circuited. As a result, the two driving-point admittances are

$$Y_{11} = \left.\frac{I_1}{V_1}\right|_{V_2=0} \qquad\qquad Y_{22} = \left.\frac{I_2}{V_2}\right|_{V_1=0} \tag{1.13}$$

whereas the two transfer admittances are

$$Y_{21} = \left.\frac{I_2}{V_1}\right|_{V_2=0} \qquad\qquad Y_{12} = \left.\frac{I_1}{V_2}\right|_{V_1=0} \tag{1.14}$$

The transmission parameters, which are often used for passive device analysis, are determined for the independent input voltage source V_1 and input current I_1 in terms of the output voltage V_2 and output current I_2. In this case, if the submatrix

$$\begin{bmatrix} m_{11} & n_{11} \\ m_{21} & n_{21} \end{bmatrix}$$

given in (1.4) is nonsingular, one can obtain

$$\begin{bmatrix} V_1 \\ I_1 \end{bmatrix} = -\begin{bmatrix} m_{11} & n_{11} \\ m_{21} & n_{21} \end{bmatrix}^{-1} \begin{bmatrix} m_{12} & n_{12} \\ m_{22} & n_{22} \end{bmatrix} \begin{bmatrix} V_2 \\ -I_2 \end{bmatrix} = [ABCD] \begin{bmatrix} V_2 \\ -I_2 \end{bmatrix} \tag{1.15}$$

where $[ABCD]$ is the forward transmission two-port network matrix. In a scalar form, (1.15) can be written as

$$V_1 = AV_2 - BI_2 \tag{1.16}$$

$$I_1 = CV_2 - DI_2 \qquad\qquad (1.17)$$

where A, B, C, and D are the transmission parameters. The voltage source V_1 and current component I_1 are determined by setting $I_2 = 0$ for the open-circuited output terminal in (1.16) and (1.17) as

$$A = \left.\frac{V_1}{V_2}\right|_{I_2=0} \qquad\qquad C = \left.\frac{I_1}{V_2}\right|_{I_2=0} \qquad\qquad (1.18)$$

where A is the reverse voltage transfer function and C is the reverse transfer admittance. Similarly, the input independent variables V_1 and I_1 are determined by setting $V_2 = 0$ when the output terminal is short circuited as

$$B = \left.\frac{V_1}{I_2}\right|_{V_2=0} \qquad\qquad D = \left.\frac{I_1}{I_2}\right|_{V_2=0} \qquad\qquad (1.19)$$

where B is the reverse transfer impedance and D is the reverse current transfer function. The reason a minus sign is associated with I_2 in (1.15) through (1.17) is that historically the input signal for transmission networks is considered as flowing to the input port, whereas the output current is flowing to the load. The direction of the current $-I_2$ entering the load is shown in Figure 1.2.

1.2   Scattering Parameters

The concept of incident and reflected voltage and current parameters can be illustrated by the single-port network shown in Figure 1.3, where the network impedance Z is connected to the signal source V_S with the internal impedance Z_S. In a common case, the terminal current I and voltage V consist of incident and reflected components (assume their root-mean-square or rms values). When the load impedance Z is equal to the conjugate of the source impedance expressed as $Z = Z_S^*$, the terminal current I becomes the incident current I_i, which is written as

$$I_i = \frac{V_S}{Z_S^* + Z_S} = \frac{V_S}{2\operatorname{Re}Z_S} \qquad\qquad (1.20)$$

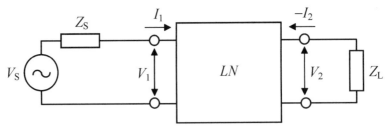

Figure 1.2 Basic diagram of loaded two-port transmission system.

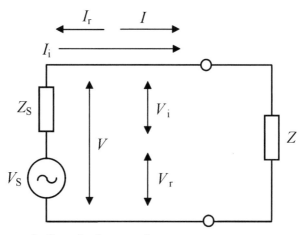

Figure 1.3 Incident and reflected voltages and currents.

The terminal voltage V, defined as the incident voltage V_i, can be determined from

$$V_i = \frac{Z_S^* V_S}{Z_S^* + Z_S} = \frac{Z_S^* V_S}{2\,\mathrm{Re}\,Z_S} \tag{1.21}$$

Consequently, the incident power, which is equal to the maximum available power from the source, can be obtained by

$$P_i = \mathrm{Re}\left(V_i I_i^*\right) = \frac{|V_S|^2}{4\,\mathrm{Re}\,Z_S} \tag{1.22}$$

The incident power can be rewritten in a normalized form using (1.21) as

$$P_i = \frac{|V_i|^2\,\mathrm{Re}\,Z_S}{|Z_S^*|^2} \tag{1.23}$$

This allows the normalized incident voltage wave a to be defined as the square root of the incident power P_i by

$$a = \sqrt{P_i} = \frac{V_i \sqrt{\mathrm{Re}\,Z_S}}{Z_S^*} \tag{1.24}$$

Similarly, the normalized reflected voltage wave b, defined as the square root of the reflected power P_r, can be written as

$$b = \sqrt{P_r} = \frac{V_r \sqrt{\mathrm{Re}\,Z_S}}{Z_S} \tag{1.25}$$

The incident power P_i can be expressed through the incident current I_i and the reflected power P_r can be expressed, respectively, through the reflected current I_r as

$$P_i = \left|I_i\right|^2 \mathrm{Re}\,Z_S \tag{1.26}$$

$$P_r = \left|I_r\right|^2 \mathrm{Re}\,Z_S \tag{1.27}$$

As a result, the normalized incident voltage wave a and reflected voltage wave b can be given by

$$a = \sqrt{P_i} = I_i\sqrt{\mathrm{Re}\,Z_S} \tag{1.28}$$

$$b = \sqrt{P_r} = I_r\sqrt{\mathrm{Re}\,Z_S} \tag{1.29}$$

The parameters a and b can also be called the *normalized incident* and *reflected current waves* or simply *normalized incident* and *reflected waves*, respectively, since the normalized current waves and the normalized voltage waves represent the same parameters.

The voltage V and current I related to the normalized incident and reflected waves a and b can be written as

$$V = V_i + V_r = \frac{Z_S^*}{\sqrt{\mathrm{Re}\,Z_S}}a + \frac{Z_S}{\sqrt{\mathrm{Re}\,Z_S}}b \tag{1.30}$$

$$I = I_i - I_r = \frac{1}{\sqrt{\mathrm{Re}\,Z_S}}a - \frac{1}{\sqrt{\mathrm{Re}\,Z_S}}b \tag{1.31}$$

where

$$a = \frac{V + Z_S I}{2\sqrt{\mathrm{Re}\,Z_S}} \qquad b = \frac{V - Z_S^* I}{2\sqrt{\mathrm{Re}\,Z_S}} \tag{1.32}$$

The source impedance Z_S is often purely real and, therefore, is used as the normalized impedance. In microwave design technique, the characteristic impedance of passive two-port networks (including transmission lines and connectors) is considered real and equal to 50Ω. This is very important for measuring S-parameters when all transmission lines, source, and load should have the same real impedance. For $Z_S = Z_S^* = Z_0$, where Z_0 is the characteristic impedance, the ratio of the normalized reflected wave and the normalized incident wave for a single-port network is called the *reflection coefficient* Γ defined as

$$\Gamma = \frac{b}{a} = \frac{V - Z_S^* I}{V + Z_S I} = \frac{V - Z_S I}{V + Z_S I} = \frac{Z - Z_S}{Z + Z_S} \tag{1.33}$$

where $Z = V/I$.

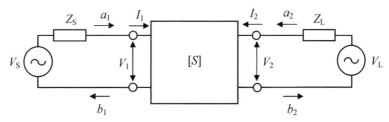

Figure 1.4   Basic diagram of S-parameter two-port network.

For the two-port network shown in Figure 1.4, the normalized reflected waves b_1 and b_2 can also be represented, respectively, by the normalized incident waves a_1 and a_2 as

$$b_1 = S_{11}a_1 + S_{12}a_2 \tag{1.34}$$

$$b_2 = S_{21}a_1 + S_{22}a_2 \tag{1.35}$$

or, in matrix form,

$$\begin{bmatrix} b_1 \\ b_2 \end{bmatrix} = \begin{bmatrix} S_{11} & S_{11} \\ S_{21} & S_{21} \end{bmatrix} \begin{bmatrix} a_1 \\ a_2 \end{bmatrix} \tag{1.36}$$

where the incident waves a_1 and a_2 and the reflected waves b_1 and b_2 for complex source and load impedances Z_S and Z_L are given by

$$a_1 = \frac{V_1 + Z_S I_1}{2\sqrt{\text{Re}\,Z_S}} \qquad a_2 = \frac{V_2 + Z_L I_2}{2\sqrt{\text{Re}\,Z_L}} \tag{1.37}$$

$$b_1 = \frac{V_1 - Z_S^* I_1}{2\sqrt{\text{Re}\,Z_S}} \qquad b_2 = \frac{V_2 - Z_L^* I_2}{2\sqrt{\text{Re}\,Z_L}} \tag{1.38}$$

where S_{11}, S_{12}, S_{21}, and S_{22} are the S-parameters of the two-port network.
From (1.36) it follows that, if $a_2 = 0$, then

$$S_{11} = \frac{b_1}{a_1}\bigg|_{a_2=0} \qquad S_{21} = \frac{b_2}{a_1}\bigg|_{a_2=0} \tag{1.39}$$

where S_{11} is the reflection coefficient and S_{21} is the transmission coefficient for ideal matching conditions at the output terminal when there is no incident power reflected from the load.
Similarly,

$$S_{12} = \frac{b_1}{a_2}\bigg|_{a_1=0} \qquad S_{22} = \frac{b_2}{a_2}\bigg|_{a_1=0} \tag{1.40}$$

where S_{12} is the transmission coefficient and S_{22} is the reflection coefficient for ideal matching conditions at the input terminal.

1.3   Conversions Between Two-Port Networks

The parameters describing the same two-port network through different two-port matrices (impedance, admittance, or transmission) can be cross-converted, and the elements of each matrix can be expressed by the elements of other matrices. For example, (1.11) and (1.12) for the admittance Y-parameters can be easily solved for the independent input voltage source V_1 and input current I_1 as

$$V_1 = -\frac{Y_{22}}{Y_{21}}V_2 - \frac{1}{Y_{21}}I_2 \tag{1.41}$$

$$I_1 = -\frac{Y_{11}Y_{22} - Y_{12}Y_{21}}{Y_{21}}V_2 - \frac{Y_{11}}{Y_{21}}I_2 \tag{1.42}$$

By comparing equivalent equations (1.16) and (1.17), and (1.41) and (1.42), the direct relationships between the transmission $ABCD$-parameters and admittance Y-parameters can be written as

$$A = -\frac{Y_{22}}{Y_{21}} \qquad\qquad B = -\frac{1}{Y_{21}} \tag{1.43}$$

$$C = -\frac{\Delta Y}{Y_{21}} \qquad\qquad D = -\frac{Y_{11}}{Y_{21}} \tag{1.44}$$

where $\Delta Y = Y_{11}Y_{22} - Y_{12}Y_{21}$.

A summary of the relationships between the impedance Z-parameters, admittance Y-parameters, and transmission $ABCD$-parameters is shown in Table 1.1, where

$$\Delta Z = Z_{11}Z_{22} - Z_{12}Z_{21}$$

To convert S-parameters to the admittance Y-parameters, it is convenient to represent (1.37) and (1.38) as

$$I_1 = \left(a_1 - b_1\right)\frac{1}{\sqrt{Z_0}} \qquad\qquad I_2 = \left(a_2 - b_2\right)\frac{1}{\sqrt{Z_0}} \tag{1.45}$$

$$V_1 = \left(a_1 + b_1\right)\sqrt{Z_0} \qquad\qquad V_2 = \left(a_2 + b_2\right)\sqrt{Z_0} \tag{1.46}$$

where it is assumed that the source and load impedances are real and equal to $Z_S = Z_L = Z_0$.

Table 1.1   Relationships Between Z-, Y-, and ABCD-Parameters

	[Z]	[Y]	[ABCD]
[Z]	$\begin{matrix} Z_{11} & Z_{12} \\ Z_{21} & Z_{22} \end{matrix}$	$\begin{matrix} \dfrac{Y_{22}}{\Delta Y} & -\dfrac{Y_{12}}{\Delta Y} \\ -\dfrac{Y_{21}}{\Delta Y} & \dfrac{Y_{11}}{\Delta Y} \end{matrix}$	$\begin{matrix} \dfrac{A}{C} & \dfrac{AD-BC}{C} \\ \dfrac{1}{C} & \dfrac{D}{C} \end{matrix}$
[Y]	$\begin{matrix} \dfrac{Z_{22}}{\Delta Z} & -\dfrac{Z_{12}}{\Delta Z} \\ -\dfrac{Z_{21}}{\Delta Z} & \dfrac{Z_{11}}{\Delta Z} \end{matrix}$	$\begin{matrix} Y_{11} & Y_{12} \\ Y_{21} & Y_{22} \end{matrix}$	$\begin{matrix} \dfrac{D}{B} & -\dfrac{AD-BC}{B} \\ -\dfrac{1}{B} & \dfrac{A}{B} \end{matrix}$
[ABCD]	$\begin{matrix} \dfrac{Z_{11}}{Z_{21}} & \dfrac{\Delta Z}{Z_{21}} \\ \dfrac{1}{Z_{21}} & \dfrac{Z_{22}}{Z_{21}} \end{matrix}$	$\begin{matrix} -\dfrac{Y_{22}}{Y_{21}} & -\dfrac{1}{Y_{21}} \\ -\dfrac{\Delta Y}{Y_{21}} & -\dfrac{Y_{11}}{Y_{21}} \end{matrix}$	$\begin{matrix} A & B \\ C & D \end{matrix}$

Substituting (1.45) and (1.46) into (1.11) and (1.12) results in

$$\frac{a_1 - b_1}{\sqrt{Z_0}} = Y_{11}\left(a_1 + b_1\right)\sqrt{Z_0} + Y_{12}\left(a_2 + b_2\right)\sqrt{Z_0} \qquad (1.47)$$

$$\frac{a_2 - b_2}{\sqrt{Z_0}} = Y_{21}\left(a_1 + b_1\right)\sqrt{Z_0} + Y_{22}\left(a_2 + b_2\right)\sqrt{Z_0} \qquad (1.48)$$

which can then be respectively converted to

$$-b_1\left(1 + Y_{11}Z_0\right) - b_2 Y_{12}Z_0 = -a_1\left(1 - Y_{11}Z_0\right) + a_2 Y_{12}Z_0 \qquad (1.49)$$

$$-b_1 Y_{21}Z_0 - b_2\left(1 + Y_{22}Z_0\right) = a_1 Y_{21}Z_0 - a_2\left(1 - Y_{22}Z_0\right) \qquad (1.50)$$

In this case, (1.49) and (1.50) can be solved for the reflected waves b_1 and b_2 as

$$b_1\left[\left(1 + Y_{11}Z_0\right)\left(1 + Y_{22}Z_0\right) - Y_{12}Y_{21}Z_0^2\right]$$
$$= a_1\left[\left(1 - Y_{11}Z_0\right)\left(1 + Y_{22}Z_0\right) + Y_{12}Y_{21}Z_0^2\right] - 2a_2 Y_{12}Z_0 \qquad (1.51)$$

$$b_2\left[\left(1 + Y_{11}Z_0\right)\left(1 + Y_{22}Z_0\right) - Y_{12}Y_{21}Z_0^2\right]$$
$$= -2a_1 Y_{21}Z_0 + a_2\left[\left(1 + Y_{11}Z_0\right)\left(1 - Y_{22}Z_0\right) + Y_{12}Y_{21}Z_0^2\right] \qquad (1.52)$$

Comparing equivalent equations (1.34) and (1.35), and (1.51) and (1.52) gives the following relationships between the scattering S-parameters and admittance Y-parameters:

$$S_{11} = \frac{\left(1 - Y_{11}Z_0\right)\left(1 + Y_{22}Z_0\right) + Y_{12}Y_{21}Z_0^2}{\left(1 + Y_{11}Z_0\right)\left(1 + Y_{22}Z_0\right) - Y_{12}Y_{21}Z_0^2} \qquad (1.53)$$

Table 1.2 Conversions between S-Parameters and Z-, Y-, and ABCD-Parameters

S-parameters through Z-, Y-, and ABCD-parameters	Z-, Y-, and ABCD-parameters through S-parameters
$S_{11} = \dfrac{(Z_{11} - Z_0)(Z_{22} + Z_0) - Z_{12}Z_{21}}{(Z_{11} + Z_0)(Z_{22} + Z_0) - Z_{12}Z_{21}}$	$Z_{11} = Z_0 \dfrac{(1 + S_{11})(1 - S_{22}) + S_{12}S_{21}}{(1 - S_{11})(1 - S_{22}) - S_{12}S_{21}}$
$S_{12} = \dfrac{2Z_{12}Z_0}{(Z_{11} + Z_0)(Z_{22} + Z_0) - Z_{12}Z_{21}}$	$Z_{12} = Z_0 \dfrac{2S_{12}}{(1 - S_{11})(1 - S_{22}) - S_{12}S_{21}}$
$S_{21} = \dfrac{2Z_{21}Z_0}{(Z_{11} + Z_0)(Z_{22} + Z_0) - Z_{12}Z_{21}}$	$Z_{21} = Z_0 \dfrac{2S_{21}}{(1 - S_{11})(1 - S_{22}) - S_{12}S_{21}}$
$S_{22} = \dfrac{(Z_{11} + Z_0)(Z_{22} - Z_0) - Z_{12}Z_{21}}{(Z_{11} + Z_0)(Z_{22} + Z_0) - Z_{12}Z_{21}}$	$Z_{22} = Z_0 \dfrac{(1 - S_{11})(1 + S_{22}) + S_{12}S_{21}}{(1 - S_{11})(1 - S_{22}) - S_{12}S_{21}}$
$S_{11} = \dfrac{(1 - Y_{11}Z_0)(1 + Y_{22}Z_0) + Y_{12}Y_{21}Z_0^2}{(1 + Y_{11}Z_0)(1 + Y_{22}Z_0) - Y_{12}Y_{21}Z_0^2}$	$Y_{11} = \dfrac{1}{Z_0} \dfrac{(1 - S_{11})(1 + S_{22}) + S_{12}S_{21}}{(1 + S_{11})(1 + S_{22}) - S_{12}S_{21}}$
$S_{12} = \dfrac{-2Y_{12}Z_0}{(1 + Y_{11}Z_0)(1 + Y_{22}Z_0) - Y_{12}Y_{21}Z_0^2}$	$Y_{12} = \dfrac{1}{Z_0} \dfrac{-2S_{12}}{(1 + S_{11})(1 + S_{22}) - S_{12}S_{21}}$
$S_{21} = \dfrac{-2Y_{21}Z_0}{(1 + Y_{11}Z_0)(1 + Y_{22}Z_0) - Y_{12}Y_{21}Z_0^2}$	$Y_{21} = \dfrac{1}{Z_0} \dfrac{-2S_{21}}{(1 + S_{11})(1 + S_{22}) - S_{12}S_{21}}$
$S_{22} = \dfrac{(1 + Y_{11}Z_0)(1 - Y_{22}Z_0) + Y_{12}Y_{21}Z_0^2}{(1 + Y_{11}Z_0)(1 + Y_{22}Z_0) - Y_{12}Y_{21}Z_0^2}$	$Y_{22} = \dfrac{1}{Z_0} \dfrac{(1 + S_{11})(1 - S_{22}) + S_{12}S_{21}}{(1 + S_{11})(1 + S_{22}) - S_{12}S_{21}}$
$S_{11} = \dfrac{AZ_0 + B - CZ_0^2 - DZ_0}{AZ_0 + B + CZ_0^2 + DZ_0}$	$A = \dfrac{(1 + S_{11})(1 - S_{22}) + S_{12}S_{21}}{2S_{21}}$
$S_{12} = \dfrac{2(AD - BC)Z_0}{AZ_0 + B + CZ_0^2 + DZ_0}$	$B = Z_0 \dfrac{(1 + S_{11})(1 + S_{22}) - S_{12}S_{21}}{2S_{21}}$
$S_{21} = \dfrac{2Z_0}{AZ_0 + B + CZ_0^2 + DZ_0}$	$C = \dfrac{1}{Z_0} \dfrac{(1 - S_{11})(1 - S_{22}) - S_{12}S_{21}}{2S_{21}}$
$S_{22} = \dfrac{-AZ_0 + B - CZ_0^2 + DZ_0}{AZ_0 + B + CZ_0^2 + DZ_0}$	$D = \dfrac{(1 - S_{11})(1 + S_{22}) + S_{12}S_{21}}{2S_{21}}$

$$S_{12} = \frac{-2Y_{12}Z_0}{(1 + Y_{11}Z_0)(1 + Y_{22}Z_0) - Y_{12}Y_{21}Z_0^2} \tag{1.54}$$

$$S_{21} = \frac{-2Y_{21}Z_0}{(1 + Y_{11}Z_0)(1 + Y_{22}Z_0) - Y_{12}Y_{21}Z_0^2} \tag{1.55}$$

$$S_{22} = \frac{(1 + Y_{11}Z_0)(1 - Y_{22}Z_0) + Y_{12}Y_{21}Z_0^2}{(1 + Y_{11}Z_0)(1 + Y_{22}Z_0) - Y_{12}Y_{21}Z_0^2} \tag{1.56}$$

The relationships among the S-parameters and the Z- and ABCD-parameters can be obtained in a similar fashion. Table 1.2 shows the conversions between S-parameters and Z-, Y-, and ABCD-parameters for the simplified case when the source impedance Z_S and the load impedance Z_L are equal to the characteristic impedance Z_0 [3].

1.4   Practical Two-Port Networks

1.4.1   Single-Element Networks

The simplest networks, which include only one element, can be constructed by a series-connected admittance Y, as shown in Figure 1.5(a), or by a parallel-connected impedance Z, as shown in Figure 1.5(b).

The two-port network consisting of the single series admittance Y can be described in a system of the admittance Y-parameters as

$$I_1 = YV_1 - YV_2 \qquad\qquad (1.57)$$

$$I_2 = -YV_1 + YV_2 \qquad\qquad (1.58)$$

or, in matrix form,

$$[Y] = \begin{bmatrix} Y & -Y \\ -Y & Y \end{bmatrix} \qquad\qquad (1.59)$$

which means that $Y_{11} = Y_{22} = Y$ and $Y_{12} = Y_{21} = -Y$. The resulting matrix is a singular matrix with $|Y| = 0$. Consequently, it is impossible to determine such a two-port network with the series admittance Y-parameters through a system of the impedance Z-parameters. However, by using the transmission $ABCD$-parameters, it can be described by

$$[ABCD] = \begin{bmatrix} 1 & 1/Y \\ 0 & 1 \end{bmatrix} \qquad\qquad (1.60)$$

Similarly, for a two-port network with the single parallel impedance Z,

$$[Z] = \begin{bmatrix} Z & Z \\ Z & Z \end{bmatrix} \qquad\qquad (1.61)$$

which means that $Z_{11} = Z_{12} = Z_{21} = Z_{22} = Z$. The resulting matrix is a singular matrix with $|Z| = 0$. In this case, it is impossible to determine such a two-port network with the parallel impedance Z-parameters through a system of the admittance

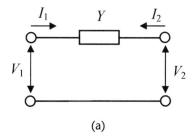
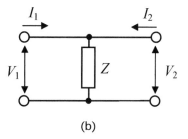

(a)   (b)

Figure 1.5   Single-element networks.

Y-parameters. By using the transmission $ABCD$-parameters, this two-port network can be described by

$$[ABCD] = \begin{bmatrix} 1 & 0 \\ 1/Z & 1 \end{bmatrix} \tag{1.62}$$

1.4.2 π- and T-Type Networks

The basic configurations of a two-port network that usually describe the electrical properties of the active devices can be represented in the form of a π-circuit, shown in Figure 1.6(a), or a T-circuit, shown in Figure 1.6(b). Here, the π-circuit includes the current source $g_m V_1$ and the T-circuit includes the voltage source $r_m I_1$.

By deriving the two loop equations using Kirchhoff's current law or applying (1.13) and (1.14) for the π-circuit, we can obtain

$$I_1 - \left(Y_1 + Y_3\right)V_1 + Y_3 V_2 = 0 \tag{1.63}$$

$$I_2 + \left(g_m - Y_3\right)V_1 + \left(Y_2 + Y_3\right)V_2 = 0 \tag{1.64}$$

Equations (1.63) and (1.64) can be rewritten as matrix equation (1.3) with

$$[M] = \begin{bmatrix} 1 & 0 \\ 0 & 1 \end{bmatrix} \quad \text{and} \quad [N] = \begin{bmatrix} -\left(Y_1 + Y_3\right) & Y_3 \\ -g_m + Y_3 & -\left(Y_2 + Y_3\right) \end{bmatrix}$$

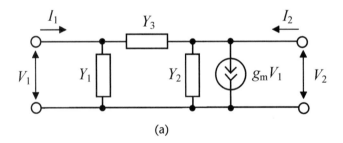

(a)

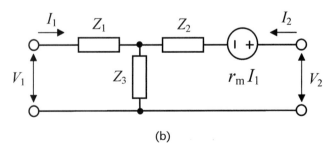

(b)

Figure 1.6 Basic diagrams of π- and T-type networks.

Since matrix $[M]$ is nonsingular, such a two-port network can be described by a system of the admittance Y-parameters as

$$[Y] = -[M]^{-1}[N] = \begin{bmatrix} Y_1 + Y_3 & -Y_3 \\ g_m - Y_3 & Y_2 + Y_3 \end{bmatrix} \tag{1.65}$$

Similarly, for a two-port network in the form of a T-circuit using Kirchhoff's voltage law or applying (1.8) and (1.9), one can obtain

$$[Z] = -[M]^{-1}[N] = \begin{bmatrix} Z_1 + Z_3 & Z_3 \\ r_m + Z_3 & Z_2 + Z_3 \end{bmatrix} \tag{1.66}$$

If $g_m = 0$ for a π-circuit and $r_m = 0$ for a T-circuit, their corresponding matrices in a system of the transmission $ABCD$-parameters can be written as follows:

For a π-Circuit:

$$[ABCD] = \begin{bmatrix} 1 + \dfrac{Y_2}{Y_3} & \dfrac{1}{Y_3} \\ Y_1 + Y_2 + \dfrac{Y_1 Y_2}{Y_3} & 1 + \dfrac{Y_1}{Y_3} \end{bmatrix} \tag{1.67}$$

For a T-Circuit:

$$[ABCD] = \begin{bmatrix} 1 + \dfrac{Z_2}{Z_3} & Z_1 + Z_2 + \dfrac{Z_1 Z_2}{Z_3} \\ \dfrac{1}{Z_3} & 1 + \dfrac{Z_1}{Z_3} \end{bmatrix} \tag{1.68}$$

Based on the appropriate relationships between impedances of a T-circuit and admittances of a π-circuit, these two circuits become equivalent with the respect to the effect on any other two-port network. For the π-circuit shown in Figure 1.7(a),

$$I_1 = Y_1 V_{13} + Y_3 V_{12} = Y_1 V_{13} + Y_3 (V_{13} - V_{23})$$
$$= (Y_1 + Y_3) V_{13} - Y_3 V_{23} \tag{1.69}$$

$$I_2 = Y_2 V_{23} - Y_3 V_{12} = Y_2 V_{23} - Y_3 (V_{13} - V_{23})$$
$$= -Y_3 V_{13} + (Y_2 + Y_3) V_{23} \tag{1.70}$$

Solving (1.69) and (1.70) for voltages V_{13} and V_{23} yields

$$V_{13} = \frac{Y_2 + Y_3}{Y_1 Y_2 + Y_1 Y_2 + Y_1 Y_2} I_1 + \frac{Y_3}{Y_1 Y_2 + Y_1 Y_2 + Y_1 Y_2} I_2 \tag{1.71}$$

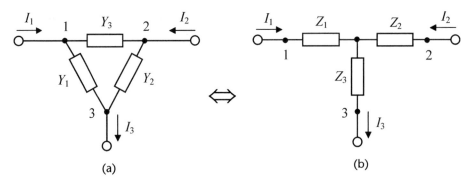

Figure 1.7 Equivalence of π- and T-circuits.

$$V_{23} = \frac{Y_3}{Y_1Y_2 + Y_1Y_2 + Y_1Y_2}I_1 + \frac{Y_1 + Y_3}{Y_1Y_2 + Y_1Y_2 + Y_1Y_2}I_2 \tag{1.72}$$

Similarly, for the T-circuit shown in Fig. 1.7(b),

$$\begin{aligned} V_{13} &= Z_1I_1 + Z_3I_3 = Z_1I_1 + Z_3(I_1 + I_2) \\ &= (Z_1 + Z_3)I_1 + Z_3I_2 \end{aligned} \tag{1.73}$$

$$\begin{aligned} V_{13} &= Z_1I_1 + Z_3I_3 = Z_1I_1 + Z_3(I_1 + I_2) \\ &= Z_3I_1 + (Z_2 + Z_3)I_2 \end{aligned} \tag{1.74}$$

and the equations for currents I_1 and I_2 can be obtained by

$$I_1 = \frac{Z_2 + Z_3}{Z_1Z_2 + Z_1Z_2 + Z_1Z_2}V_{13} - \frac{Z_3}{Z_1Z_2 + Z_1Z_2 + Z_1Z_2}V_{23} \tag{1.75}$$

$$I_2 = -\frac{Z_3}{Z_1Z_2 + Z_1Z_2 + Z_1Z_2}V_{13} + \frac{Z_1 + Z_3}{Z_1Z_2 + Z_1Z_2 + Z_1Z_2}V_{23} \tag{1.76}$$

To establish a T-to-π transformation, it is necessary to equate the coefficients for V_{13} and V_{23} in (1.75) and (1.76) to the corresponding coefficients in (1.69) and

Table 1.3 Relationships Between π- and T-Circuit Parameters

T- to π-transformation	π- to T-transformation
$Y_1 = \dfrac{Z_2}{Z_1Z_2 + Z_2Z_3 + Z_1Z_3}$	$Z_1 = \dfrac{Y_2}{Y_1Y_2 + Y_2Y_3 + Y_1Y_3}$
$Y_2 = \dfrac{Z_1}{Z_1Z_2 + Z_2Z_3 + Z_1Z_3}$	$Z_2 = \dfrac{Y_1}{Y_1Y_2 + Y_2Y_3 + Y_1Y_3}$
$Y_3 = \dfrac{Z_3}{Z_1Z_2 + Z_2Z_3 + Z_1Z_3}$	$Z_3 = \dfrac{Y_3}{Y_1Y_2 + Y_2Y_3 + Y_1Y_3}$

(1.70). Similarly, to establish a π-to-T transformation, it is necessary to equate the coefficients for I_1 and I_2 in (1.73) and (1.74) to the corresponding coefficients in (1.71) and (1.72). The resulting relationships between admittances for a π-circuit and impedances for a T-circuit are given in Table 1.3.

1.5   Lumped Elements

Generally, passive hybrid or integrated circuits are designed based on lumped elements, distributed elements, or a combination of both types of elements. Distributed elements represent any sections of the transmission lines of different lengths, types, and characteristic impedances. The basic lumped elements are inductors and capacitors that are small in size in comparison with the transmission-line wavelength λ, and usually their linear dimensions are less than $\lambda/10$ or even $\lambda/16$. In applications where lumped elements are used, their basic advantages are small physical size and low production costs. However, their main drawbacks are a lower quality factor and reduced power-handling capability compared with distributed elements.

1.5.1   Inductors

Inductors are lumped elements that store energy in a magnetic field. The lumped inductors can be implemented using several different configurations such as a short section of a strip conductor or wire, a single loop, or a spiral. The printed high-impedance microstrip-section inductor is usually used for low inductance values, typically less than 2 nH, and often meandered to reduce the component size. The printed microstrip single-loop inductors are not very popular due to their limited inductance per unit area. The approximate expression for the microstrip short-section inductance in free space is given by

$$L(\text{nH}) = 0.2 \times 10^{-3} l \left[\ln\left(\frac{l}{W+t} \right) + 1.193 + \frac{W+t}{3l} \right] K_{\text{g}} \qquad (1.77)$$

where the conductor length l, conductor width W, and conductor thickness t are in microns, and the term K_{g} accounts for the presence of a ground plane defined as

$$K_{\text{g}} = 0.57 - 0.145 \ln \frac{W}{h} \quad \text{for} \quad \frac{W}{h} > 0.05 \qquad (1.78)$$

where h is the spacing from the ground plane [4, 5].

Spiral inductors can have a circular configuration, a rectangular (square) configuration [shown in Figure 1.8(a)], or an octagonal configuration [shown in Figure 1.8(b)], if the technology allows 45° routing. The circular geometry is superior in electrical performance, whereas the rectangular shapes are easy to lay out and fabricate. Printed inductors are based on using thin-film or thick-film Si or GaAs fabrication processes, and the inner conductor is pulled out to connect with other circuitry through a bondwire, an air bridge, or by using multilevel crossover metal. The general expression for a spiral inductor, which is also valid for its planar

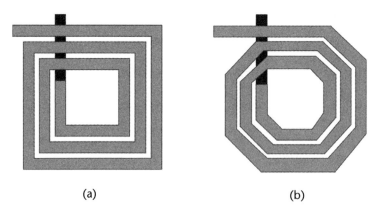

(a)   (b)

Figure 1.8 Spiral inductor layouts.

integration with an accuracy of around 3%, is based on a Wheeler formula and can be obtained as

$$L(\text{nH}) = \frac{K_1 n^2 d_{\text{avg}}}{1 + K_2 \rho}$$ (1.79)

where

n = number of turns

$d_{\text{avg}} = (d_{\text{out}} + d_{\text{in}})/2$ = average diameter

$\rho = (d_{\text{out}} + d_{\text{in}})/(d_{\text{out}} - d_{\text{in}})$ = fill ratio

d_{out} = outer diameter (μm)

d_{in} = inner diameter (μm)

and the coefficients K_1 and K_2 are layout dependent, where square: $K_1 = 2.34$, $K_2 = 2.75$; hexagonal: $K_1 = 2.33$, $K_2 = 3.82$; and octagonal: $K_1 = 2.25$, $K_2 = 3.55$ [6, 7].

In contrast to the capacitors, high-quality inductors cannot be readily available in a standard complementary metal-oxide-semiconductor (CMOS) technology. Therefore, it is necessary to use special techniques to improve the inductor electrical performance. By using a standard CMOS technology with only two metal layers and a heavily doped substrate, the spiral inductor will have a large series resistance compared with three-four metal layer technologies, and the substrate losses become a very important factor due to the relatively low resistivity of silicon. A major source of substrate losses is the capacitive coupling when current is flowing not only through the metal strip, but also through the silicon substrate. Another important source of substrate losses is the inductive coupling when, due to the planar inductor structure, the magnetic field penetrates deeply into the silicon substrate, inducing current loops and related losses. However, the latter effects are particularly important for large-area inductors and can be overcome by using silicon micromachining techniques [8].

The simplified equivalent circuit for the CMOS spiral microstrip inductor is shown in Figure 1.9, where L_s models the self and mutual inductances, R_s is the series coil resistance, C_{ox} is the parasitic oxide capacitance from the metal layer to the substrate, R_{si} is the resistance of the conductive silicon substrate, C_{si} is the silicon

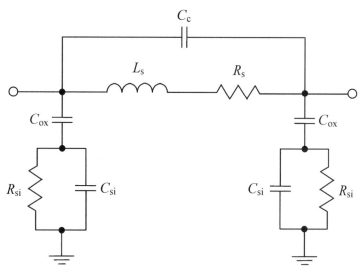

Figure 1.9   Equivalent circuit of a square spiral inductor.

substrate parasitic capacitance, and C_c is the parasitic coupling capacitance [9]. The parasitic silicon substrate capacitance C_{si} is sufficiently small and in most cases it can be neglected. Such a model shows an accurate agreement between simulated and measured data within 10% across a variety of inductor geometries and substrate dopings up to 20 GHz [10]. At frequencies well below the inductor self-resonant radian frequency ω_{SRF}, the coupling capacitance C_c between metal segments due to fringing fields in both the dielectric and air regions can also be neglected since the relative dielectric constant of the oxide is sufficiently small [11]. In this case, if one side of the inductor is grounded, the self-resonant radian frequency of the spiral inductor can be approximately calculated from

$$\omega_{SRF} = \frac{1}{\sqrt{L_s C_{ox}}} \sqrt{\frac{L_s - R_s^2 C_{ox}}{L_s - R_{si}^2 C_{ox}}} \tag{1.80}$$

At frequencies higher than the self-resonant frequency ω_{SRF}, the inductor exhibits a capacitive behavior. The self-resonant frequency ω_{SRF} is limited mainly by the parasitic oxide capacitance C_{ox}, which is inversely proportional to the oxide thickness between the metal layer and substrate. The frequency at which the inductor quality factor Q is maximal can be obtained as

$$\omega_Q = \frac{1}{\sqrt{L_s C_{ox}}} \sqrt{\frac{R_s}{2R_{si}}} \left(\sqrt{1 + \frac{4R_{si}}{3R_s}} - 1 \right)^{0.5} \tag{1.81}$$

The inductor metal conductor series resistance R_s can be easily calculated at low frequencies as the product of the sheet resistance and the number of squares of the metal trace. However, at high frequencies, the skin effect and other magnetic field effects will cause a nonuniform current distribution in the inductor profile. In this case, a simple increase in a diameter of the inductor metal turn does not necessarily reduce correspondingly the inductor series resistance. For example, for the same inductance value, the difference in resistance between two inductors, when one of them has a two times wider metal strip, is a factor of only 1.35 [12]. Moreover, at very high frequencies, the largest contribution to the series resistance does not come from the longer outer turns, but from the inner turns. This phenomenon is a result of the generation of circular eddy currents in the inner conductors, whose direction is such that they oppose the original change in magnetic field. On the inner side of the inner turn, coil current and eddy current flow in the same direction, so the current density is larger than average. On the outer side, both currents cancel, and the current density is smaller than average. As a result, the current in the inner turn is pushed to the inside of the conductor.

In hybrid or monolithic applications, bondwires are used to interconnect different components such as lumped elements, planar transmission lines, solid-state devices, and integrated circuits. These bondwires, which are usually made of gold or aluminium, have 0.5- to 1.0-mil diameters, and their lengths are electrically shorter compared with the operating wavelength. To characterize the electrical behavior of the bondwires, simple formulas in terms of their inductances and series resistances can be used. As a first-order approximation, the parasitic capacitance associated with bondwires can be neglected. In this case, the bondwire inductance can be estimated as follows when $l \gg d$, where l is the bondwire length in microns and d is the bondwire diameter in microns:

$$L(\mathrm{nH}) = 0.2 \times 10^{-3} l \left(\ln \frac{4l}{d} + 0.5 \frac{d}{l} - 1 + C \right) \tag{1.82}$$

where $C = \tanh(4\delta/d)/4$ is the frequency-dependent correction factor, which is a function of the bondwire's diameter and its material's skin depth δ [6, 13].

1.5.2   Capacitors

Capacitors are lumped elements that store energy due to an electric field between two electrodes (or plates) when a voltage is applied across them. In this case, charge of equal magnitude but opposite sign accumulates on the opposing capacitor plates. The capacitance depends on the area of the plates, separation, and dielectric material between them. The basic structure of a chip capacitor, shown in Figure 1.10(a), consists of two parallel plates, each of area $A = W \times l$, which are separated by a dielectric material of thickness d and permittivity $\varepsilon_0 \varepsilon_r$, where ε_0 is the free-space permittivity (8.85×10^{-12} F/m) and ε_r is the relative dielectric constant.

Chip capacitors are usually used in hybrid integrated circuits when relatively high capacitance values are required. In the parallel-plate configuration, the equation for calculating the capacitance is commonly written as

$$C(\mathrm{pF}) = 8.85 \times 10^{-3} \varepsilon_r \frac{Wl}{d} \tag{1.83}$$

where W, l, and d are dimensions in millimeters. Generally, the low-frequency bypass capacitor values are expressed in microfarads and nanofarads, high-frequency

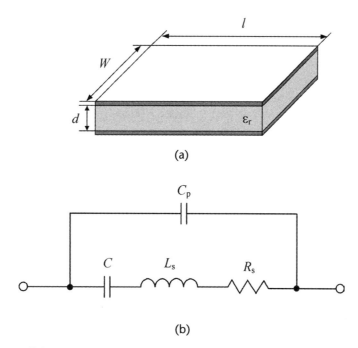

Figure 1.10   Parallel capacitor topology and its equivalent circuit.

blocking and tuning capacitors are expressed in picofarads, and parasitic or fring-ing capacitances are written in femtofarads. The basic formula given by (1.83) can also be applied to capacitors based on a multilayer technique [5]. The lumped-element equivalent circuit of a capacitor is shown in Figure 1.10(b), where L_s is the series plate inductance, R_s is the series contact and plate resistance, and C_p is the parasitic parallel capacitance. When $C \gg C_p$, the radian frequency ω_{SRF}, at which the reactances of series elements C and L_s become equal, is called the *capacitor self-resonant frequency*, and the capacitor impedance is equal to the resistance R_s.

For monolithic applications where relatively low capacitances (typically less than 0.5 pF) are required, planar series capacitances in the form of microstrip or inter-digital configurations can be used. These capacitors are simply formed by gaps in the center conductor of the microstrip lines, and they do not require any dielectric films. The gap capacitor shown in Figure 1.11(a) can be equivalently represented by a series coupling capacitance and two parallel fringing capacitances [14]. The interdigital capacitor is a multifinger periodic structure, as shown in Figure 1.11(b), where the capacitance occurs across the narrow gap between thin-film transmission-line conductors [15]. These gaps are essentially very long and folded to use a small amount of area. In this case, it is important to keep the size of the capacitor very small relative to the wavelength, so that it can be treated as a lumped element. A larger total width-to-length ratio results in the desired greater shunt capacitance and lower series inductance. An approximate expression for the total capacitance of an interdigital structure with $s = W$ and length l less than a quarter wavelength can be given by

$$C(\text{pF}) = \left(\varepsilon_r + 1\right)l\left[(N - 3)A_1 + A_2\right] \tag{1.84}$$

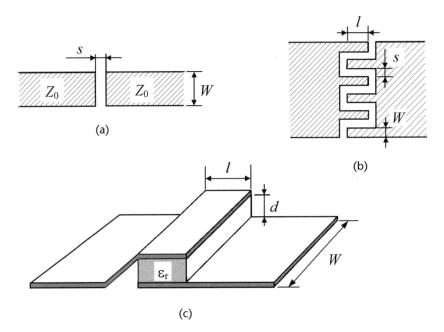

Figure 1.11   Different series capacitor topologies.

where N is the number of fingers and

$$A_1(\text{pF/}\mu\text{m}) = 4.409 \ \tanh\left[0.55\left(\frac{h}{W}\right)^{0.45}\right] \times 10^{-6} \qquad (1.85)$$

$$A_2(\text{pF/}\mu\text{m}) = 9.92 \ \tanh\left[0.52\left(\frac{h}{W}\right)^{0.5}\right] \times 10^{-6} \qquad (1.86)$$

where h is the spacing from the ground plane.

Series planar capacitors with larger values, which are called metal–insulator–metal (MIM) capacitors, can be realized by using an additional thin dielectric layer (typically less than 0.5 µm) between two metal plates, as shown in Figure 1.11(c) [5]. The bottom plate of the capacitor uses a thin unplated metal, and typically the dielectric material is silicon nitride (Si_3N_4) for integrated circuits on GaAs and silicon dioxide (SiO_2) for integrated circuits on Si. The top plate uses a thick plated conductor to reduce the loss in the capacitor. These capacitors are used to achieve higher capacitance values in small areas (10 pF and greater), with typical tolerances from 10% to 15%. The capacitance can be calculated according to (1.83).

1.6   Transmission Lines

Transmission lines are widely used in matching circuits of power amplifiers, in hybrid couplers, or in power combiners and dividers. When the propagated signal wavelength is compared to its physical dimension, the transmission line can be considered to be a two-port network with distributed parameters, where the voltages and currents vary in magnitude and phase over length.

1.6.1   Basic Parameters

Schematically, a transmission line is often represented as a two-wire line, as shown in Figure 1.12(a), where its electrical parameters are distributed along its length. The physical properties of a transmission line are determined by four basic parameters:

1. The series inductance L due to the self-inductive phenomena of two conductors
2. The shunt capacitance C in view of the proximity between two conductors
3. The series resistance R due to the finite conductivity of the conductors
4. The shunt conductance G, which is related to the dielectric losses in the material.

As a result, a transmission line of length Δx represents a lumped-element circuit, as shown in Figure 1.12(b), where ΔL, ΔC, ΔR, and ΔG are the series inductance, shunt capacitance, series resistance, and shunt conductance per unit length, respectively. If all of these elements are distributed uniformly along the transmission line, and their values do not depend on the chosen position of Δx, this transmission line is called the *uniform transmission line* [16]. Any finite length of the uniform transmission line can be viewed as a cascade of section length Δx.

To define the distribution of the voltages and currents along the uniform transmission line, it is necessary to write the differential equations using Kirchhoff's

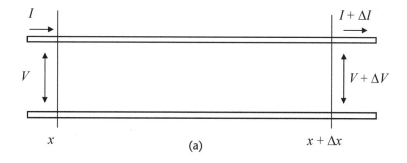

(a)

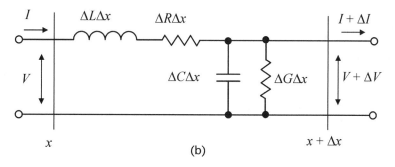

(b)

Figure 1.12 Transmission-line schematics.

voltage law for instantaneous values of the voltages and currents in the line section of length Δx, distant x from its beginning. For the sinusoidal steady-state condition, the telegrapher equations for $V(x)$ and $I(x)$ are given by

$$\frac{d^2V(x)}{dx^2} - \gamma^2 V(x) = 0 \tag{1.87}$$

$$\frac{d^2I(x)}{dx^2} - \gamma^2 I(x) = 0 \tag{1.88}$$

where $\gamma = \alpha + j\beta = \sqrt{(\Delta R + j\omega\Delta L)(\Delta G + j\omega\Delta C)}$ is the complex propagation constant (which is a function of frequency), α is the attenuation constant, and β is the phase constant. The general solutions of (1.87) and (1.88) for voltage and current of the traveling wave in the transmission line can be written as

$$V(x) = A_1 \exp(-\gamma x) + A_2 \exp(\gamma x) \tag{1.89}$$

$$I(x) = \frac{A_1}{Z_0} \exp(-\gamma x) - \frac{A_2}{Z_0} \exp(\gamma x) \tag{1.90}$$

where $Z_0 = \sqrt{(\Delta R + j\omega\Delta L)/(\Delta G + j\omega\Delta C)}$ is the characteristic impedance of the transmission line; $V_i = A_1 \exp(-\gamma x)$ and $V_r = A_2 \exp(\gamma x)$ represent the incident voltage and the reflected voltage, respectively; and $I_i = A_1 \exp(-\gamma x)/Z_0$ and $I_r = A_2 \exp(\gamma x)/Z_0$ are the incident current and the reflected current, respectively. From (1.89) and (1.90), it follows that the characteristic impedance of transmission line Z_0 represents the ratio of the incident (reflected) voltage to the incident (reflected) current at any position on the line as

$$Z_0 = \frac{V_i(x)}{I_i(x)} = \frac{V_r(x)}{I_r(x)} \tag{1.91}$$

For a lossless transmission line, when $R = G = 0$ and the voltage and current do not change with position, the attenuation constant $\alpha = 0$, the propagation constant $\gamma = j\beta = j\omega\sqrt{\Delta L\Delta C}$, and the phase constant $\beta = \omega\sqrt{\Delta L\Delta C}$. Consequently, the characteristic impedance is reduced to $Z_0 = \sqrt{L/C}$ and represents a real number. The wavelength is defined as $\lambda = 2\pi/\beta = 2\pi/\omega\sqrt{\Delta L\Delta C}$, and the phase velocity as $v_p = \omega/\beta = 1/\sqrt{\Delta L\Delta C}$.

Figure 1.13 represents a transmission line of the characteristic impedance Z_0 terminated with a load Z_L. In this case, the constants A_1 and A_2 are determined at the position $x = l$ by

$$V(l) = A_1 \exp(-\gamma l) + A_2 \exp(\gamma l) \tag{1.92}$$

$$I(l) = \frac{A_1}{Z_0} \exp(-\gamma l) - \frac{A_2}{Z_0} \exp(\gamma l) \tag{1.93}$$

and are equal to

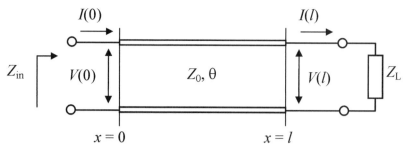

Figure 1.13   Loaded transmission line.

$$A_1 = \frac{V(l) + Z_0 I(l)}{2} \, \exp(\gamma l) \qquad (1.94)$$

$$A_2 = \frac{V(l) - Z_0 I(l)}{2} \, \exp(-\gamma l) \qquad (1.95)$$

As a result, wave equations for voltage $V(x)$ and current $I(x)$ can be rewritten as

$$V(x) = \frac{V(l) + Z_0 I(l)}{2} \, \exp[\gamma(l - x)] + \frac{V(l) - Z_0 I(l)}{2} \, \exp[-\gamma(l - x)] \quad (1.96)$$

$$I(x) = \frac{V(l) + Z_0 I(l)}{2Z_0} \, \exp[\gamma(l - x)] - \frac{V(l) - Z_0 I(l)}{2Z_0} \, \exp[-\gamma(l - x)] \quad (1.97)$$

which allows their determination at any position on the transmission line.

The voltage and current amplitudes at $x = 0$ as functions of the voltage and current amplitudes at $x = l$ can be determined from (1.96) and (1.97) as

$$V(0) = \frac{V(l) + Z_0 I(l)}{2} \, \exp(\gamma l) + \frac{V(l) - Z_0 I(l)}{2} \, \exp(-\gamma l) \qquad (1.98)$$

$$I(0) = \frac{V(l) + Z_0 I(l)}{2Z_0} \, \exp(\gamma l) - \frac{V(l) - Z_0 I(l)}{2Z_0} \, \exp(-\gamma l) \qquad (1.99)$$

By using the ratios $\cosh x = [\exp(x) + \exp(-x)]/2$ and $\sinh x = [\exp(x) - \exp(-x)]/2$, (1.98) and (1.99) can be rewritten in the form

$$V(0) = V(l)\cosh(\gamma l) + Z_0 I(l)\sinh(\gamma l) \qquad (1.100)$$

$$I(0) = \frac{V(l)}{Z_0} \, \sinh(\gamma l) + I(l)\cosh(\gamma l) \qquad (1.101)$$

which represents the transmission equations of the symmetrical reciprocal two-port network expressed through the $ABCD$-parameters when $AD - BC = 1$ and $A = D$. Consequently, the transmission $ABCD$-matrix of the lossless transmission line with $\alpha = 0$ can be defined as

$$[ABCD] = \begin{bmatrix} \cos\theta & jZ_0\sin\theta \\ \dfrac{j\sin\theta}{Z_0} & \cos\theta \end{bmatrix} \qquad (1.102)$$

Using the formulas to transform $ABCD$-parameters into S-parameters yields

$$[S] = \begin{bmatrix} 0 & \exp(-j\theta) \\ \exp(-j\theta) & 0 \end{bmatrix} \qquad (1.103)$$

where $\theta = \beta l$ is the electrical length of the transmission line.

In the case of the loaded lossless transmission line, the reflection coefficient Γ is defined as the ratio between the reflected voltage wave and the incident voltage wave given at a position x as

$$\Gamma(x) = \frac{V_r}{V_i} = \frac{A_2}{A_1}\exp(2j\beta x) \qquad (1.104)$$

By taking into account (1.94) and (1.95), the reflection coefficient for $x = l$ can be defined as

$$\Gamma = \frac{Z - Z_0}{Z + Z_0} \qquad (1.105)$$

where Γ represents the load reflection coefficient and $Z = Z_L = V(l)/I(l)$. If the load is mismatched, only part of the available power from the source is delivered to the load. This power loss is called the *return loss* (RL), and it is calculated in decibels as

$$RL = -20\log_{10}|\Gamma| \qquad (1.106)$$

For a matched load when $\Gamma = 0$, a return loss is of ∞ dB. A total reflection with $\Gamma = 1$ means a return loss of 0 dB when all incident power is reflected.

According to the general solution for voltage at a position x in the transmission line,

$$V(x) = V_i(x) + V_r(x) = V_i[1 + \Gamma(x)] \qquad (1.107)$$

Hence, the maximum amplitude (when the incident and reflected waves are in phase) is

$$V_{max}(x) = |V_i|\big[1 + |\Gamma(x)|\big] \qquad (1.108)$$

and the minimum amplitude (when these two waves are out of phase) is

$$V_{min}(x) = |V_i|\big[1 - |\Gamma(x)|\big] \qquad (1.109)$$

The ratio of V_{max} to V_{min}, which is a function of the reflection coefficient Γ, represents the voltage standing wave ratio (VSWR). The VSWR is a measure of mismatch and can be written as

$$\text{VSWR} = \frac{V_{max}}{V_{min}} = \frac{1 + |\Gamma|}{1 - |\Gamma|} \tag{1.110}$$

which can change from 1 to ∞ (where VSWR = 1 implies a matched load). For a load impedance with zero imaginary part when $Z_L = R_L$, the VSWR can be calculated as VSWR = R_L/Z_0 when $R_L \geq Z_0$ and VSWR = Z_0/R_L when $Z_0 \geq R_L$.

From (1.100) and (1.101), it follows that the input impedance of the loaded loss-less transmission line can be obtained as

$$Z_{in} = \frac{V(0)}{I(0)} = Z_0 \frac{Z_L + jZ_0 \tan(\theta)}{Z_0 + jZ_L \tan(\theta)} \tag{1.111}$$

which gives an important dependence between the input impedance, the transmission-line parameters (electrical length and characteristic impedance), and the arbitrary load impedance.

1.6.2   Microstrip Line

Planar transmission lines as an evolution of the coaxial and parallel-wire lines are compact and readily adaptable to hybrid and monolithic integrated circuit fabrication technologies at RF and microwave frequencies [17]. In a microstrip line, the grounded metallization surface covers only one side of dielectric substrate, as shown in Figure 1.14. Such a configuration is equivalent to a pair-wire system for the image of the conductor in the ground plane, which produces the required symmetry [18]. In this case, the electric and magnetic field lines are located in both the dielectric region between the strip conductor and the ground plane and in the air region above the substrate. As a result, the electromagnetic wave propagated along a microstrip line is not a pure TEM, since the phase velocities in these two regions are not the same. However, in a quasistatic approximation, which gives sufficiently accurate

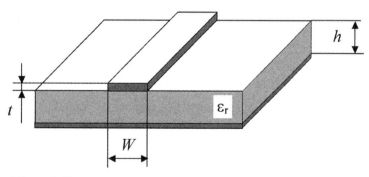

Figure 1.14   Microstrip-line structure.

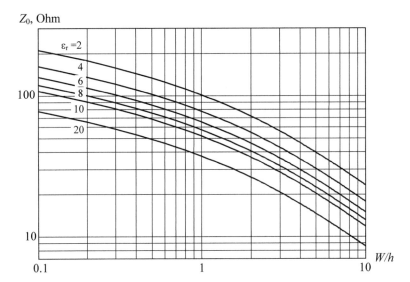

Figure 1.15 Microstrip-line characteristic impedance versus W/h.

results as long as the height of the dielectric substrate is very small compared with the wavelength, it is possible to obtain the explicit analytical expressions for its electrical characteristics. Since a microstrip line is an open structure, it has a major fabrication advantage over the stripline due to simplicity of practical realization, interconnection, and adjustments.

The exact expression for the characteristic impedance of a lossless microstrip line even with finite strip thickness is complicated [19, 20]. However, in practice, it is possible to use a sufficiently simple formula to estimate the characteristic impedance Z_0 of a microstrip line with zero strip thickness [21]:

$$Z_0 = \frac{120\pi}{\sqrt{\varepsilon_r}} \frac{h}{W} \frac{1}{1 + 1.735\varepsilon_r^{-0.0724}(W/h)^{-0.836}} \qquad (1.112)$$

Figure 1.15 shows the characteristic impedance Z_0 of a microstrip line with zero strip thickness as a function of the normalized strip width W/h for various ε_r. Conductor loss is a result of several factors related to the metallic material composing the ground plane and walls, among which are conductivity, skin effect, and surface ruggedness. For most microstrip lines (except some kinds of semiconductor substrate such as silicon), the conductor loss is much more significant than the dielectric loss. The conductor losses increase with increasing characteristic impedance due to greater resistance of narrow strips.

1.6.3   Coplanar Waveguide

A coplanar waveguide (CPW) is similar in structure to a slotline, the only difference being a third conductor centered in the slot region. The center strip conductor and two outer grounded conductors lie in the same plane on the substrate surface, as shown in Figure 1.16 [22, 23]. A coplanar configuration has some advantages such

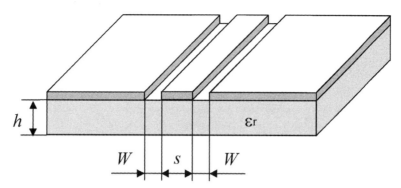

Figure 1.16 Coplanar waveguide structure.

as low dispersion, ease of attaching shunt and series circuit components, no need for via holes, and simple realization of short-circuited ends, which makes a coplanar waveguide very suitable for hybrid and monolithic integrated circuits. In contrast to the microstrip and stripline, the coplanar waveguide has shielding between adjacent lines that creates better isolation between them.

However, like microstrip and striplines, the coplanar waveguide can be also described by a quasi-TEM approximation for both numerical and analytical calculations. Because of the high dielectric constant of the substrate, most of the RF energy is stored in the dielectric, and the loading effect of the grounded cover is negligible if it is more than two slot widths away from the surface. Similarly, the thickness of the dielectric substrate with higher relative dielectric constants is not so critical, and practically it should be one or two times the width W of the slots. Figure 1.17 shows the characteristic impedance Z_0 of a coplanar waveguide as a function of the parameter $s/(s + 2W)$ for various ε_r according to the approximate expression of the characteristic impedance Z_0 for zero metal thickness, which is satisfactorily accurate for a wide range of substrate thicknesses, given in [24].

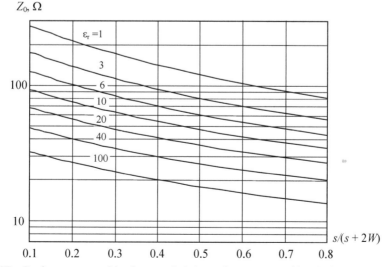

Figure 1.17 Coplanar waveguide characteristic impedance versus $s/(s + 2W)$.

1.7   Noise Figure

Electrical circuits have several primary noise sources. *Thermal* or *white noise* is created by random motion of charge carriers due to thermal excitation, being always found in any conducting medium whose temperature is above absolute zero whatever the nature of the conduction process or the nature of the mobile charge carriers [25]. This random motion of carriers creates a fluctuating voltage on the terminals of each resistive element that increases with temperature. However, if the average value of such a voltage is zero, then the noise power on its terminal is not zero being proportional to the resistance of the conductor and to its absolute temperature. The resistor as a thermal noise source can be represented by either of the noise sources shown in Figure 1.18.

The noise voltage source and noise current source can be described by Nyquist equations through their mean-square noise voltage and noise current values, respectively, as

$$\overline{e_n^2} = 4kTR\Delta f \tag{1.113}$$

$$\overline{i_n^2} = \frac{4kT\Delta f}{R} \tag{1.114}$$

where $k = 1.38 \times 10^{-23}$ J/K is the Boltzmann constant, T is the absolute temperature, and $kT = 4 \times 10^{-21}$ W/Hz $= -174$ dBm/Hz at ambient temperature $T = 290$ K. The thermal noise is proportional to the frequency bandwidth Δf, and it can be represented by the voltage source in series with resistor R, or by the current source in parallel to the resistor R. The maximum noise power can be delivered to the load when $R = R_L$, where R_L is the load resistance, being equal to $kT\Delta f$. Hence, the noise power density when the noise power is normalized by Δf is independent of frequency and is considered to be white noise. The rms noise voltage and current are proportional to the square root of the frequency bandwidth Δf.

Shot noise is associated with the carrier injection through the device *p-n* junction, being generated by the movement of individual electrons within the current flow. In each forward-biased junction, there is a potential barrier that can be overcome by the carriers with higher thermal energy. Such a process is random and mean-square noise current can be given by

$$\overline{i_n^2} = 2qI\Delta f \tag{1.115}$$

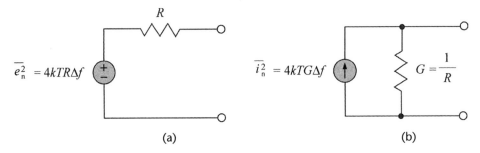

$\overline{e_n^2} = 4kTR\Delta f$ (a)

$\overline{i_n^2} = 4kTG\Delta f$ $G = \dfrac{1}{R}$ (b)

Figure 1.18   Equivalent circuits to represent thermal noise sources.

where q is the electron charge and I is the direct current flowing through the p-n junction. The shot noise depends on the thermal energy of the carriers near the potential barrier and its power density is independent of frequency. It has essentially a flat spectral distribution and can be treated as the thermal or white type of noise with current source $\overline{i_n^2}$ connected in parallel to the small-signal junction resistance. In a voltage noise representation, when the noise voltage source is connected in series with such a resistor, it can be written as

$$\overline{e_n^2} = 2kTr\Delta f \tag{1.116}$$

where $r = kT/qI$ is the junction resistance.

It is well known that any linear noisy two-port network can be represented as a noise-free two-port part with noise sources at the input and the output connected in a different way [26, 27]. For example, the noisy linear two-port network with internal noise sources shown in Figure 1.19(a) can be redrawn, either in the impedance form with external series voltage noise sources [Figure 1.19(b)] or in the admittance form with external parallel current noise sources [Figure 1.19(c)].

However, to fully describe the noise properties of the two-port network at fixed frequency, sometimes it is convenient to represent it through the noise-free two-port part and the noise sources equivalently located at the input. Such a circuit is equivalent to the configurations with noise sources located at the input and the output [28]. In this case, it is enough to use four parameters: the noise spectral densities of both noise sources and the real and imaginary parts of its correlation spectral density. These four parameters can be defined by measurements at the two-port network terminals. The two-port network current and voltage amplitudes are related to each other through a system of two linear algebraic equations. By taking into account the noise sources at the input and the output, these equations in the impedance and admittance forms can be respectively written as

$$V_1 = Z_{11}I_1 + Z_{12}I_2 - V_{n1} \tag{1.117}$$

$$V_2 = Z_{21}I_1 + Z_{22}I_2 - V_{n2} \tag{1.118}$$

and

$$I_1 = Y_{11}V_1 + Y_{12}V_2 - I_{n1} \tag{1.119}$$

$$I_2 = Y_{21}V_1 + Y_{22}V_2 - I_{n2} \tag{1.120}$$

where the voltage and current noise amplitudes represent the Fourier transforms of noise fluctuations.

The equivalent two-port network with voltage and current noise sources located at its input is shown in Figure 1.20(a), where [Y] is the two-port network admittance matrix, and the ratios between current and voltage amplitudes can be written as

$$I_1 = Y_{11}(V_1 + V_{ni}) + Y_{12}V_2 - I_{ni} \tag{1.121}$$

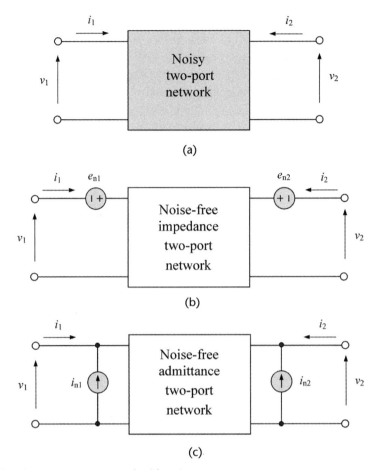

Figure 1.19 Linear two-port network with noise sources.

$$I_2 = Y_{21}\left(V_1 + V_{ni}\right) + Y_{22}V_2 \tag{1.122}$$

From comparison of (1.119) and (1.120) with (1.121) and (1.122), it follows that

$$V_{ni} = -\frac{I_{n2}}{Y_{21}} \tag{1.123}$$

$$I_{ni} = I_{n1} - \frac{Y_{11}}{Y_{21}}I_{n2} \tag{1.124}$$

representing the relationships between the current noise sources at the input and the output corresponding to the circuit shown in Figure 1.19(c) and the voltage and current noise sources at the input only corresponding to the circuit shown in Figure 1.20(a). In this case, (1.123) and (1.124) are valid only if $Y_{21} \neq 0$, which always takes place in practice. Similar equations can be written for the circuit with the series noise voltage source followed by a parallel noise current source shown in Figure 1.20(b) in terms of impedance Z-parameters to represent the relationships between the voltage noise sources at the input and the output corresponding to the circuit shown in Figure 1.19(b). The use of voltage and current noise sources at the input enables the combination of all internal two-port network noise sources.

To evaluate the quality of the two-port network, it is important to know the amount of noise added to a signal passing through it. Usually this can be done by introducing an important parameter such as *noise figure* or *noise factor*. The noise figure of the two-port network is intended to be an indication of its noisiness. The lower the noise figure, the less noise that is contributed by the two-port network. The noise figure is defined as

$$F = \frac{S_{in}/N_{in}}{S_{out}/N_{out}} \tag{1.125}$$

where S_{in}/N_{in} is the signal-to-noise ratio available at the input and S_{out}/N_{out} is the signal-to-noise ratio available at the output.

For a two-port network characterized by the available power gain G_A, the noise figure can be rewritten as

$$F = \frac{S_{in}/N_{in}}{G_A S_{in}/G_A(N_{in} + N_{add})} = 1 + \frac{N_{add}}{N_{in}} \tag{1.126}$$

where N_{add} is the additional noise power added by the two-port network referred to the input. From (1.126) it follows that the noise figure depends on the source impedance Z_S shown in Figure 1.21(a), but not on the circuit connected to the output of the two-port network.

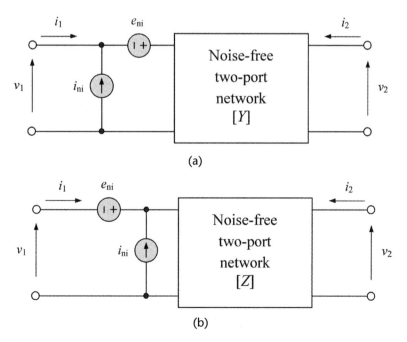

(a)

(b)

Figure 1.20   Linear two-port network with noise sources at input.

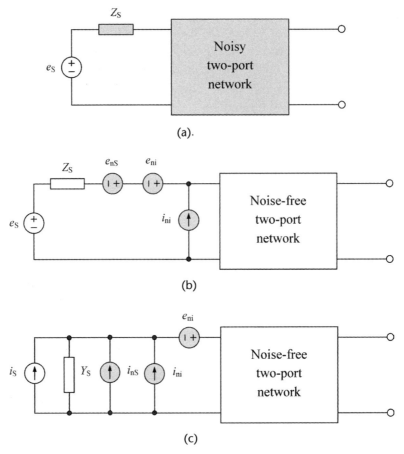

Figure 1.21 Linear two-port networks used to calculate noise figures.

Hence, if the two-port network is driven from the source with impedance $Z_S = R_S + jX_S$, the noise figure F of this two-port network in terms of the model shown in Figure 1.21(b) with input voltage and current noise sources and a noise-free two-port network can be obtained by

$$F = 1 + \frac{\overline{|e_n + Z_S i_n|^2}}{4kTR_S \Delta f}$$

$$= 1 + \frac{R_n + |Z_S|^2 G_n + 2\sqrt{R_n G_n}\, \mathrm{Re}(CZ_S)}{R_S} \tag{1.127}$$

where

$$R_n = \frac{\overline{e_n^2}}{4kT\Delta f} \tag{1.128}$$

is the equivalent input-referred noise resistance corresponding to the noise voltage source, where $\overline{e_n^2} = \overline{e_{nS}^2} + \overline{e_{ni}^2}$;

$$G_n = \frac{\overline{i_n^2}}{4kT\Delta f} \tag{1.129}$$

is the equivalent input-referred noise conductance corresponding to the noise current source, where $\overline{i_n^2} = \overline{i_{ni}^2}$; and

$$C = \frac{\overline{i_n e_n^*}}{\sqrt{\overline{i_n^2} \, \overline{e_n^2}}} \tag{1.130}$$

is the correlation coefficient representing a complex number less than or equal to unity in magnitude [27]. Here, G_n and R_n generally do not represent the particular circuit immittances but depend on the bias level resulting in a dependence of the noise figure on the operating bias point of the active device.

As the source impedance Z_S is varied over all values with positive R_S, the noise figure F has a minimum value of

$$F_{min} = 1 + 2\sqrt{R_n G_n}\left[\sqrt{1 - (\operatorname{Im} C)^2} + \operatorname{Re} C\right] \tag{1.131}$$

which occurs for the optimum source impedance $Z_{Sopt} = R_{Sopt} + jX_{Sopt}$ given by

$$\left|Z_{Sopt}\right|^2 = \frac{R_n}{G_n} \tag{1.132}$$

$$X_{Sopt} = \sqrt{\frac{R_n}{G_n}} \operatorname{Im} C \tag{1.133}$$

As a result, the noise figure F for the input impedance Z_S, which is not optimum, can be expressed in terms of F_{min} as

$$\begin{aligned}
F &= F_{min} + \left|Z_S - Z_{Sopt}\right|^2 \frac{G_n}{R_S} \\
&= F_{min} + \left[\left(R_S - R_{Sopt}\right)^2 + \left(X_S - X_{Sopt}\right)^2\right]\frac{G_n}{R_S}
\end{aligned} \tag{1.134}$$

Similarly, the noise figure F can be equivalently expressed using a model shown in Figure 1.21(c) with source admittance $Y_S = G_S + jB_S$ as

$$\begin{aligned}
F &= F_{min} + \left|Y_S - Y_{Sopt}\right|^2 \frac{R_n}{G_S} \\
&= F_{min} + \left[\left(G_S - G_{Sopt}\right)^2 + \left(B_S - B_{Sopt}\right)^2\right]\frac{R_n}{G_S}
\end{aligned} \tag{1.135}$$

where F_{min} is the minimum noise figure of the two-port network, which can be realized with respect to the source admittance Y_S; $Y_{Sopt} = G_{Sopt} + jB_{Sopt}$ is the optimal source admittance; and R_n is the equivalent noise resistance, which measures how rapidly the noise figure degrades when the source admittance Y_S deviates from its optimum value Y_{Sopt} [29]. Since the admittance Y_S is generally complex, then its real and imaginary parts can be controlled independently. To obtain the minimum value of the noise figure, the two matching conditions of $G_S = G_{Sopt}$ and $B_S = B_{Sopt}$ must be satisfied.

In a multistage transmitter system, the input signal travels through a cascade of many different components, each of which may degrade the signal-to-noise ratio to some degree. For a cascade of two stages having available gains G_{A1} and G_{A2} and noise figures F_1 and F_2, using (1.126) results in the output-to-input noise power ratio N_{out}/N_{in} written as

$$\frac{N_{out}}{N_{in}} = G_{A2}\left[G_{A1}\left(1 + \frac{N_{add1}}{N_{in}}\right) + \frac{N_{add2}}{N_{in}}\right] = G_{A1}G_{A2}\left(F_1 + \frac{F_2 - 1}{G_{A1}}\right) \quad (1.136)$$

where N_{add1} and N_{add2} are the additional noise powers added by the first and second stages, respectively. Consequently, an overall noise figure $F_{1,2}$ for a two-stage system based on (1.125) can be given by

$$F_{1,2} = F_1 + \frac{1}{G_{A1}}(F_2 - 1) \quad (1.137)$$

Equation (1.137) can be generalized to a multistage transmitter system with n stages as

$$F_{1,n} = F_1 + \frac{F_2 - 1}{G_{A1}} + \cdots + \frac{F_n - 1}{G_{A1}G_{A2}\cdots G_{A(n-1)}} \quad (1.138)$$

which means that the noise figure of the first stage has the predominant effect on the overall noise figure, unless G_{A1} is small or F_2 is large [30].

References

[1]   L. O. Chua, C. A. Desoer, and E. S. Kuh, *Linear and Nonlinear Circuits*, New York: McGraw-Hill, 1987.

[2]   D. M. Pozar, *Microwave Engineering*, New York: John Wiley & Sons, 2004.

[3]   D. A. Frickey, "Conversions Between S, Z, Y, h, ABCD, and T Parameters Which Are Valid for Complex Source and Load Impedances," *IEEE Trans. Microwave Theory Tech.*, Vol. MTT-42, pp. 205–211, Feb. 1994.

[4]   E. F. Terman, *Radio Engineer's Handbook*, New York: McGraw-Hill, 1945.

[5]   I. J. Bahl, *Lumped Elements for RF and Microwave Circuits*, Norwood, MA: Artech House, 2003.

[6]   H. A. Wheeler, "Simple Inductance Formulas for Radio Coils," *Proc. IRE*, Vol. 16, pp. 1398–1400, Oct. 1928.

[7]   S. S. Mohan et al., "Simple Accurate Expressions for Planar Spiral Inductances," *IEEE J. Solid-State Circuits*, Vol. SC-34, pp. 1419–1424, Oct. 1999.

[8]   J. M. Lopez-Villegas et al., "Improvement of the Quality Factor of RF Integrated Inductors by Layout Optimization," *IEEE Trans. Microwave Theory Tech.*, Vol. MTT-48, pp. 76–83, Jan. 2000.

[9]   J. R. Long and M. A. Copeland, "The Modeling, Characterization, and Design of Monolithic Inductors for Silicon RF ICs," *IEEE J. Solid-State Circuits*, Vol. SC-32, pp. 357–369, Mar. 1997.

[10] N. A. Talwalkar, C. P. Yue, and S. S. Wong, "Analysis and Synthesis of On-Chip Spiral Inductors," *IEEE Trans. Electron Devices*, Vol. ED-52, pp. 176–182, Feb. 2005.

[11] N. M. Nguyen and R. G. Meyer, "Si IC-Compatible Inductors and *LC* Passive Filters," *IEEE J. Solid-State Circuits*, Vol. SC-25, pp. 1028–1031, Aug. 1990.

[12] J. Craninckx and M. S. J. Steyaert, "A 1.8-GHz Low-Phase-Noise CMOS VCO Using Optimized Hollow Spiral Inductors," *IEEE J. Solid-State Circuits*, Vol. SC-32, pp. 736–744, May 1997.

[13] S. L. March, "Simple Equations Characterize Bond Wires," *Microwaves & RF*, Vol. 30, pp. 105–110, Nov. 1991.

[14] P. Benedek and P. Silvester, "Equivalent Capacitances of Microstrip Gaps and Steps," *IEEE Trans. Microwave Theory Tech.*, Vol. MTT-20, pp. 729–733, Nov. 1972.

[15] G. D. Alley, "Interdigital Capacitors and Their Applications in Lumped Element Microwave Integrated Circuits," *IEEE Trans. Microwave Theory Tech.*, Vol. MTT-18, pp. 1028–1033, Dec. 1970.

[16] S. A. Schelkunoff, "The Impedance Concept and Its Application to Problems of Reflection, Refraction, Shielding and Power Absorption," *Bell Syst. Tech. J.*, Vol. 17, pp. 17–48, Jan. 1938.

[17] R. M. Barrett, "Microwave Printed Circuits—The Early Years," *IEEE Trans. Microwave Theory Tech.*, Vol. MTT-32, pp. 983–990, Sep. 1984.

[18] D. D. Grieg and H. F. Engelmann, "Microstrip—A New Transmission Technique for the Kilomegacycle Range," *Proc. IRE*, Vol. 40, pp. 1644–1650, Dec. 1952.

[19] E. O. Hammerstad, "Equations for Microstrip Circuit Design," *Proc. 5th Europ. Microwave Conf.*, 1975, pp. 268–272.

[20] I. J. Bahl and R. Garg, "Simple and Accurate Formulas for Microstrip with Finite Strip Thickness," *Proc. IEEE*, Vol. 65, pp. 1611–1612, Nov. 1977.

[21] R. S. Carson, *High-Frequency Amplifiers*. New York: John Wiley & Sons, 1975.

[22] C. P. Weng, "Coplanar Waveguide: A Surface Strip Transmission Line Suitable for Nonreciprocal Gyromagnetic Device Applications," *IEEE Trans. Microwave Theory Tech.*, Vol. MTT-17, pp. 1087–1090, Dec. 1969.

[23] R. N. Simons, *Coplanar Waveguide Circuits, Components, and Systems*, New York: John Wiley & Sons, 2001.

[24] G. Ghione and C. Naldli, "Analytical Formulas for Coplanar Lines in Hybrid and Monolithic," *Electronics Lett.*, Vol. 20, pp. 179–181, Feb. 1984.

[25] A. van der Ziel, *Noise*, Englewood Cliffs: Prentice-Hall, 1954.

[26] H. C. Montgomery, "Transistor Noise in Circuit Applications," *Proc. IRE*, Vol. 40, pp. 1461–1471, 1952.

[27] H. Rothe and W. Dahlke, "Theory of Noise Fourpoles," *Proc. IRE*, Vol. 44, pp. 811–818, June 1956.

[28] A. G. T. Becking, H. Groendijk, and K. S. Knol, "The Noise Factor of Four-Terminal Networks," *Philips Res. Rep.*, Vol. 10, pp. 349–357, 1955.

[29] IRE Subcommittee 7.9 on Noise (H. A. Haus, Chairman), "Representation of Noise in Linear Twoports," *Proc. IRE*, Vol. 48, pp. 69–74, Jan. 1960.

[30] H. T. Friis, "Noise Figures of Radio Receivers," *Proc. IRE*, Vol. 32, pp. 419–422, July 1944.

CHAPTER 2

Power Amplifier Design Fundamentals

Power amplifier design requires accurate active device modeling, effective impedance matching (depending on the technical requirements and operating conditions), stability during operation, and ease of practical implementation. The quality of a power amplifier design is evaluated by its ability to achieve maximum power gain across the required frequency bandwidth under stable operating conditions with minimum amplifier stages, and the requirements for linearity or high efficiency can be considered where it is needed. For stable operation, it is necessary to evaluate the operating frequency domains in which the active device may be potentially unstable.

2.1   Main Characteristics

Power amplifier design aims for maximum power gain and efficiency for a given value of output power with a predictable degree of stability. To extract the maximum power from a source generator, it is a well-known fact that the external load should have a vector value that is a conjugate of the internal impedance of the source [1]. The power delivered from a generator to a load when matched on this basis is called the available power of the generator [2]. In this case, the power gain of a four-terminal network is defined as the ratio of the power delivered to the load connected at the output terminals to the power available from the generator connected to the input terminals, usually measured in decibels, and this ratio is called the *power gain* irrespective of whether it is greater than or less than 1 [3, 4].

Figure 2.1 shows the basic block schematic for a single-stage power amplifier circuit, which includes an active device, an input matching circuit to match with the source impedance, and an output matching circuit to match with the load impedance. Generally, the two-port active device is characterized by a system of the immittance W-parameters, that is, any system of the impedance Z-parameters or admittance Y-parameters [5, 6]. The input and output matching circuits transform the source and load immittances W_S and W_L into specified values between ports

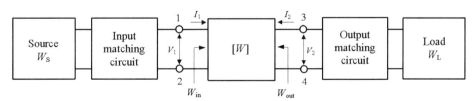

Figure 2.1   Block schematic of single-stage power amplifier.

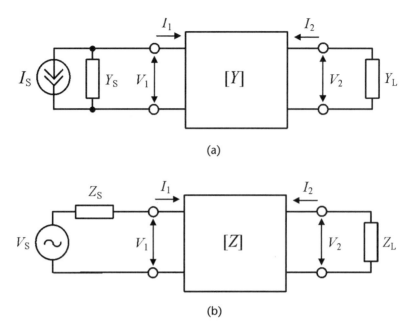

Figure 2.2 Two-port loaded amplifier networks.

1–2 and 3–4, respectively, by means of which the optimal design operation mode of the power amplifier is realized.

The given technical requirements and the convenience of the design realization (using the appropriate types of the active devices and circuit schematics) determine the choice of system for the admittance Y-parameters or impedance Z-parameters. For the given input and output voltages, let the active device be characterized by a matrix of Y-parameters. In this case, the source of the input signal is represented by the current source with an internal admittance Y_S, and a load is characterized by the load admittance Y_L, as shown in Figure 2.2(a). If a two-port active device is described by a system of Z-parameters, the source of the input signal is represented by the voltage source with an internal impedance Z_S, whereas the load is characterized by the load impedance Z_L, as shown in Figure 2.2(b). In both cases, the admittances Y_S and Y_L and the impedances Z_S and Z_L are seen looking toward the source and load through the input and output matching circuits.

To calculate the electrical characteristics of a power amplifier, first consider a system of the admittance Y-parameters. The active device in this case is described by the following system of two equations:

$$I_1 = Y_{11}V_1 + Y_{12}V_2$$
$$I_2 = Y_{21}V_1 + Y_{22}V_2 \tag{2.1}$$

Depending on impedance matching at the input and output device ports, several definitions of the amplifier power gain in terms of Y-parameters can be derived:

- *Operating power gain* ($G_P = P_L/P_{in}$) is the ratio of power dissipated in the active load G_L to power delivered to the input port of the active device with admittance Y_{in}. This gain is independent of G_S but is strongly dependent on G_L.

- *Available power gain* ($G_A = P_{out}/P_S$) is the ratio of power available at the output port of the active device with admittance Y_{out} to power available from the source G_S. This power gain depends on G_S but not G_L.
- *Transducer power gain* ($G_T = P_L/P_S$) is the ratio of power dissipated in the active load G_L to power available from the source G_S. This power gain assumes complex-conjugate impedance matching at the input and output ports of the active device being dependent on both G_S and G_L.
- *Maximum available gain* (*MAG*) is the theoretical power gain of the active device when its reverse transfer admittance Y_{12} is set equal to zero. It represents a theoretical limit on the gain that can be achieved with the given active device, assuming complex-conjugate impedance matching of the input and output ports of the active device with the source and load, respectively.

In practice, to characterize an amplifier circuit gain property, generally two types of power gain are used: operating power gain G_P and transducer power gain G_T. The operating power gain G_P is used when we need to calculate the power at the input port of the device, which is necessary to provide the given power delivered to the load. However, to analyze the stability conditions, it is important to know both the value of the source impedance and the value of the load impedance. Therefore, in this case it is preferable to use the transducer power gain G_T, which must be maximized within the restrictions imposed by the stability consideration.

First, consider the evaluation of G_P in terms of the two-port network Y-parameters. If V_1 is the amplitude at the input port of the active device, then

$$P_{in} = 0.5V_1^2 \operatorname{Re} Y_{in} \tag{2.2}$$

where $Y_{in} = I_1/V_1$ is the input admittance (between input ports 1–2) of a two-port network loaded on the admittance Y_L. Given that $I_2 = -Y_L V_2$, the expression for Y_{in} is defined from (2.1) as

$$Y_{in} = Y_{11} - \frac{Y_{12}Y_{21}}{Y_{22} + Y_L} \tag{2.3}$$

The output power dissipated in a load is obtained by

$$P_L = 0.5V_2^2 \operatorname{Re} Y_L \tag{2.4}$$

As a result, the operating power gain G_P can be written as

$$G_P = \frac{P_L}{P_{in}} = \frac{|Y_{21}|^2 \operatorname{Re} Y_L}{|Y_{22} + Y_L|^2 \operatorname{Re} Y_{in}} \tag{2.5}$$

The operating power gain G_P does not depend on the source parameters and characterizes only the effectiveness of the power delivery from the input port of the active device to the load. This gain helps to evaluate the gain property of a multistage amplifier when the overall operating power gain $G_{P(total)}$ is equal to the product of each stage G_P.

The transducer power gain G_T includes an assumption of the complex-conjugate matching of the load and the source. This means that $Y_{in} = Y_S^*$, where Y_S^* is the source admittance conjugately matched to the input port of the active device Y_{in}.

If $I_S = Y_S V_1 + I_1$, then the expression for the source current I_S using (2.1) can be defined by

$$I_S = \frac{\left(Y_{11} + Y_S\right)\left(Y_{22} + Y_L\right) - Y_{12}Y_{21}}{Y_{22} + Y_L} V_1 \tag{2.6}$$

From (2.4) and (2.6) it follows that the transducer power gain G_T can be written as

$$G_T = \frac{P_L}{P_S} = \frac{4\left|Y_{21}\right|^2 \operatorname{Re} Y_S \operatorname{Re} Y_L}{\left|\left(Y_{11} + Y_S\right)\left(Y_{22} + Y_L\right) - Y_{12}Y_{21}\right|^2} \tag{2.7}$$

Similarly, the operating and transducer power gains G_P and G_T and the input and output impedances Z_{in} and Z_{out} can be expressed through the same analytical forms of a system of Z-parameters. Thus, by using the immittance W-parameters, they can be generalized as

$$W_{in} = W_{11} - \frac{W_{12}W_{21}}{W_{22} + W_L} \tag{2.8}$$

$$W_{out} = W_{22} - \frac{W_{12}W_{21}}{W_{11} + W_S} \tag{2.9}$$

$$G_P = \frac{\left|W_{21}\right|^2 \operatorname{Re} W_L}{\left|W_{22} + W_L\right|^2 \operatorname{Re} W_{in}} \tag{2.10}$$

$$G_T = \frac{4\left|W_{21}\right|^2 \operatorname{Re} W_S \operatorname{Re} W_L}{\left|\left(W_{11} + W_S\right)\left(W_{22} + W_L\right) - W_{12}W_{21}\right|^2} \tag{2.11}$$

where W_{ij} $(i, j = 1, 2)$ are the immittance two-port parameters of the active device equivalent circuit.

From (2.8) and (2.9) it follows that if $W_{12} = 0$, then $W_{in} = W_{11}$ and $W_{out} = W_{22}$. As a result, in the case of the complex-conjugate matching at the input and output ports of the active device, the expression for MAG is obtained from (2.10) or (2.11) as

$$MAG = \frac{\left|W_{21}\right|^2}{4 \operatorname{Re} W_{11} \operatorname{Re} W_{22}} \tag{2.12}$$

the magnitude of which depends only on the active device immittance parameters.

The bipolar transistor, simplified, small-signal π-hybrid equivalent circuit shown in Figure 2.3 provides an example for a conjugately matched bipolar power amplifier. The impedance Z-parameters of the equivalent circuit of the bipolar transistor in a common-emitter configuration can be written as

$$Z_{11} = r_b + \frac{1}{g_m + j\omega C_\pi} \qquad Z_{12} = \frac{1}{g_m + j\omega C_\pi}$$

$$Z_{21} = -\frac{1}{j\omega C_c}\frac{g_m - j\omega C_c}{g_m + j\omega C_\pi} \qquad Z_{22} = \left(1 + \frac{C_\pi}{C_c}\right)\frac{1}{g_m + j\omega C_\pi} \tag{2.13}$$

where

g_m = device transconductance

r_b = series base resistance

C_π = base-emitter capacitance including both diffusion and junction components

C_c = feedback collector capacitance.

By setting the device feedback impedance Z_{12} to zero and complex-conjugate matching the conditions at the input as $R_S = \mathrm{Re}Z_{in}$ and $L_{in} = -\mathrm{Im}Z_{in}/\omega$ and at the output as $R_L = \mathrm{Re}Z_{out}$ and $L_{out} = -\mathrm{Im}Z_{out}/\omega$, the small-signal transducer power gain G_T can be calculated by

$$G_T = \left(\frac{f_T}{f}\right)^2 \frac{1}{8\pi f_T r_b C_c} \tag{2.14}$$

where $f_T = g_m/2\pi C\pi$ is the device transition frequency. Equation (2.14) gives a well-known expression for maximum operating frequency of the bipolar junction transistor (BJT) device when the maximum available gain is equal to unity:

$$f_{max} = \sqrt{\frac{f_T}{8\pi r_b C_c}} \tag{2.15}$$

Figure 2.4 shows the simplified circuit schematic for a conjugately matched field-effect transistor (FET) power amplifier. The admittance Y-parameters of the

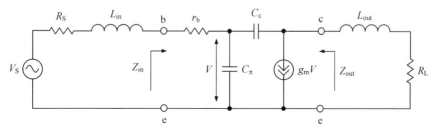

Figure 2.3 Simplified equivalent circuit of matched bipolar power amplifier.

small-signal equivalent circuit of any FET device in a common-source configuration can be written as

$$Y_{11} = \frac{j\omega C_{gs}}{1 + j\omega C_{gs}R_{gs}} + j\omega C_{gd} \qquad Y_{12} = -j\omega C_{gd}$$

$$Y_{21} = \frac{g_m}{1 + j\omega C_{gs}R_{gs}} - j\omega C_{gd} \qquad Y_{22} = \frac{1}{R_{ds}} + j\omega\left(C_{ds} + C_{gd}\right) \qquad (2.16)$$

where

g_m = device transconductance

R_{gs} = gate-source resistance

C_{gs} = gate-source capacitance

C_{gd} = feedback gate-drain capacitance

C_{ds} = drain-source capacitance

R_{ds} = differential drain-source resistance.

Since the value of the gate-drain capacitance C_{gd} is usually relatively small, the effect of feedback admittance Y_{12} can be neglected in a simplified case. Then, it is necessary to set $R_S = R_{gs}$ and $L_{in} = 1/\omega^2 C_{gs}$ for input matching, while $R_L = R_{ds}$ and $L_{out} = 1/\omega^2 C_{ds}$ for output matching. Hence, the small-signal transducer power gain G_T can approximately be calculated by

$$G_T\left(C_{gd} = 0\right) = MAG = \left(\frac{f_T}{f}\right)^2 \frac{R_{ds}}{4R_{gs}} \qquad (2.17)$$

where $f_T = g_m/2\pi C_{gs}$ is the device transition frequency and MAG is the maximum available gain representing a theoretical limit for the power gain that can be achieved under complex-conjugate matching conditions. Equation (2.17) gives a well-known expression for maximum operating frequency of the FET device when the maximum available gain is equal to unity:

$$f_{max} = \frac{f_T}{2}\sqrt{\frac{R_{ds}}{R_{gs}}} \qquad (2.18)$$

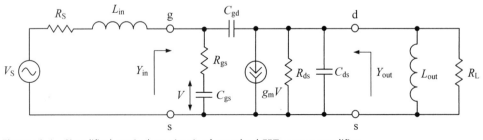

Figure 2.4   Simplified equivalent circuit of matched FET power amplifier.

From (2.14) and (2.17) it follows that the small-signal power gain of a conjugately matched power amplifier for any type of active device drops off as $1/f^2$ or 6 dB per octave. Therefore, if a power gain G_T is known at the transition frequency f_T, then $G_T(f)$ can readily be predicted at a certain frequency f by

$$G_T(f) = G_T(f_T)\left(\frac{f_T}{f}\right)^2 \tag{2.19}$$

Note that the previous analysis is based on the linear small-signal consideration when the nonlinear device current source as a function of both input and output voltages can be generally characterized by the linear transconductance g_m as a function of the input voltage, and the output differential resistance R_{ds} is a function of the output voltage. This is a result of a Taylor series expansion of the output current as a function of the input and output voltages with maintaining only the dc and linear components. Such an approach helps to explain and derive the maximum achievable power amplifier parameters in a linear approximation. In this case, the active device is operated in a Class A mode when one-half of the dc power is dissipated in the device, while the other half is transformed to the fundamental-frequency output power flowing into the load, resulting in a maximum ideal collector efficiency of 50%. The device output resistance R_{out} remains constant and can be calculated as a ratio of the dc supply voltage to the dc current flowing through the active device. In a nonlinear case, for a complex-conjugate matching procedure, the device output immittance under large-signal consideration should be calculated using a Fourier series analysis of the output current and voltage fundamental components. This means that, unlike a linear Class A mode, the active device is operated in a device linear region for only part of the entire period, and its output resistance is defined as a ratio of the fundamental-frequency output voltage to the fundamental-frequency output current. This is not a physical resistance resulting in a power loss inside the device, but an equivalent resistance that must be used in conjugate matching procedures. In this case, the complex-conjugate matching is valid and necessary, first, to compensate for the reactive part of the device output impedance and, second, to provide a proper load resistance resulting in a maximum power gain for a given supply voltage and required output power delivered to the load. Note that this is not the maximum available small-signal power gain that can be achieved in a linear operation mode, but a maximum achievable large-signal power gain that can be achieved for a particular operation mode with a certain conduction angle. Of course, the maximum large-signal power gain is smaller than the small-signal power gain for the same input power, since the output power in a nonlinear operation mode also includes the powers at the harmonic components of the fundamental frequency and the large-signal transconductance is lower than the small-signal one.

Therefore, it makes more practical sense not to introduce separately the concepts of the gain match with respect to the linear power amplifiers and the power match in nonlinear power amplifiers since the maximum large-signal power gain as a function of the conduction angle corresponds to the maximum fundamental-frequency output power delivered to the load due to large-signal conjugate output matching. It is very important to provide conjugate matching at both input and output device ports to achieve maximum power gain in a large-signal mode. In Class A mode, the maximum small-signal power gain ideally remains constant regardless of the output power level.

The transistor characterization in a large-signal mode can be done based on an equivalent quasi-harmonic nonlinear approximation under the condition of sinusoidal port voltages [7]. In this case, the large-signal impedances are generally determined in the following manner. The designer tunes the load network (often by trial and error) to maximize the output power to the required level using a particular transistor at a specified frequency and supply voltage. Then, the transistor is removed from the circuit and the impedance seen by the collector is measured at the carrier frequency. The complex conjugate of the measured impedance then represents the equivalent large-signal output impedance of the transistor at that frequency, supply voltage, and output power. A similar design process is used to measure the input impedance of the transistor and to maximize the power-added efficiency of the power amplifier. It is especially important with power transistors in the microwave region either to characterize the transistors in the packaging configuration in which they will be utilized in the circuit application, or to accurately know the package parameters so that suitable corrections can be made if the transistors are mounted in a different environment [8].

2.2   Impedance Matching

Impedance matching is necessary to provide maximum delivery to the load of the RF power available from the source. This means that, when the electrical signal propagates in the circuit, a portion of this signal will be reflected at the interface between the sections with different impedances. Therefore, it is necessary to establish the conditions that allow the entire electrical signal to be transmitted without any reflection. To determine the optimum value for load impedance Z_L at which the power delivered to the load is maximized, consider the equivalent circuit shown in Figure 2.5(a).

The power delivered to the load can be defined as

$$P = \frac{1}{2}V_{in}^2 \operatorname{Re}\left(\frac{1}{Z_L}\right) = \frac{1}{2}V_S^2 \left|\frac{Z_L}{Z_S + Z_L}\right|^2 \operatorname{Re}\left(\frac{1}{Z_L}\right) \qquad (2.20)$$

where

$Z_S = R_S + jX_S$ = source impedance

$Z_L = R_L + jX_L$ = load impedance

V_S = source voltage amplitude

V_{in} = load voltage amplitude.

Substituting the real and imaginary parts of the source and load impedances Z_S and Z_L into (2.20) yields

$$P = \frac{1}{2}V_S^2 \frac{R_L}{\left(R_S + R_L\right)^2 + \left(X_S + X_L\right)^2} \qquad (2.21)$$

Assume the source impedance Z_S is fixed and it is necessary to vary the real and imaginary parts of the load impedance Z_L until maximum power is delivered to the load. To maximize the output power, we must apply the following analytical conditions in the form of derivatives with respect to the output power:

$$\frac{\partial P}{\partial R_L} = 0 \qquad \frac{\partial P}{\partial X_L} = 0 \tag{2.22}$$

Then, applying these conditions to (2.21), the following system of two equations can be obtained with some simplifications:

$$R_S^2 - R_L^2 + \left(X_L + X_S\right)^2 = 0 \tag{2.23}$$

$$X_L(X_L + X_S) = 0 \tag{2.24}$$

By solving (2.23) and (2.24) simultaneously for R_S and X_S, we can obtain

$$R_S = R_L \qquad X_L = -X_S \tag{2.25}$$

or, in an impedance form,

$$Z_L = Z_S^* \tag{2.26}$$

where * denotes the complex-conjugate value [3, 9].

Equation (2.26) is called the *impedance conjugate matching condition*, and its fulfillment results in a maximum power delivered to the load for fixed source impedance. It should be noted that the term *impedance* was introduced by Oliver Lodge in 1889, and it referred to the ratio *V/I* in the special circuit comprised of a resistance and an inductance, where *I* and *V* are the amplitudes of an alternating current and the driving force that produced it [10].

The *admittance conjugate matching condition*, applied to the equivalent circuit shown in Figure 2.5(b), is written as

$$Y_L = Y_S^* \tag{2.27}$$

which can be readily obtained in the same way. Thus, the conjugate matching conditions in a common case can be determined through the immittance *W*-parameters, which represent any system of the impedance *Z*-parameters or admittance *Y*-parameters, in the form of

$$W_L = W_S^* \tag{2.28}$$

where W_S is the source immittance and W_L is the load immittance. The term *immittance* was introduced by Bode to refer to a complex number, which may be either the impedance or the admittance of a system [11].

For a single-stage power amplifier, the matching circuit is connected between the source and the input of an active device, as shown in Figure 2.6(a), and between

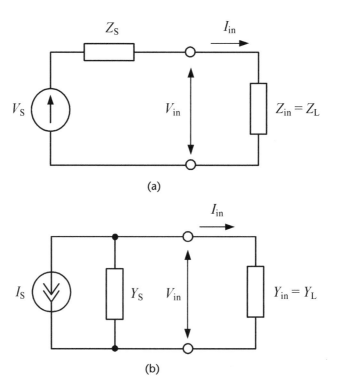

Figure 2.5   Equivalent circuits with (a) voltage and (b) current sources.

the output of an active device and the load, as shown in Figure 2.6(b). In the latter case, the main objective is to properly transform the load immittance W_L to the optimum device output immittance W_{out}, the value of which is properly determined by the supply voltage, the output power, the saturation voltage of the active device, and the selected class of the active device operation to maximize the operating efficiency and output power of the power amplifier. For a multistage power amplifier, the load represents an input circuit for the next stage. Therefore, the matching circuit is connected between the output of the active device of the preceding amplifier stage and the input of the active device of the succeeding stage of the power amplifier, as shown in Figure 2.6(c).

Note that (2.28) is given in a general immittance form without indication of whether it is used in a small-signal or large-signal application. In the latter case, this only means that the device immittance W-parameters are fundamentally averaged over large-signal swings across the device equivalent circuit parameters and that the conjugate matching principle is valid in both the small-signal application and the large-signal application where the optimum equivalent device output resistance (or conductance) at the fundamental frequency is matched to the load resistance (or conductance). In addition, the effect of the device output reactive elements is eliminated by the conjugate reactance of the load network. In addition, the matching circuits should be designed to realize the required voltage and current waveforms at the device output, to provide the stabile operating conditions and to satisfy the requirements for the power amplifier amplitude and phase characteristics. The losses in the matching circuits must be as small as possible to deliver the output power to

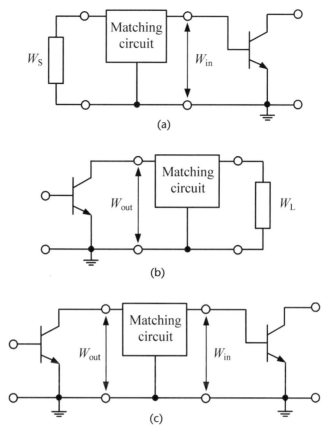

Figure 2.6   Matching circuit arrangements.

the load with maximum efficiency. Finally, it is desirable for the matching circuit to be easy to tune.

Generally, matching networks take advantage of various configurations that can be used to connect a generating system efficiently to its useful load. However, to obtain high transmission efficiency, these networks should be properly designed. The lumped matching circuits shown in Figure 2.7 in the form of an (a) L-transformer, (b) π-transformer, or (c) T-transformer have proved for a long time to be effective for power amplifier design [1]. The simplest and most popular matching network is the matching circuit in the form of the L-transformer. The transforming properties of this matching circuit can be analyzed by using the equivalent transformation of a parallel into a series representation of RX circuit.

Consider the parallel RX circuit shown in Figure 2.8(a), where R_1 is the real (resistive) part and X_1 is the imaginary (reactive) part of the circuit impedance $Z_1 = jX_1R_1/(R_1 + jX_1)$, and the series RX circuit shown in Figure 2.8(b), where R_2 is the resistive part and X_2 is the reactive part of the circuit impedance $Z_2 = R_2 + jX_2$. These two circuits, series and parallel, can be considered equivalent at some frequency if $Z_1 = Z_2$, resulting in

$$R_2 + jX_2 = \frac{R_1X_1^2}{R_1^2 + X_1^2} + j\frac{R_1^2X_1}{R_1^2 + X_1^2} \tag{2.29}$$

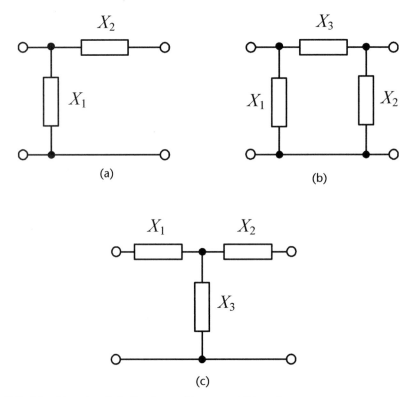

Figure 2.7 Matching circuits in the form of *L*-, *π*-, and *T*-transformers.

Equation (2.29) can be rearranged by two separate equations for real and imaginary parts as

$$R_1 = R_2(1 + Q^2) \tag{2.30}$$

$$X_1 = X_2(1 + Q^{-2}) \tag{2.31}$$

where $Q = R_1/|X_1| = |X_2|/R_2$ is the circuit quality factor, which is equal for both the series and parallel RX circuits.

Consequently, if the reactive impedance $X_1 = -X_2(1 + Q^{-2})$ is connected in parallel to the series circuit R_2X_2, it allows the reactive impedance (or reactance) of the series circuit to be compensated. In this case, the input impedance of such a

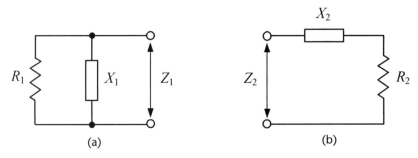

Figure 2.8 Impedance parallel and series equivalent circuits.

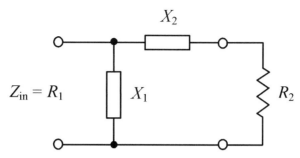

Figure 2.9 Input impedance of loaded two-port network.

two-port network shown in Figure 2.9 will be resistive only and equal to R_1. Consequently, to transform the resistance R_1 into the other resistance R_2 at the given frequency, it is sufficient to connect between them a two-port L-transformer with the opposite signs of the reactances X_1 and X_2, the parameters of which can be easily calculated from the following simple equations:

$$|X_1| = \frac{R_1}{Q} \tag{2.32}$$

$$|X_2| = R_2 Q \tag{2.33}$$

where

$$Q = \sqrt{\frac{R_1}{R_2} - 1} \tag{2.34}$$

is the circuit (loaded) quality factor expressed through the resistances to be matched. Thus, to design a matching circuit with fixed resistances to be matched, first we need to calculate the circuit quality factor Q according to (2.34) and then define the reactive elements according to (2.32) and (2.33).

Due to the opposite signs of the reactances X_1 and X_2, two possible circuit configurations (one in the form of a lowpass filter section and the other in the form of a highpass filter section) with the same transforming properties can be realized, as shown in Figure 2.10 together with the design equations.

The matching circuits in the form of a π-transformer [Figure 2.10(a)] and a T-transformer [Figure 2.10(a)] can be realized by appropriate connection of two L-transformers, as shown in Figure 2.11. For each L-transformer, the resistances R_1 and R_2 are transformed to some intermediate resistance R_0 with the value of $R_0 < (R_1, R_2)$ for a π-transformer and the value of $R_0 > (R_1, R_2)$ for a T-transformer. The value of R_0 is not fixed and can be chosen arbitrarily depending on the frequency bandwidth. This means that, compared to the simple L-transformer with fixed parameters for the same ratio of R_2/R_1, the parameters of a π-transformer or a T-transformer can be different. However, they provide narrower frequency bandwidths due to higher quality factors because the intermediate resistance R_0 is either greater or smaller than each of the resistances R_1 and R_2. By taking into account

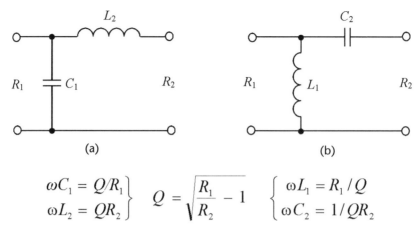

$$\left. \begin{array}{l} \omega C_1 = Q/R_1 \\ \omega L_2 = QR_2 \end{array} \right\} \qquad Q = \sqrt{\dfrac{R_1}{R_2} - 1} \qquad \left\{ \begin{array}{l} \omega L_1 = R_1/Q \\ \omega C_2 = 1/QR_2 \end{array} \right.$$

Figure 2.10 *L*-type matching circuits and relevant equations.

the two possible circuit configurations of the *L*-transformer shown in Figure 2.10, there is the possibility that different circuit configurations can be designed for the two-port impedance transformers shown in Figure 2.11(a), where $X_3 = X_3' + X_3''$, and in Figure 2.11(b), where $X_3 = X_3' X_3'' / (X_3' + X_3'')$.

Some of the most widely used two-port *LC*-type π- and *T*-transformers, together with the design formulas, are discussed in [12, 13]. The π-transformers are usually used as output matching circuits of high-power amplifiers in the Class B operation mode when it is necessary to achieve a sinusoidal drain (or collector) voltage waveform by appropriate harmonic suppression. For the π-transformer with shunt capacitors used as an interstage matching circuit, the input and output capacitances

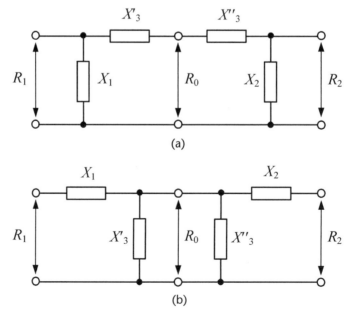

Figure 2.11 Matching circuits developed by connecting two *L*-transformers.

of the transistors can be easily included in the matching circuit elements. Besides, a π-transformer can be directly used as the load network for a high-efficiency Class E mode with proper calculation of its design parameters [14]. The T-transformers are usually used in high-power amplifiers as input, interstage, and output matching circuits, especially the matching network with series inductor and shunt and series capacitors. By using such a T-transformer for output matching of a power amplifier, it is easy to realize a high-efficiency Class F operating mode, because the series inductor connected to the drain (or collector) of the active device creates open-circuit harmonic impedance conditions [14].

Now let us demonstrate a lumped matching network technique for the design of a 150W MOSFET power amplifier with a supply voltage of 50V, operating in a frequency bandwidth of 132 to 174 MHz and providing a power gain greater than 10 dB. These requirements can be satisfied using a silicon n-channel enhancement-mode vertical-double diffused metal oxide semiconductor FET (VDMOSFET) device designed for power amplification in the VHF range. In this case, the center bandwidth frequency is equal to fc = $\sqrt{132 \times 174}$ = 152 MHz. For this frequency, the manufacturer states the following values of the input and output impedances: Z_{in} = $(0.9 - j1.2)\Omega$ and Z_{out} = $(1.8 + j2.1)\Omega$. Both Z_{in} and Z_{out} represent the series combination of an input or output resistance with a capacitive or inductive reactance, respectively. To cover the required frequency bandwidth, the low-Q matching circuits chosen should be such that they allow reduction of the in-band amplitude ripple and improvement of the input VSWR. The value of a quality factor for a 3-dB bandwidth level must be less than Q = 152/(174 - 132) = 3.6. As a result, it is very convenient to design input and output matching circuits using simple L-transformers in the form of lowpass and highpass filter sections with a constant value of Q [15].

To match the input series capacitive impedance to the standard 50Ω source impedance in a sufficiently wide frequency bandwidth, it is preferable to use three filter sections, as shown in Figure 2.12. From the negative reactive part of the input impedance Z_{in}, it follows that the input capacitance at the operating frequency of 152 MHz is equal to approximately 873 pF. To compensate at the center bandwidth frequency for this capacitive reactance, it is sufficient to connect an inductance of 1.3 nH in series to it. Now, when the device input capacitive reactance is compensated, in order to simplify the matching design procedure, it is best to cascade L-transformers with equal values for Q. Although equal Q values are not absolutely necessary, this provides a convenient guide for both analytical calculation of the matching circuit parameters and the Smith chart graphical design.

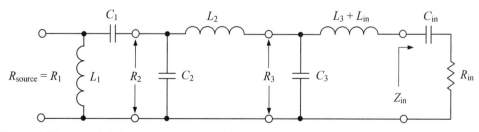

Figure 2.12   Complete broadband input matching circuit.

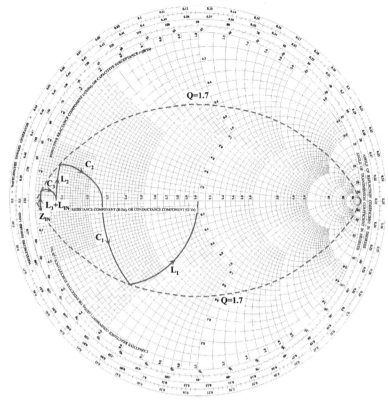

Figure 2.13 Smith chart with elements from Figure 2.12.

In this case, the following ratio can be written for the input matching circuit:

$$\frac{R_1}{R_2} = \frac{R_2}{R_3} = \frac{R_3}{R_{in}} \tag{2.35}$$

resulting in $R_2 = 13\Omega$ and $R_3 = 3.5\Omega$ for $R_{source} = R_1 = 50\Omega$ and $R_{in} = 0.9\Omega$. Consequently, a quality factor of each L-transformer is equal to $Q = 1.7$ according to (2.34). The elements of the input matching circuit using the formulas given in Figure 2.10 can be calculated as $L_1 = 31$ nH, $C_1 = 47$ pF, $L_2 = 6.2$ nH, $C_2 = 137$ pF, $L_3 = 1.6$ nH, and $C_3 = 509$ pF.

This equal-Q approach significantly simplifies the matching circuit design using the Smith chart [16]. When calculating a Q value, it is necessary to plot a circle of equal Q values on the Smith chart. Then, each element of the input matching circuit can be readily determined, as shown in Figure 2.13. Each trace for the series inductance must be plotted until the intersection point with the Q-circle, whereas each trace for the parallel capacitance should be plotted until it intersects with a horizontal real axis.

In practice, to simplify power amplifier designs at microwave frequencies, simple matching circuits are very often used, including an L-transformer with a series transmission line as the basic matching section. It is convenient to analyze

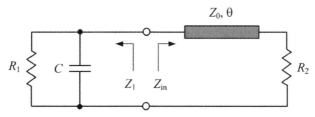

Figure 2.14   *L*-transformer with series transmission line.

the transforming properties of this matching circuit by substituting the equivalent transformation of the parallel RX circuit into the series one. For example, R_1 is the resistance and $X_1 = -1/\omega C$ is the reactance of the impedance $Z_1 = jR_1X_1/(R_1 + jX_1)$ for a parallel RC circuit, and $R_{in} = \mathrm{Re}Z_{in}$ is the resistance and $X_{in} = \mathrm{Im}Z_{in}$ is the reactance of the impedance $Z_{in} = R_{in} + jX_{in}$ for the series transmission-line circuit shown in Figure 2.14.

For complex-conjugate matching when $Z_1 = Z_{in}^*$, we obtain

$$\frac{R_1X_1^2}{R_1^2 + X_1^2} + j\frac{R_1^2X_1}{R_1^2 + X_1^2} = R_{in} - jX_{in} \tag{2.36}$$

The solution of (2.36) can be written in the form of two expressions for real and imaginary impedance parts as

$$R_1 = R_{in}(1 + Q^2) \tag{2.37}$$

$$X_1 = -X_{in}(1 + Q^{-2}) \tag{2.38}$$

where $Q = R_1/|X_1| = X_{in}/R_{in}$ is the quality factor equal for both parallel capacitive and series transmission-line circuits.

By using (1.111) from Chapter 1, which defines the dependence between the input impedance Z_{in}, the transmission-line parameters (electrical length θ and characteristic impedance Z_0), and the load impedance $R_L = R_2$, the real and imaginary parts of the input impedance Z_{in} can be written as

$$R_{in} = Z_0^2 R_2 \frac{1 + \tan^2\theta}{Z_0^2 + \left(R_2\tan\theta\right)^2} \tag{2.39}$$

$$X_{in} = Z_0\tan\theta\frac{Z_0^2 - R_2^2}{Z_0^2 + \left(R_2\tan\theta\right)^2} \tag{2.40}$$

From (2.40) it follows that an inductive input impedance (necessary to compensate for the capacitive parallel component) is provided when $Z_0 > R_2$ for $\theta < \pi/2$ and $Z_0 < R_2$ for $\pi/2 < \theta < \pi$. As a result, to transform the resistance R_1 into the other resistance R_2 at the given frequency, it is necessary to connect a two-port lowpass *L*-transformer (including a parallel capacitor and a series transmission line) between them. When one parameter (usually the characteristic impedance Z_0) is known, the matching circuit parameters can be calculated from the following two equations:

$$C = \frac{Q}{\omega R_1} \tag{2.41}$$

$$\sin 2\theta = \frac{2Q}{\dfrac{Z_0}{R_2} - \dfrac{R_2}{Z_0}} \tag{2.42}$$

where

$$Q = \sqrt{\frac{R_1}{R_2}\left[\cos^2\theta + \left(\frac{R_2}{Z_0}\right)^2 \sin^2\theta\right] - 1} \tag{2.43}$$

is the circuit (loaded) quality factor defined as a function of the resistances R_1 and R_2 and the parameters of the transmission line (characteristic impedance Z_0 and electrical length θ).

It follows from (2.42) and (2.43) that the electrical length θ can be calculated as a result of the numerical solution of a transcendental equation with one unknown parameter. However, it is more convenient to combine these two equations and to rewrite them in the implicit form of

$$\frac{R_1}{R_2} = \frac{1 + \left(\dfrac{Z_0}{R_2} - \dfrac{R_2}{Z_0}\right)^2 \sin^2\theta\cos^2\theta}{\cos^2\theta + \left(\dfrac{R_2}{Z_0}\right)^2 \sin^2\theta} \tag{2.44}$$

2.3 Gain and Stability

In early RF vacuum-tube transmitters, it was observed that the tubes and associated circuits may have damped or undamped oscillations depending on the circuit losses, the feedback coupling, the grid and anode potentials, and the reactance or tuning of the parasitic circuits [17, 18]. Various parasitic oscillator circuits such as the tuned-grid-tuned-anode circuit with capacitive feedback or Hartley, Colpitts, or Meissner oscillators can be realized at high frequencies, which potentially can be eliminated by adding a small resistor close to the grid or anode connections of the tubes for damping the circuits. Inductively coupled rather than capacitively coupled input and output circuits should be used wherever possible.

According to the immittance approach to the stability analysis of the active non-reciprocal two-port network, it is necessary and sufficient for its unconditional stability if the following system of equations can be satisfied for the given active device:

$$\text{Re}\left[W_S(\omega) + W_{\text{in}}(\omega)\right] > 0 \tag{2.45}$$

$$Im\left[W_S(\omega) + W_{in}(\omega) \right] = 0 \qquad (2.46)$$

or

$$Re\left[W_L(\omega) + W_{out}(\omega) \right] > 0 \qquad (2.47)$$

$$Im\left[W_L(\omega) + W_{out}(\omega) \right] = 0 \qquad (2.48)$$

where ReW_S and ReW_L are considered to be greater than zero [19, 20]. The active two-port network can be treated as unstable or potentially unstable in the case of the opposite signs in (2.45) and (2.47).

According to (2.48) and (2.49), the value of ReW_{in} depends on the load immittance W_L, whereas the variation of the source immittance W_S leads to the change of ReW_{out}. Therefore, if ReW_{out} has a negative value, the active two-port network will be potentially unstable in certain limits of the values of immittance W_S. Consequently, for unconditionally stable amplifier operation mode, it is necessary to satisfy the following condition:

$$Re\left[W_{out}(\omega) \right]\Big|_{min} > 0 \qquad (2.49)$$

Analyzing (2.49) in extremum, the minimum positive value of ReW_{out} for a given constant value of ReW_S can be derived by solving equation $\partial ReW_{out}/\partial ImW_S = 0$ as

$$Re\,W_{out} = Re\,W_{22} - \frac{\left| W_{12}W_{21}\right| + Re\left(W_{12}W_{21}\right)}{2\,Re\left(W_{11} + W_S\right)} \qquad (2.50)$$

The second term on the right-hand side of (2.50) as a function of ReW_S can take a maximum negative value when $ReW_S = 0$. Consequently, the requirement of a positive minimum value of ReW_{out} given by (2.49) can be written as

$$2\,Re\,W_{11}\,Re\,W_{22} - \left| W_{12}W_{21}\right| - Re\left(W_{12}W_{21}\right) > 0 \qquad (2.51)$$

A similar result can be obtained by optimizing the immittance W_{in} as a function of W_L. From (2.51) it follows that under such a condition the active device is unconditionally stable at any frequencies where this condition is fulfilled, regardless of any values of the source and load immittances W_S and W_L. By normalizing (2.51), a special relationship between the device immittance parameters called the *device stability factor* can be derived as

$$K = \frac{2\,Re\,W_{11}\,Re\,W_{22} - Re\left(W_{12}W_{21}\right)}{\left| W_{12}\,W_{21}\right|} \qquad (2.52)$$

which shows a stability margin indicating how far from a zero value are the real parts in (2.45) and (2.47) being positive [20].

Note that the applicability of (2.51) and (2.52) is restricted to the following requirements:

$$\text{Re}\big[W_{11}(\omega)\big] > 0 \qquad \text{Re}\big[W_{22}(\omega)\big] > 0 \qquad (2.53)$$

A comparison of (2.51) and (2.52) shows that the active device is unconditionally stable if $K > 1$ and potentially unstable if $K < 1$. From (2.53) and the condition

$$\big|\text{Re}\big(W_{12}W_{21}\big)\big| \le \big|W_{12}W_{21}\big| \qquad (2.54)$$

it follows that the smallest possible value for the device's stability factor can be defined as $K = -1$, which means that the values of K can be arranged only in the interval $[-1, \infty)$.

When the active device is potentially unstable, an improvement of the power amplifier stability can be provided with the appropriate choice of the source and load immittances W_S and W_L. In this case, the circuit stability factor K_T is defined in the same way as the device stability factor K, but by taking into account $\text{Re}W_S$ and $\text{Re}W_L$ along with the device's W-parameters. In this case, the circuit stability factor is given by

$$K_T = \frac{2\text{Re}\big(W_{11} + W_S\big)\text{Re}\big(W_{22} + W_L\big) - \text{Re}\big(W_{12}W_{21}\big)}{\big|W_{12}\ W_{21}\big|} \qquad (2.55)$$

If the circuit stability factor $K_T \ge 1$, the power amplifier is unconditionally stable. However, the power amplifier becomes potentially unstable if $K_T < 1$. The value of $K_T = 1$ corresponds to the border of the circuit unconditional stability. The values of the circuit stability factor K_T and device stability factor K become equal if $\text{Re}W_S = \text{Re}W_L = 0$.

For the device stability factor $K > 1$, the operating power gain G_P has to be maximized [21, 22]. By analyzing (2.10) in extremum, it is possible to derive the optimum values $\text{Re}W_L^o$ and $\text{Im}W_L^o$, at which the operating power gain G_P is maximal, by solving the following system of two equations:

$$\frac{\partial G_P}{\partial\,\text{Re}\,W_L} = 0 \qquad \frac{\partial G_P}{\partial\,\text{Im}\,W_L} = 0 \qquad (2.56)$$

As a result, the optimum values $\text{Re}W_L^o$ and $\text{Im}W_L^o$ depend on the immittance parameters of the active device and the device stability factor according to

$$\text{Re}\,W_L^o = \frac{\big|W_{12}W_{21}\big|}{2\,\text{Re}\,W_{11}}\sqrt{K^2 - 1} \qquad (2.57)$$

$$\text{Im}\,W_L^o = \frac{\text{Im}\big(W_{12}W_{21}\big)}{2\,\text{Re}\,W_{11}} - \text{Im}\,W_{22} \qquad (2.58)$$

Substituting the obtained values of $\operatorname{Re} W_{L}^{o}$ and $\operatorname{Im} W_{L}^{o}$ into (2.10) yields an expression for calculating the maximum value of G_{Pmax} written as

$$G_{\mathrm{Pmax}} = \left|\frac{W_{21}}{W_{12}}\right| / \left(K + \sqrt{K^{2} - 1}\right) \tag{2.59}$$

from which it follows that G_{Pmax} can be achieved only if $K > 1$, whereas $G_{\mathrm{Pmax}}^{A} = \left|W_{21}/W_{12}\right|$ when $K = 1$.

If the source is conjugately matched with the input of the active device, the following conditions must be satisfied:

$$\operatorname{Re} W_{S} = \operatorname{Re} W_{\mathrm{in}} \qquad \operatorname{Im} W_{S} + \operatorname{Im} W_{\mathrm{in}} = 0 \tag{2.60}$$

Then, by substituting these expressions into (2.8), the optimum values $\operatorname{Re} W_{S}^{o}$ and $\operatorname{Im} W_{S}^{o}$ as functions of the immittance parameters of the active device and the device stability factor can be derived as

$$\operatorname{Re} W_{S}^{o} = \frac{\left|W_{12} W_{21}\right|}{2 \operatorname{Re} W_{22}} \sqrt{K^{2} - 1} \tag{2.61}$$

$$\operatorname{Im} W_{S}^{o} = \frac{\operatorname{Im}\left(W_{12} W_{21}\right)}{2 \operatorname{Re} W_{22}} - \operatorname{Im} W_{11} \tag{2.62}$$

A comparison of (2.61) and (2.62) with (2.57) and (2.58), respectively, shows that these expressions are identical. Consequently, the power amplifier with an unconditionally stable active device provides a maximum power gain operation only if the input and output of the active device are conjugately matched with the source and load impedances, respectively. For the lossless input matching circuit when the power available at the source is equal to the power delivered to the input port of the active device (i.e., when $P_{S} = P_{\mathrm{in}}$), the maximum operating power gain is equal to the maximum transducer power gain (i.e., $G_{\mathrm{Pmax}} = G_{\mathrm{Tmax}}$).

Domains of the device's potential instability include the operating frequency ranges where the active device stability factor is equal to $K < 1$. Within the bandwidth of such a frequency domain, parasitic oscillations can occur, defined by internal positive feedback and the operating conditions of the active device. The instabilities may not be self-sustaining, induced by the RF drive power but remaining on its removal. One of the most serious cases of power amplifier instability can occur when the load impedance varies. Under these conditions, the transistor may be destroyed almost instantaneously. However, even it is not destroyed, the instability can result in a tremendously increased level of spurious emissions in the output spectrum of the power amplifier. Generally, the following classification can be applied to linear instabilities [23]:

- Low-frequency oscillations produced by thermal feedback effects
- Oscillations due to internal feedback

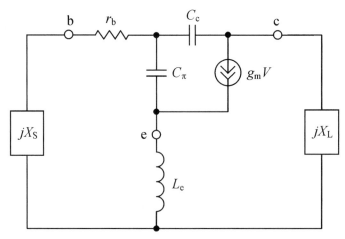

Figure 2.15 Simplified bipolar π-hybrid equivalent circuit with emitter lead inductance.

- Negative resistance or conductance-induced instabilities due to transit-time effects, avalanche multiplication, and so forth.
- Oscillations due to external feedback as a result of insufficient decoupling of the dc supply and so forth.

Therefore, it is very important to determine the effect of the device feedback parameters on the origin of the parasitic self-oscillations and to establish possible circuit configurations of the parasitic oscillators. Based on the simplified bipolar equivalent circuit shown in Figure 2.15, the device stability factor can be expressed through the parameters of the transistor equivalent circuit as

$$K = 2r_b g_m \frac{1 + \dfrac{g_m}{\omega_T C_c}}{\sqrt{1 + \left(\dfrac{g_m}{\omega C_c}\right)^2}} \tag{2.63}$$

where $\omega_T = 2\pi f_T$ [12, 13].

At very low frequencies, the bipolar transistors are potentially stable and the fact that $K \to 0$ when $f \to 0$ in (2.63) can be explained by simplifying the bipolar equivalent circuit. In practice, at low frequencies, it is necessary to take into account the dynamic base-emitter resistance r_π and Early collector-emitter resistance r_{ce}, the presence of which substantially increase the value of the device stability factor. This gives only one unstable frequency domain with $K < 1$ and low-boundary frequency f_{p1}. However, an additional region of possible low-frequency oscillations can occur due to thermal feedback where the collector junction temperature becomes frequently dependent, and the common-base configuration is especially affected by this [24].

Equating the device stability factor K with unity allows us to determine the high-boundary frequency of a frequency domain of the bipolar transistor potential instability as

$$f_{p2} = \frac{g_m}{2\pi C_c} / \sqrt{(2r_b g_m)^2 \left(1 + \frac{g_m}{\omega_T C_c}\right)^2 - 1} \qquad (2.64)$$

When $r_b g_m > 1$ and $g_m \gg \omega_T C_c$, (2.64) reduces to

$$f_{p2} \approx \frac{1}{4\pi r_b C_\pi} \qquad (2.65)$$

At higher frequencies, the presence of parasitic reactive intrinsic transistor parameters and package parasitics can be of great importance in view of power amplifier stability. The parasitic series emitter lead inductance L_e shown in Figure 2.15 has a major effect on the stability factor of the device. The presence of L_e leads to the appearance of the second frequency domain of potential instability at higher frequencies. The circuit analysis shows that the second frequency domain of potential instability can be realized only under the particular ratios between the normalized parameters $\omega_T L_e / r_b$ and $\omega_T r_b C_c$ [12, 13]. For example, the second domain does not occur for any values of L_e when $\omega_T r_b C_c \geq 0.25$.

An appearance of the second frequency domain of the device's potential instability is a result of the corresponding changes in the device feedback phase conditions and takes place only under a simultaneous effect of the collector capacitance C_c and emitter lead inductance L_e. If the effect of one of these factors is lacking, the active device is characterized by only the first domain of its potential instability.

Figure 2.16 shows the potentially realizable equivalent circuits of the parasitic oscillators. If the value of a series-emitter inductance L_e is negligible, the parasitic oscillations can occur only when the values of the source and load reactances are positive, that is, when $\mathrm{Im}Z_S = jX_S > 0$ and $\mathrm{Im}Z_L = jX_L > 0$. In this case, the parasitic oscillator shown in Figure 2.16(a) represents the inductive three-point circuit, where the inductive elements L_S and L_L in combination with the collector capacitance C_c form a Hartley oscillator. From a practical point of view, the more the value of the collector dc-feed inductance exceeds the value of the base-bias inductance, the more likely it is that low-frequency parasitic oscillators can be created. It was observed that a very low inductance, even a short between the emitter and the base, can produce very strong and dangerous oscillations that may easily destroy a transistor [23]. Therefore, it is recommended that the value of the base choke inductance be increased and the value of the collector choke inductance be decreased.

The presence of L_e leads to narrowing of the first frequency domain of the potential instability, which is limited to the high-boundary frequency f_{p2}, and can contribute to the appearance of the second frequency domain of the potential instability at higher frequencies. The parasitic oscillator that corresponds to the first frequency domain of the device potential instability can be realized only if the source and load reactances are inductive, that is, if $\mathrm{Im}Z_S = jX_S > 0$ and $\mathrm{Im}Z_L = jX_L > 0$, with the equivalent circuit of such a parasitic oscillator shown in Figure 2.16(b). The parasitic oscillator corresponding to the second frequency domain of the device potential instability can be realized only if the source reactance is capacitive and the load reactance is inductive, that is, if $\mathrm{Im}Z_S = -jX_S < 0$ and $\mathrm{Im}Z_L = jX_L > 0$, with the equivalent circuit shown in Figure 2.16(c). The series emitter

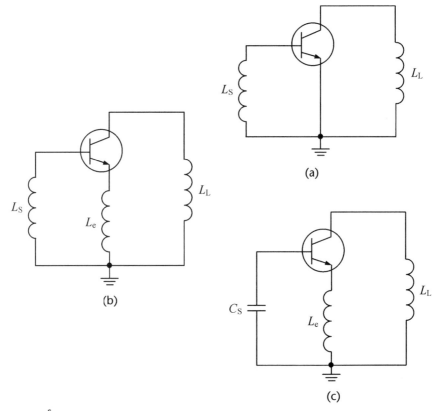

Figure 2.16 Equivalent circuits of parasitic bipolar oscillators.

inductance L_e is an element of fundamental importance for the parasitic oscillator that corresponds to the second frequency domain of the device potential instability. It changes the circuit phase conditions so it becomes possible to establish the oscillation phase-balance condition at high frequencies. However, if it is possible to eliminate the parasitic oscillations at high frequencies by other means, increasing L_e will result in narrowing of the low-frequency domain of potential instability, thus making the power amplifier potentially more stable, although at the expense of reduced power gain.

Similar analysis of the MOSFET power amplifier also shows two frequency domains of MOSFET potential instability due to the internal feedback gate-drain capacitance C_{gd} and series source inductance L_s [13]. Because of the very high gate-leakage resistance, the value of the low-boundary frequency f_{p1} is sufficiently small. For usually available conditions for power MOSFET devices when $g_m R_{ds} = 10 \div 30$ and $C_{gd}/C_{gs} = 0.1 \div 0.2$, the high-boundary frequency f_{p2} can be approximately calculated from

$$f_{p2} \approx \frac{1}{4\pi R_{gs} C_{gs}} \tag{2.66}$$

Note that power MOSFET devices have a substantially higher value of $g_m R_{ds}$ at small values of the drain current than at high values. Consequently, for small drain current, MOSFET devices are characterized by a wider domain of potential instability. This domain is significantly wider than the same first domain of the potential instability of the bipolar transistor. The series source inductance L_s contributes to the appearance of the second frequency domain of the device potential instability. The potentially realizable equivalent circuits of the MOSFET parasitic oscillators are the same as for the bipolar transistor, as shown in Figure 2.16 [13].

Thus, to prevent the parasitic oscillations and to provide a stable operating mode for any power amplifier, it is necessary to take into consideration the following common requirements:

- Use an active device with stability factor $K > 1$.
- If it is impossible to choose an active device with $K > 1$, it is necessary to provide the circuit stability factor $K_T > 1$ by choosing the appropriate real parts of the source and load immittances.
- Disrupt the equivalent circuits of the possible parasitic oscillators.
- Choose the proper reactive parameters of the matching circuit elements adjacent to the input and output ports of the active device, which are necessary to avoid the self-oscillation conditions.

Generally, the parasitic oscillations can arise at any frequency within the potential instability domains for particular values of the source and load immittances W_S and W_L. The frequency dependencies of W_S and W_L are very complicated and very often cannot be predicted exactly, especially in multistage power amplifiers. Therefore, it is very difficult to propose a unified approach for providing a stable operating mode to power amplifiers with different circuit configurations and operating frequencies. In practice, the parasitic oscillations can arise close to the operating frequencies due to the internal positive feedback inside the transistor and at the frequencies sufficiently far from the operating frequencies due to the external positive feedback created by the surface-mounted elements. As a result, stability analyses of power amplifiers must include methods for preventing parasitic oscillations in different frequency ranges.

2.4   Basic Classes of Operation

As established at the end of the 1910s, amplifier efficiency can reach quite high values when suitable adjustments are made to the grid and anode voltages [25]. With a resistive load, the anode current is in phase with the grid voltage, whereas it leads with the capacitive load and it lags with the inductive load. On the assumption that the anode current and anode voltage both have sinusoidal variations, the maximum possible output of the amplifying device would be just half the dc supply power, resulting in an anode efficiency of 50%. However, by using a pulsed-shaped anode current, it is possible to achieve anode efficiency considerably in excess of 50%, potentially as high as 90%, by choosing the proper operating conditions. By applying the proper negative bias voltage to the grid terminal to provide the pulsed anode current of different width with the angle θ, the anode current becomes equal to zero, where the double angle 2θ represents a conduction angle of the amplifying device [26]. In this case, a theoretical anode efficiency approaches 100% when the conduction angle, during which the anode current flows, reduces to zero, with starting efficiency of 50% corresponding to the conduction angle of 360° or 100% duty ratio.

Generally, power amplifiers can be classified in three classes according to their mode of operation: *linear mode*, in which the amplifier's operation is confined to the substantially linear portion of the active device characteristic curve; *critical mode*, in which the anode current ceases to flow, but operation extends beyond the linear portion up to the saturation and cutoff regions; and *nonlinear mode*, in which the anode current ceases to flow during a portion of each cycle, with a duration that depends on the grid bias [27]. When high efficiency is required, power amplifiers of the third class are employed since the presence of harmonics contributes to the attainment of high efficiencies. To suppress harmonics of the fundamental frequency to deliver a sinusoidal signal to the load, a parallel resonant circuit can be used in the load network, which bypasses harmonics through a low-impedance path and, by virtue of its resonance to the fundamental, receives energy at that frequency. At the very beginning of the 1930s, power amplifiers operating in the first two classes with a 100% duty ratio were called Class A power amplifiers, whereas power amplifiers operating in the third class with a 50% duty ratio were considered Class B power amplifiers [28].

To analytically determine the operating classes of a power amplifier, consider the simple resistive stage shown in Figure 2.17, where L_{ch} is the ideal choke inductor with zero series resistance and infinite reactance at the operating frequency, C_b is the dc-blocking capacitor with infinite value having zero reactance at the operating frequency, and R_L is the load resistor. The dc supply voltage V_{cc} is applied to both plates of the dc-blocking capacitor, being constant during the entire signal period. The active device behaves as an ideal voltage- or current-controlled current source having zero saturation resistance.

Let us assume the input signal to be in a cosine form of

$$v_{in} = V_b + V_{in}\cos\omega t \qquad (2.67)$$

where V_b is the input dc bias voltage. The operating point must be fixed at the middle point of the linear part of the device transfer characteristic with $V_{in} \le V_b - V_p$, where V_p is the device pinch-off voltage. Usually, to simplify an analysis of the power amplifier operation, the device transfer characteristic is represented by a piecewise-linear approximation. As a result, the output current is cosinusoidal:

$$i = I_q + I\cos\omega t \qquad (2.68)$$

with the quiescent current I_q greater or equal to the collector current amplitude I. In this case, the output collector current contains only two components—dc and cosine—and the averaged current magnitude is equal to a quiescent current or dc component I_q.

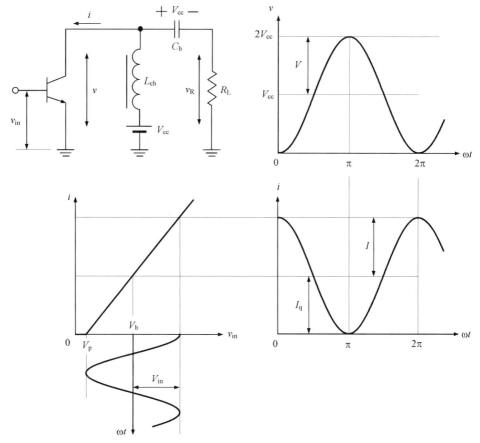

Figure 2.17   Voltage and current waveforms in Class A operation.

The output voltage v across the device collector represents a sum of the dc supply voltage V_{cc} and cosine voltage v_R across the load resistor R_L. Consequently, greater output current i results in greater voltage v_R across the load resistor R_L and smaller output voltage v across the device output. Thus, for a purely real load impedance $Z_L = R_L$, the collector voltage v is shifted by 180° relative to the input voltage v_{in} and can be written as

$$v = V_{cc} + V\cos(\omega t + 180°) = V_{cc} - V\cos\omega t \qquad (2.69)$$

where V is the output voltage amplitude.

Substituting (2.68) into (2.69) yields

$$v = V_{cc} - \left(i - I_q\right)R_L \qquad (2.70)$$

where $R_L = V/I$, and (2.70) can be rewritten as

$$i = \left(I_q + \frac{V_{cc}}{R_L}\right) - \frac{v}{R_L} \qquad (2.71)$$

which determines a linear dependence of the collector current versus collector voltage. Such a combination of the cosine collector voltage and current waveforms is known as a Class A operating mode. In practice, because of the device's inherent nonlinearities, it is necessary to connect a parallel LC circuit with resonant frequency equal to the operating frequency to suppress any possible harmonic components.

Circuit theory prescribes that the collector efficiency η can be written as

$$\eta = \frac{P}{P_0} = \frac{1}{2}\frac{I}{I_q}\frac{V}{V_{cc}} = \frac{1}{2}\frac{I}{I_q}\xi \qquad (2.72)$$

where $P_0 = I_q V_{cc}$ is the dc output power, $P = 0.5IV$ is the power delivered to the load resistance R_L at the fundamental frequency f_0, and $\xi = V/V_{cc}$ is the collector voltage peak factor.

Then, by assuming the ideal conditions of zero saturation voltage when $\xi = 1$ and maximum output current amplitude when $I/I_q = 1$, from (2.72) it follows that the maximum collector efficiency in a Class A operating mode is equal to $\eta = 50\%$. However, as also follows from (2.72), increasing the value of I/I_q can further increase the collector efficiency. This leads to a step-by-step nonlinear transformation of the current cosine waveform to its pulsed waveform when the amplitude of the collector current exceeds zero value during only a part of the entire signal period. In this case, an active device is operated in the active region followed by the operation in the pinch-off region when the collector current is zero, as shown in Figure 2.18.

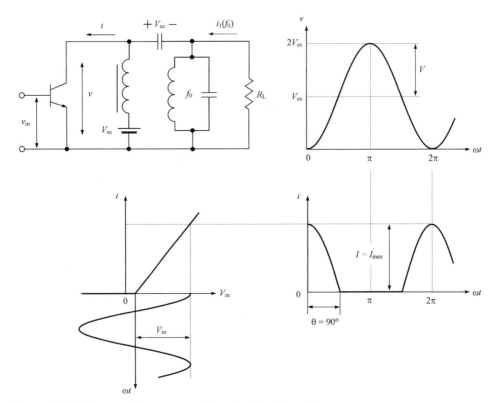

Figure 2.18 Voltage and current waveforms in Class B operation.

As a result, the frequency spectrum at the device output will generally contain the second-, third-, and higher-order harmonics of the fundamental frequency. However, due to the high quality factor of the parallel resonant LC circuit, only the fundamental-frequency signal flows to the load, while the short-circuit conditions are fulfilled for higher-order harmonic components. Therefore, ideally the collector voltage represents a purely sinusoidal waveform with the voltage amplitude $V \le V_{cc}$.

Analytically such an operation can be written as

$$i(\omega t) = \begin{cases} I_q + I\cos\omega t & -\theta \le \omega t < \theta \\ 0 & \theta \le \omega t < 2\pi - \theta \end{cases} \tag{2.73}$$

where the conduction angle 2θ is the angle of a current flow indicating the part of the RF current cycle, for which device conduction occurs and determines the moment when output current $i(\omega t)$ takes a zero value [29]. At this moment

$$i(\theta) = I_q + I\cos\theta = 0 \tag{2.74}$$

and half the conduction angle θ can be calculated by

$$\cos\theta = -\frac{I_q}{I} \tag{2.75}$$

As a result, the basic definitions for the nonlinear operating modes of a power amplifier through half the conduction angle θ can be introduced as follows:

- When $\theta > 90°$, then $\cos\theta < 0$ and $I_q > 0$, corresponding to Class AB operation.
- When $\theta = 90°$, then $\cos\theta = 0$ and $I_q = 0$, corresponding to Class B operation.
- When $\theta < 90°$, then $\cos\theta > 0$ and $I_q < 0$, corresponding to Class C operation.

The periodic pulsed output current $i(\omega t)$ can be written as a Fourier-series expansion

$$i(\omega t) = I_0 + I_1\cos\omega t + I_2\cos 2\omega t + I_3\cos 3\omega t + \cdots \tag{2.76}$$

where the dc and fundamental-frequency components can be obtained, respectively, by

$$I_0 = \frac{1}{2\pi}\int_{-\theta}^{\theta} I(\cos\omega t - \cos\theta)d\omega t = I\gamma_0 \tag{2.77}$$

$$I_1 = \frac{1}{\pi}\int_{-\theta}^{\theta} I(\cos\omega t - \cos\theta)\cos\omega t\, d\omega t = I\gamma_1 \tag{2.78}$$

where

$$\gamma_0 = \frac{1}{\pi}(\sin\theta - \theta\cos\theta) \tag{2.79}$$

$$\gamma_1 = \frac{1}{\pi}(\theta - \sin\theta\cos\theta) \tag{2.80}$$

are the current coefficients for the dc and fundamental-frequency components, respectively [30, 31].

From (2.77) it follows that the dc current component is a function of θ in the operating modes where $\theta < 180°$, in contrast to a Class A operating mode where $\theta = 180°$ and the dc current is equal to the quiescent current during the entire period.

The collector efficiency of a power amplifier with a shunt resonant circuit, biased to operate in the nonlinear modes, can be obtained by

$$\eta = \frac{P_1}{P_0} = \frac{1}{2}\frac{I_1}{I_0}\xi = \frac{1}{2}\frac{\gamma_1}{\gamma_0}\xi \tag{2.81}$$

which is a function of θ only, where

$$\frac{\gamma_1}{\gamma_0} = \frac{\theta - \sin\theta\cos\theta}{\sin\theta - \theta\cos\theta} \tag{2.82}$$

The Class B power amplifiers had been defined as those which operate with a negative grid bias such that the anode current is practically zero with no excitation grid voltage, and in which the output power is proportional to the square of the excitation voltage [32]. If $\xi = 1$ and $\theta = 90°$, then from (2.81) and (2.82) it follows that the maximum collector efficiency in the Class B operating mode is equal to

$$\eta = \frac{\pi}{4} \cong 78.5\%$$

The fundamental-frequency power delivered to the load $P_L = P_1$ is defined as

$$P_1 = \frac{VI_1}{2} = \frac{VI\gamma_1(\theta)}{2} \tag{2.83}$$

showing its direct dependence on the conduction angle 2θ. This means that reduction in θ results in lower γ_1, and, to increase the fundamental-frequency power P_1, it is necessary to increase the current amplitude I. Since the current amplitude I is determined by the input voltage amplitude V_{in}, the input power P_{in} must be increased. The collector efficiency also increases with a reduced value for θ and becomes maximum when $\theta = 0°$, where the ratio γ_1/γ_0 is maximal, as follows from 2.82. For example, the collector efficiency η increases from 78.5% to 92% when θ is reduced from 90° to 60°. However, it requires the input voltage amplitude V_{in} to be increased by 2.5 times, resulting in lower values of the power-added efficiency (PAE), which is defined as

$$PAE = \frac{P_1 - P_{in}}{P_0} = \frac{P_1}{P_0}\left(1 - \frac{1}{G_P}\right) \tag{2.84}$$

where $G_P = P_1/P_{in}$ is the operating power gain.

Class C power amplifiers had been defined as those that operate with a negative grid bias more than sufficient to reduce the anode current to zero with no excitation grid voltage, and in which the output power varies as the square of the anode voltage between limits [32]. The main distinction between Class B and Class C amplifiers is in the duration of the output current pulses, which are shorter for Class C amplifiers when the active device is biased beyond the pinch-off point. To achieve the maximum anode efficiency in Class C amplifiers, the active device should be biased (negative) considerably lower than the pinch-off point to provide sufficiently low conduction angles [33].

To obtain an acceptable trade-off between a high power gain and a high power-added efficiency in different situations, the conduction angle should be chosen within the range of $120° \leq 2\theta \leq 190°$. If it is necessary to provide high collector efficiency for the active device having a high-gain capability, then a Class C operating mode must be chosen with θ close to 60°. However, when the input power is limited and power gain is not sufficient, a Class AB operating mode with small quiescent current when θ is slightly greater than 90° is recommended. In the latter case, the linearity of the power amplifier can be significantly improved. From (2.82) it follows that the ratio of the fundamental-frequency component of the anode current to the dc current is a function of θ only, which means that, if the operating angle is maintained constant, the fundamental component of the anode current will replicate linearly to the variation of the dc current, thus providing linear operation of Class C power amplifiers when dc current is directly proportional to the grid voltage [34].

2.5   Nonlinear Active Device Models

Generally, for an accurate power amplifier simulation and matching circuit design for different operating frequencies and output power levels, it is necessary to represent an active device in the form of a nonlinear equivalent circuit. This circuit can adequately describe the small- and large-signal electrical behavior of the power amplifier up to the device transition frequency f_T and higher to its maximum frequency f_{max} that allows a sufficient number of harmonic components to be taken into account. Accurate device modeling is extremely important to develop monolithic integrated circuits. Better approximations of the final design can only be achieved if the nonlinear device behavior is described accurately.

2.5.1   LDMOSFETs

Figure 2.19(a) shows the cross section of the physical structure of a laterally-diffused metal oxide semiconductor FET (LDMOSFET) device where a heavily doped p^+-sinker is inserted between the top source and p^+-substrate (source grounding) for low resistivity to provide high-current flow between the drain and source terminals [35]. The lightly doped p-epilayer and n-drift layer are required to provide sufficient distance between regions to prevent latch-up (forward-biased p-n diodes) and for drain-source breakdown protection. The parasitic gate-drain capacitance is directly related to the overlap of the gate oxide onto the heavily doped n^+-source region. To describe accurately the nonlinear properties of the large-size MOSFET device, it is necessary to consider its two-dimensional gate-distributed nature along both the channel length and channel width, resulting in lower values for the intrinsic series gate and shunt gate-source resistances. Figure 2.19(b) shows the nonlinear MOSFET equivalent circuit with extrinsic parasitic elements, which can properly describe the nonlinear behavior of both VDMOSFET and LDMOSFET devices [36, 37].

The nonlinear current source $i(v_{gs}, v_{ds}, \tau)$ as a function of the input gate-source and output drain-source voltages incorporating a self-heating effect can be described accurately enough using hyperbolic functions [37, 38]. A careful analytical description of the transition from the quadratic region to the linear region of the device transfer characteristic provides for a more accurate prediction of intermodulation distortion [39]. The overall channel carrier transit time τ also includes the effect of the transcapacitance required for charge conservation. The drain-source capacitance C_{ds} and gate-drain capacitance C_{gd} are considered to be the junction capacitances and they strongly depend on the drain-source voltage. The extrinsic parasitic elements are represented by the gate and drain bondwire inductances L_g and L_d, source inductance L_s, source and drain bulk and ohmic resistances R_s and R_d, and gate contact and ohmic resistance R_g. The effect of the gate-source channel resistance

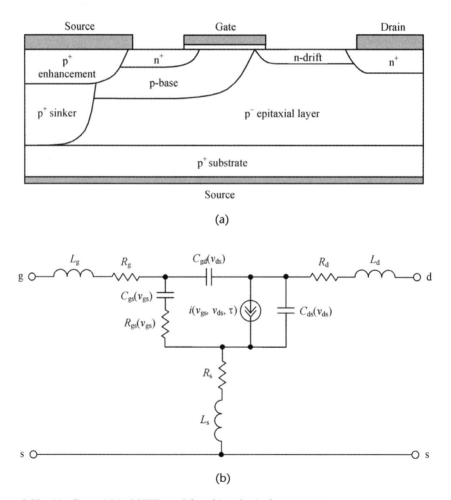

Figure 2.19   Nonlinear LDMOSFET model and its physical structure.

R_{gs} becomes significant at higher frequencies close to the transition frequency $f_T = g_m/2\pi C_{gs}$, where g_m is the device transconductance. To account for the self-heating effect and substrate losses, a special four-port thermal circuit and a series combination of the resistance and capacitance between the external drain and source terminals can be included [40].

An empirical nonlinear model developed for silicon LDMOS transistors, which is single-piece and continuously differentiable, can be written as

$$I_{ds} = \beta V_{gst}^{VGexp}\left(1 + \lambda V_{ds}\right)\tanh\left(\frac{\alpha V_{ds}}{V_{gst}}\right)\left[1 + K_1 \exp\left(V_{BReff1}\right)\right]$$

$$+ I_{ss}\exp\left(\frac{V_{ds} - V_{BR}}{V_T}\right) \tag{2.85}$$

where

$$V_{gst} = V_{st}\ln\left[1 + \exp\left(\frac{V_{gst2}}{V_{st}}\right)\right]$$

$$V_{gst2} = V_{gst1} - \frac{1}{2}\left(V_{gst1} + \sqrt{\left(V_{gst1} - V_K\right)^2 + \Delta^2} - \sqrt{V_K^2 + \Delta^2}\right)$$

$$V_{gst1} = V_{gs} - V_{th0} - \gamma V_{ds}$$

$$V_{BReff1} = \frac{V_{ds} - V_{BReff}}{K_2} + M_3\frac{V_{ds}}{V_{BReff}}$$

$$V_{BReff} = \frac{V_{BR}}{2}\left[1 + \tanh\left(M_1 - V_{gst}M_2\right)\right]$$

where

λ = drain current slope parameter

β = transconductance parameter

V_{th0} = forward threshold voltage

V_{st} = subthreshold slope coefficient

V_T = temperature voltage

I_{ss} = forward diode leakage current

V_{BR} = breakdown voltage

K_1, K_2, M_1, M_2, M_3 = breakdown parameters

$V_K, V_{Gexp}, \Delta, \gamma$ = gate-source voltage parameters [38].

The gate-source capacitance C_{gs} can be analytically described as a function of the gate-source voltage since it is practically independent on the drain-source voltage. It is equal to the oxide capacitance C_{ox} in the accumulation region, significantly reduces in the depletion region, slightly decreases and reaches its minimum in the weak-inversion region, then significantly increases in the moderate-inversion region, and becomes practically constant in the strong-inversion or saturation region, as

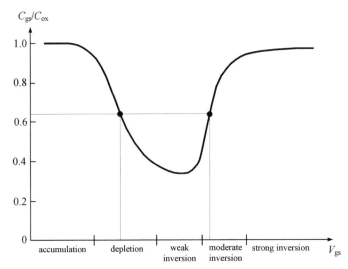

Figure 2.20 Gate-source capacitance versus gate-source voltage.

shown in Figure 2.20 [41]. The approximation function for the gate-source capacitance C_{gs} as the dependence of V_{gs} can be derived by using two components containing the hyperbolic functions as

$$C_{gs} = C_{gs1} + C_{gs2}\left\{1 + \tanh\left[C_{gs6}\left(V_{gs} + C_{gs3}\right)\right]\right\}$$
$$+ C_{gs4}\left[1 - \tanh\left(C_{gs5}V_{gs}\right)\right] \tag{2.86}$$

where C_{gs1}, C_{gs2}, C_{gs3}, C_{gs4}, C_{gs5}, and C_{gs6} are the approximation parameters [38].

The gate-source resistance R_{gs} is determined by the effect of the channel inertia in responding to rapid changes of the time varying gate-source voltage, and varies in such a manner that the charging time $\tau_g = R_{gs}C_{gs}$ remains approximately constant. Thus, the increase of R_{gs} in the velocity saturation region, when the channel conductivity decreases, is partially compensated for by the decrease of C_{gs} due to nonuniform channel charge distribution [42]. The effect of R_{gs} becomes significant at higher frequencies close to the transition frequency f_T of the MOSFET and can be neglected when designing RF circuits that operate below 2 GHz, as used for commercial wireless applications [43, 44].

2.5.2 GaAs MESFETs and GaN HEMTs

Adequate representation for metal-semiconductor field effect transistors (MESFETs) and high electron mobility transfers (HEMTs) in a frequency range up to at least 25 GHz can be provided using the nonlinear model shown in Figure 2.21(a), which is very similar to a nonlinear MOSFET model [45, 46]. The intrinsic model is described by the channel charging resistance R_{gs}, which represents the resistive path for the charging of the gate-source capacitance C_{gs}, feedback gate-drain capacitance C_{gd}, and drain-source capacitance C_{ds} with the gate-source diode to model the forward conduction current $i_{gs}(v_{gs})$ and gate-drain diode to account for the gate-drain avalanche current $i_{gd}(v_{gs}, v_{ds})$, which can occur at large-signal operating conditions. The gate-source capacitance C_{gs} and gate-drain capacitance C_{gd} represent the charge depletion region and can be treated as the voltage-dependent Schottky-barrier diode capacitances, being the nonlinear functions of the gate-source voltage v_{gs} and drain-source voltage v_{ds}. For negative gate-source voltage and small drain-source voltage, these capacitances are practically equal.

However, when the drain-source voltage is increased beyond the current saturation point, the gate-drain capacitance C_{gd} is much more heavily back-biased than the gate-source capacitance C_{gs}. Therefore, the gate-source capacitance C_{gs} is significantly more important and usually dominates the input impedance of the MESFET or HEMT device. The influence of the drain-source capacitance C_{ds} on the device behavior is insignificant and its value is bias independent. The capacitance C_{dsd} and resistance R_{dsd} model the dispersion of the MESFET or HEMT current-voltage characteristics due to a trapping effect in the device channel, which leads to discrepancies between the dc and S-parameter measurements at higher frequencies [47, 48]. A large-signal model for monolithic power amplifier design must be accurate for all operating conditions. In addition, the model parameters should be easily extractable and the model must be as simple as possible.

Various nonlinear MESFET and HEMT models with different complexity are available, and each one can be considered sufficiently accurate for a particular application. For example, although the Materka model does not fulfill charge

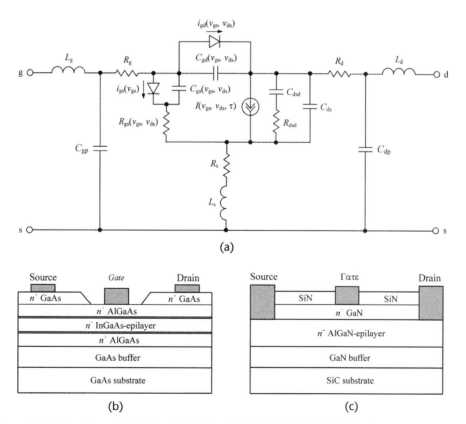

Figure 2.21 Nonlinear MESFET and HEMT model with HEMT physical structures.

conservation, it seems to be an acceptable compromise between accuracy and model simplicity for MESFETs, but not for HEMTs, where it is preferable to use the Angelov model [49, 50]. For example, it can be used to predict the large-signal behavior of pHEMT devices using high-power, high-efficiency 60-GHz monolithic microwave integrated circuits (MMICs) [51]. By using three additional terms of the gate power-series function in the Angelov model, better accuracy can be achieved for large-signal modeling of AlGaN/GaN HEMT devices on a SiC substrate [52]. This model can also be improved by incorporating two additional analytical expressions to model device behavior in the saturation region [53].

Figure 2.21(b) shows the cross section of the physical structure of an InGaAs/AlGaAs HEMT device, in which an undoped InGaAs n-epilayer is used as a channel and two heavily n-doped AlGaAs layers with a high energetic barrier for holes are necessary to maximize high electron mobility in the channel. In this case, spacing between the AlGaAs layer and InGaAs channel is optimized to achieve high breakdown voltage. An example of the physical structure of a AlGaN/GaN HEMT device is shown in Figure 2.21(c), where an undoped AlGaN n-epilayer is used as a channel, an n-type doped GaN layer can suppress dispersion in the device current-voltage characteristics, and a SiN passivation layer with optimized parameters contributes to a lower-trap device structure [54]. The thermal conductivity of GaN HEMT devices is improved by using a SiC substrate. Note that use of GaN-based technology results in a higher breakdown voltage, wider bandwidth, and higher efficiency for power amplifiers due to the high charge density and the ability to operate at higher voltages for GaN HEMT devices, which are characterized by lower output capacitance and on-resistance [55, 56].

The basic electrical properties of a MESFET or HEMT device can be characterized by the admittance Y-parameters expressed through the device intrinsic small-signal equivalent circuit as

$$Y_{11} = \frac{j\omega C_{gs}}{1 + j\omega C_{gs} R_{gs}} + j\omega C_{gd} \qquad Y_{12} = -j\omega C_{gd}$$
$$Y_{21} = \frac{g_m \exp(-j\omega\tau)}{1 + j\omega C_{gs} R_{gs}} - j\omega C_{gd} \qquad Y_{22} = \frac{1}{R_{ds}} + j\omega\left(C_{ds} + C_{gd}\right) \tag{2.87}$$

where g_m is the device transconductance and R_{ds} is the differential drain-source resistance [37]. In this case, the dispersion effect, which is important at higher frequencies and modeled by C_{dsd} and R_{dsd}, cannot be taken into account.

By separating the Y-parameters given in (2.87) into their real and imaginary parts, the elements of the small-signal equivalent circuit can be analytically determined as

$$C_{gd} = -\frac{\text{Im}\, Y_{12}}{\omega} \tag{2.88}$$

$$C_{gs} = \frac{\text{Im}\, Y_{11} - \omega C_{gd}}{\omega}\left[1 + \left(\frac{\text{Re}\, Y_{11}}{\text{Im}\, Y_{11} - \omega C_{gd}}\right)^2\right] \tag{2.89}$$

$$R_{gs} = \frac{\text{Re}\,Y_{11}}{\left(\text{Im}\,Y_{11} - \omega C_{gd}\right)^2 + \left(\text{Re}\,Y_{11}\right)^2} \tag{2.90}$$

$$g_m = \sqrt{\left(\text{Re}\,Y_{21}\right)^2 + \left(\text{Im}\,Y_{21} + \omega C_{gd}\right)^2}\sqrt{1 + \left(\omega C_{gs}R_{gs}\right)^2} \tag{2.91}$$

$$\tau = \frac{1}{\omega}\sin^{-1}\left(\frac{-\omega C_{gd} - \text{Im}\,Y_{21} - \omega C_{gs}R_{gs}\,\text{Re}\,Y_{21}}{g_m}\right) \tag{2.92}$$

$$C_{ds} = \frac{\text{Im}\,Y_{22} - \omega C_{gd}}{\omega} \tag{2.93}$$

$$R_{ds} = \frac{1}{\text{Re}\,Y_{22}} \tag{2.94}$$

which are valid for a wide frequency range up to the transition frequency f_T [57]. Assuming that all extrinsic parasitic elements are known, the only problem is then to determine the admittance Y-parameters of the intrinsic two-port network from on-bias experimental data [58]. Consecutive steps shown in Figure 2.22 can represent such a determination procedure [59]:

- Measure the S-parameters of the extrinsic device.
- Transform the S-parameters to the impedance Z-parameters with subtraction of the series inductances L_g and L_d.
- Transform the impedance Z-parameters to the admittance Y-parameters with subtraction of parallel capacitances C_{gp} and C_{dp}.
- Transform the admittance Y-parameters to the impedance Z-parameters with subtraction of series resistances R_g, R_s, R_d, and inductance L_s.
- Transform the impedance Z-parameters to the admittance Y-parameters of the intrinsic device two-port network.

A simple and accurate nonlinear Angelov model is capable of modeling the drain current-voltage characteristics and its derivatives, as well as the gate-source and gate-drain capacitances, for different submicron gate-length HEMT devices and commercially available MESFETs. The drain current source is described by using the hyperbolic functions as

$$I_{ds} = I_{pk}\left(1 + \tanh\psi\right)\left(1 + \lambda V_{ds}\right)\tanh\alpha V_{ds} \tag{2.95}$$

where I_{pk} is the drain current at maximum transconductance with the contribution from the output conductance subtracted, λ is the channel-length modulation parameter, and $\alpha = \alpha_0 + \alpha_1\tanh\psi$ is the saturation voltage parameter, where α_0 is the saturation voltage parameter at pinch-off and α_1 is the saturation voltage

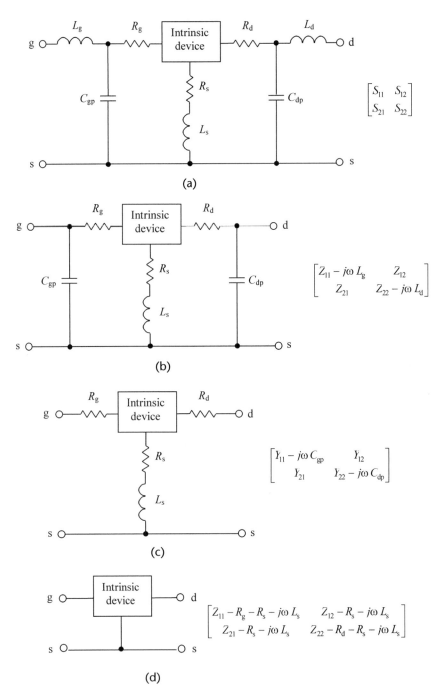

Figure 2.22 Method for extracting device intrinsic *Z*-parameters.

parameter at $V_{gs} > 0$. The parameter ψ is a power-series function centered at V_{pk} with the bias voltage V_{gs} as a variable,

$$\psi = P_1\left(V_{gs} - V_{pk}\right) + P_2\left(V_{gs} - V_{pk}\right)^2 + P_3\left(V_{gs} - V_{pk}\right)^3 + \cdots \qquad (2.96)$$

where V_{pk} is the gate voltage for maximum transconductance g_{mpk}. The model parameters as a first approximation can be easily obtained from the experimental $I_{ds}(V_{gs}, V_{ds})$ curves at a saturated channel condition when all higher terms in ψ are assumed to be zero and λ is the slope of the I_{ds}-V_{ds} characteristic.

The same hyperbolic functions can be used to model the intrinsic device capacitances C_{gs} and C_{ds}. When an accuracy of 5% to 10% is sufficient, the gate-source capacitance C_{gs} and gate-drain capacitance C_{gd} can be described by

$$C_{gs} = C_{gs0}\left[1 + \tanh\left(P_{1gsg}V_{gs}\right)\right]\left[1 + \tanh\left(P_{1gsd}V_{ds}\right)\right] \qquad (2.97)$$

$$C_{gd} = C_{gd0}\left[1 + \tanh\left(P_{1gdg}V_{gs}\right)\right]\left[1 - \tanh\left(P_{1gdd}V_{ds} + P_{1cc}V_{gs}V_{ds}\right)\right] \qquad (2.98)$$

where the product $P_{1cc}V_{gs}V_{ds}$ reflects the cross-coupling of V_{gs} and V_{ds} on C_{gd} and the coefficients P_{1gsg}, P_{1gsd}, P_{1gdg}, and P_{1gdd} are the fitting parameters.

2.5.3   Low-Voltage and High-Voltage HBTs

Figure 2.23(a) shows the modified Gummel-Poon nonlinear model of the bipolar transistor with extrinsic parasitic elements [60, 61]. Such a hybrid-π equivalent circuit can model the nonlinear electrical behavior of bipolar junction transistors (BJTs), in particularly heterojunction bipolar transistor (HBT) devices, with sufficient accuracy up to about 20 GHz. The intrinsic model is described by the dynamic diode resistance r_π, the total base-emitter junction capacitance and base charging diffusion capacitance C_π, the base-collector diode required to account for the nonlinear effects at the saturation, the internal collector-base junction capacitance C_{ci}, the external distributed collector-base capacitance C_{co}, the collector-emitter capacitance C_{ce}, and the nonlinear current source $i(v_{be}, v_{ce})$. The lateral and base semiconductor resistances underneath the base contact and the base semiconductor resistance underneath the emitter are combined into a base-spreading resistance r_b. The extrinsic parasitic elements are represented by the base bondwire inductance L_b, emitter ohmic resistance r_e, emitter inductance L_e, collector ohmic resistance r_c, and collector bondwire inductance L_c. To increase the usable operating frequency range of the device up to 50 GHz, it is necessary to include the collector current delay time τ in the collector current source as $g_m \exp(-j\omega\tau)$. The more complicated models, such as VBIC, HICUM, or MEXTRAM, include the effects of self-heating of a bipolar transistor; take into account the parasitic p-n-p transistor formed by the base, collector, and substrate regions; provide an improved description of depletion capacitances at large forward bias; and take into account avalanche and tunneling currents and other nonlinear effects corresponding to distributed high-frequency effects [62].

Figure 2.23(b) shows the modified version of a bipolar transistor equivalent circuit, where $C_c = C_{co} + C_{ci}$, $r_{b1} = r_b C_{ci}/C_c$, and $r_{b2} = r_b C_{co}/C_c$ [63]. Such an equivalent circuit becomes possible due to an equivalent π- to T-transformation of the elements r_b, C_{co}, and C_{ci} and a condition $r_b \ll (C_{ci} + C_{co})/\omega C_{ci} C_{co}$, which is usually fulfilled over a frequency range close to the device maximum frequency f_{max}. Then, from a comparison of the transistor nonlinear models, for a bipolar transistor in Figure 4.13(b), for a MOSFET device in Figure 4.9(b), and for a MESFET device in Figure

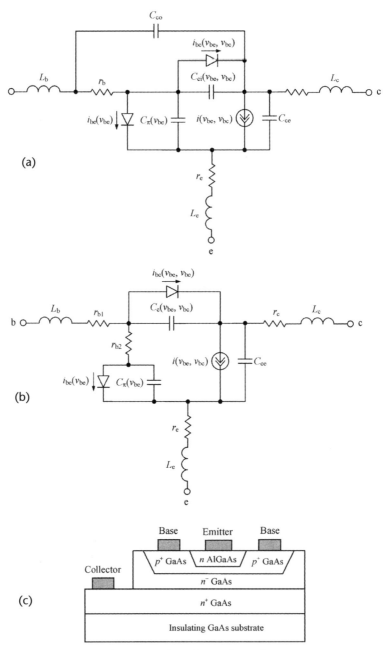

Figure 2.23 Nonlinear BJT and HBT models and HBT physical structure.

4.11(a), it is easy to detect the circuit similarity of all these equivalent circuits, which means that the basic circuit design procedure is very similar for any type of bipolar or field-effect transistor. The main difference is in the device physics and values of the model parameters. However, techniques for representation of the input and output impedances, stability analysis based on feedback effect, derivation of power gain, and efficiency are very similar.

The cross section of a physical structure of an AlGaAs/GaAs HBT device is shown in Figure 2.23(c), with a heavily p-doped base to reduce base resistance and a lightly n-doped emitter to minimize emitter capacitance [64]. The lightly n-doped collector region allows the collector-base junction to sustain relatively high voltages without breaking down. The forward-bias emitter-injection efficiency is very high since the wider-bandgap AlGaAs emitter injects electrons into the GaAs p-base at lower energy level, but the holes are prevented from flowing into the emitter by a high energy barrier, thus resulting in the ability to decrease base length and base-width modulation, and increase frequency response. By using a wide bandgap InGaP layer instead of an AlGaAs one, the device performance over temperature can be improved [65]. The high-linearity power performance in a Class AB condition at the backoff power level, the ruggedness under mismatch and overdrive conditions, and the long lifetime of InGaP/GaAs HBT technology makes it very attractive for 28V power amplifier applications [66].

The growth process used for a high-voltage HBT device is identical to the process used for the conventional low-voltage HBT device, which is widely used in handset power amplifiers, except for changes to the collector because of the higher voltage operating requirements. The epitaxial growth process starts with a highly doped n-type collector layer and a lightly n-doped collector drift region, then followed by a heavily doped p-type base layer and an InGaP emitter layer, and finishes with an InGaAs cap layer [67]. As a result, the high-voltage HBT devices exhibit collector-base breakdown voltages higher than 70V.

The bipolar transistor intrinsic Y-parameters can be written as

$$Y_{11} = \frac{1}{r_\pi} + j\omega\left(C_\pi + C_{ci}\right) \qquad Y_{12} = -j\omega C_{ci}$$
$$Y_{21} = g_m \exp\left(-j\omega\tau\right) - j\omega C_{ci} \qquad Y_{22} = \frac{1}{r_{ce}} + j\omega C_{ci} \tag{2.99}$$

where r_{ce} is the output Early resistance that models the effect of the base-width modulation on the transistor characteristics due to variations in the collector-base depletion region.

After separating the Y-parameters given in (2.99) into their real and imaginary parts, the elements of the intrinsic small-signal equivalent circuit can be determined analytically as [68]

$$C_\pi = \frac{\text{Im}\left(Y_{11} + Y_{12}\right)}{\omega} \tag{2.100}$$

$$r_\pi = \frac{1}{\text{Re}\,Y_{11}} \tag{2.101}$$

$$C_{ci} = -\frac{\operatorname{Im} Y_{12}}{\omega} \qquad (2.102)$$

$$g_m = \sqrt{\left(\operatorname{Re} Y_{21}\right)^2 + \left(\operatorname{Im} Y_{21} + \operatorname{Im} Y_{12}\right)^2} \qquad (2.103)$$

$$\tau_\pi = \frac{1}{\omega}\cos^{-1}\frac{\operatorname{Re} Y_{21} + \operatorname{Re} Y_{12}}{\sqrt{\left(\operatorname{Re} Y_{21}\right)^2 + \left(\operatorname{Im} Y_{21} + \operatorname{Im} Y_{12}\right)^2}} \qquad (2.104)$$

$$r_{ce} = \frac{1}{\operatorname{Re} Y_{22}} \qquad (2.105)$$

A simple nonlinear HBT model for computer-aided simulations can be based on representation of the collector current source through the power series and diffusion capacitances through the hyperbolic functions [69]. To equivalently represent the input impedance of a bipolar transistor, we need to take into account that C_{ce} is usually much smaller than C_c. As a result, the equivalent output capacitance can be defined as $C_{out} \cong C_c$. The input equivalent resistance R_{in} can approximately be represented by the base resistance r_b, while the input equivalent capacitance can be defined as $C_{in} \cong C_\pi + C_c$. The feedback effect of the collector capacitance C_c through C_{co} and C_{ci} is sufficiently high when load variations are directly transferred to the device input with a significant extent.

2.6 DC Biasing

The simplest way to provide a proper dc biasing condition for a power MOSFET device in Class A or Class AB operation is to use the potentiometer-type voltage divider for the gate bias and choke inductor in the drain circuit, as shown in Figure 2.24(a). However, in this case, any variations of the ambient temperature or bias voltage will lead to variations of quiescent current and, as a result, to appropriate variations of the output power, linearity, drain efficiency, and gain of the power amplifier. The threshold voltage V_{th} of the MOSFET transistor varies with temperature T linearly with the approximate velocity of $\Delta V_{th}/\Delta T \cong -2$ mV/°C. But simply adding a diode (or diode-connected MOSFET) in series to the variable resistor allows the quiescent current variation to be reduced substantially over temperature. A bias circuit corresponding to this stabilizing condition is shown in Figure 2.24(b). For a high value of V_{th}, several diodes can be connected in series. The reason to use such a simple bias circuit for power MOSFET biasing is that its dc gate current is equal to the gate leakage current only.

In contrast to MOSFET devices, where it may be possible to choose the optimum operating point with a practically zero temperature coefficient or to be limited to just an additional diode only, bipolar transistors require the more complicated approach of dc biasing depending on a class of operation. For example, in Class AB operation, the bias circuit has to deliver a dc voltage, which is slightly adjustable approximately within limits of 0.7V to 0.8V using a wide range of current values to stabilize the base current of the RF bipolar transistor. In addition, it is necessary to

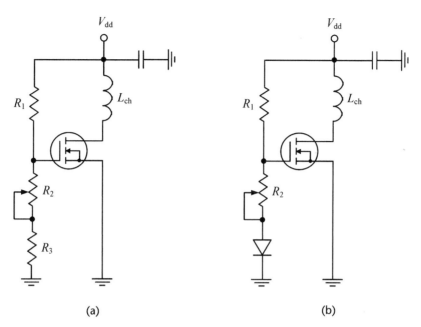

Figure 2.24 MOSFETs with simple bias circuits.

provide an operating mode for the power amplifier with temperature compensation (collector current stabilization over temperature) and minimum possible reference current (dc current from the reference dc voltage supply). One of the simplest versions of such a bias circuit with silicon diode temperature compensation is shown in Figure 2.25(a). In this bias circuit, each silicon diode can be replaced by the *n-p-n* diode-connected transistor, the collector and the base of which are directly connected between each other.

A better temperature-compensating result can be achieved using the same transistors for RF and dc paths, only with reduced area sizes for bias circuit devices. Such

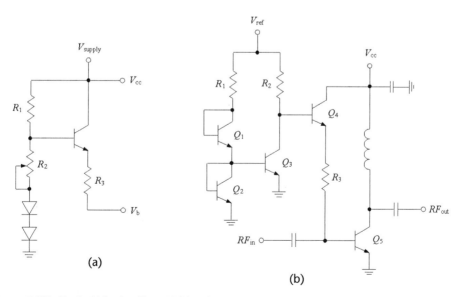

Figure 2.25 Typical bipolar Class AB bias circuits.

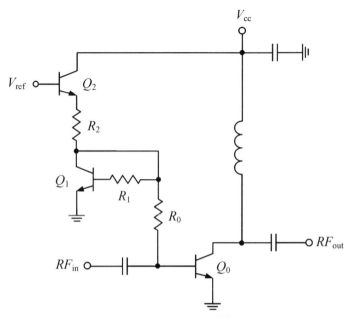

Figure 2.26 Bipolar power amplifier stage with emitter-follower bias circuit.

an approach is usually used in monolithic integrated circuit design when transistor cells with different area sizes are used for both RF power device and bias circuit transistors. Figure 2.25(b) shows the temperature and supply independent bias circuit, which is composed of the current driving transistor Q_4 and the compensation circuit including the diode-connected transistors Q_1 and Q_2, compensating transistor Q_3, and resistors R_1 and R_2 [70]. By providing the same values for R_1 and R_2, the InGaP/GaAs MMIC power amplifier for wideband code division multiple access (WCDMA) applications provides the quiescent current variations of only 6% for the temperature range of $-30°C$ to $90°C$ and 8.5% for the supply voltage range of 2.9V to 3.1V, with variations in the power gain of less than ±0.8 dB at the output power of 28 dBm.

Figure 2.26 shows the emitter-follower bias circuit that provides temperature compensation and minimizing reference current requirements [71]. The emitter follower bias circuit requires only several tens of microamperes of reference current, whereas the current-mirror bias circuit requires a few milliamperes of reference current. Both the current-mirror and emitter-follower bias circuits have similar current-voltage behavior but, for the same circuit parameters (R_0, R_1, and R_2) and device areas for Q_0, Q_1, and Q_2, the emitter follower bias circuit is less sensitive to the reference voltage variations compared with a current-mirror bias circuit. Variations in the collector supply voltage V_{cc} for limits of 3.0V to 5.0V have no effect on the quiescent current set by the reference voltage V_{ref}.

2.7 Impedance Transformers and Power Combiners

The transmission-line transformers and combiners can provide very wide operating bandwidths and operate up to frequencies of 3 GHz and higher [72, 73]. They are widely used in matching networks for antennas and power amplifiers in the HF and VHF bands, as well as in mixer circuits, and their low losses make them especially useful in high-power circuits [74, 75]. Typical structures for transmission-line transformers consist of parallel wires, coaxial cables, or bifilar twisted wire pairs. In the latter case, the characteristic impedance can easily be determined by the wire diameter, the insulation thickness, and, to some extent, by the twisting pitch [76, 77]. For coaxial cable transformers with correctly chosen characteristic impedance, the theoretical high-frequency bandwidth limit is reached when the cable length comes on the order of a half wavelength, with the overall achievable bandwidth being about a decade. By introducing low-loss high-permeability ferrites alongside a good-quality semirigid coaxial or symmetrical strip cable, the low frequency limit can be significantly improved, providing bandwidths of several or more decades.

The concept of a broadband impedance transformer consisting of a pair of interconnected transmission lines was first disclosed and described by Guanella [78, 79]. Figure 2.27(a) shows the Guanella transformer system with a transmission-line character achieved by an arrangement comprising one pair of cylindrical coils that is wound in the same sense and spaced a certain distance apart by an intervening dielectric. In this case, one cylindrical coil is located inside the insulating cylinder and the other coil is located on the outside of this cylinder. For the currents flowing through both windings in opposite directions, the corresponding flux in the coil axis

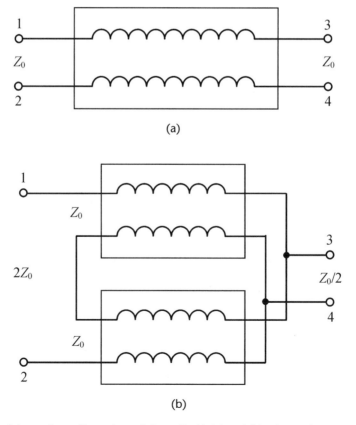

Figure 2.27   Schematic configurations of Guanella (a) 1:1 and (b) 4:1 transformers.

is negligibly small. However, for the currents flowing in the same direction through both coils, the latter may be assumed to be connected in parallel, and a coil pair represents a considerable inductance for such currents and acts like a choke coil. With terminal 4 being grounded, such a 1:1 transformer provides matching of a balanced source to an unbalanced load and is called a *balun* (*ba*lanced-to-*un*balanced transformer). In this case, if terminal 2 is grounded, it simply represents a delay line. In a particular case, when terminals 2 and 3 are grounded, the transformer performs as a phase inverter. The series-parallel connection of a plurality of coil pairs can produce a match between unequal source and load resistances. Figure 2.27(b) shows a 4:1 impedance (2:1 voltage) transmission-line transformer where the two pairs of cylindrical transmission-line coils are connected in series at the input and in parallel at the output. For the characteristic impedance Z_0 of each transmission line, this results in two times higher impedance $2Z_0$ at the input and two times lower impedance $Z_0/2$ at the output. By grounding terminal 4, such a 4:1 impedance transformer provides impedance matching of the balanced source to the unbalanced load. In this case, when terminal 2 is grounded, it performs as a 4:1 *unun* (*un*balanced-to-*un*balanced transformer). With a series-parallel connection of n coil pairs with the characteristic impedance Z_0 each, the input impedance is equal to nZ_0 and the output impedance is equal to Z_0/n. Because a Guanella transformer adds voltages that have equal delays through the transmission lines, such a technique results in the so called *equal-delay* transmission-line transformers.

The simplest transmission-line transformer represents a quarterwave transmission line whose characteristic impedance is chosen to give the correct impedance transformation. However, this transformer provides a narrowband performance valid only around frequencies for which the transmission line is odd multiples of a quarter-wavelength. If a ferrite sleeve is added to the transmission line, common-mode currents flowing in both transmission-line inner and outer conductors in phase and in the same direction are suppressed and the load may be balanced and floating above ground [80, 81]. If the characteristic impedance of the transmission line is equal to the terminating impedances, the transmission is inherently broadband. If not, there will be a dip in the response at the frequency, at which the transmission line is a quarter-wavelength long.

A coaxial cable transformer, the physical configuration and equivalent circuit representation of which are shown in Figures 2.28(a) and (b), respectively, consists of the coaxial line arranged inside the ferrite core or wound around the ferrite core. Due to its practical configuration, the coaxial cable transformer takes a position between the lumped and distributed systems. Therefore, at lower frequencies its equivalent circuit represents a conventional low-frequency transformer [Figure 2.28(c)], whereas at higher frequencies it is a transmission line with the characteristic impedance Z_0 [Figure 2.28(d)]. The advantage of such a transformer is that the parasitic interturn capacitance determines its characteristic impedance, whereas in the conventional wire-wound transformer with discrete windings, this parasitic capacitance negatively contributes to the transformer frequency performance.

When $R_S = R_L = Z_0$, the transmission line can be considered a transformer with a 1:1 impedance transformation. To avoid any resonant phenomena, especially for complex loads, which can contribute to significant output power variations, as a general rule, the length l of the transmission line is kept to no more than an eighth

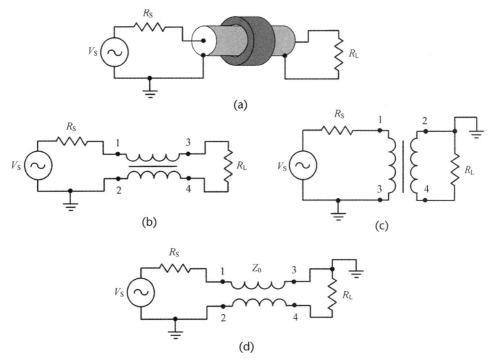

Figure 2.28 Schematic configurations of coaxial cable transformer.

of a wavelength λ_{min} at the highest operating frequency, that is, $l \le \lambda_{min}/8$, where λ_{min} is the minimum wavelength in the transmission line corresponding to the high operating frequency f_{max}.

The low-frequency bandwidth limit of a coaxial cable transformer is determined by the effect of the magnetizing inductance L_m of the outer surface of the outer conductor according to the equivalent low-frequency transformer model shown in Figure 2.29(a), where the transmission line is represented by the ideal 1:1 transformer [75]. The resistance R_0 represents the losses of the transmission line. An approximation to the magnetizing inductance can be made by considering the outer surface of the coaxial cable to be the same as that of a straight wire (or linear conductor) that, at higher frequencies where the skin effect causes the current to be concentrated on the outer surface, would have the self-inductance defined by

$$L_m = 2l\left[\ln\left(\frac{2l}{r}\right) - 1\right]nH \qquad (2.106)$$

where l is the length of the coaxial cable in centimeters and r is the radius of the outer surface of the outer conductor in centimeters [75].

High permeability of core materials results in shorter transmission lines. If a toroid is used for the core, the magnetizing inductance L_m is obtained by

$$L_m = 4\pi n^2 \mu \frac{A_e}{L_e}nH \qquad (2.107)$$

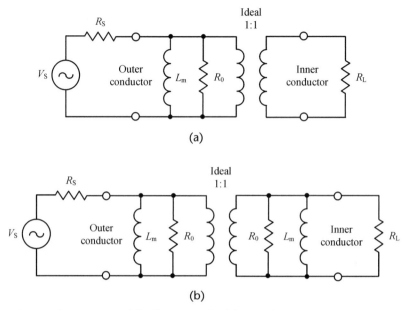

Figure 2.29 Low-frequency models of 1:1 coaxial cable transformer.

where

   n = number of turns

   μ = core permeability

   A_e = effective cross-sectional area of the core (cm^2)

   L_e = average magnetic path length (cm) [82].

By considering the transformer equivalent circuit shown in Figure 2.29(a), the ratio between the power delivered to the load P_L and power available at the source $P_S = V_S^2/8R_S$, when $R_S = R_L$, can be obtained from

$$\frac{P_L}{P_S} = \frac{\left(2\omega L_m\right)^2}{R_S^2 + \left(2\omega L_m\right)^2} \tag{2.108}$$

which gives the minimum operating frequency f_{min} for a given magnetizing inductance L_m as

$$f_{min} \geq \frac{R_S}{4\pi L_m} \tag{2.109}$$

when taking into account the maximum decrease of the output power by 3 dB.

A similar low-frequency model for a coaxial cable transformer using twisted or parallel wires is shown in Figure 2.29(b) [75]. Here, the model is symmetrical because both conductors are exposed to any magnetic material and therefore contribute identically to the losses and low-frequency performance of the transformer.

An approach using the transmission line based on a single bifilar wound coil to realize a broadband 1:4 impedance transformation was introduced by Ruthroff [83, 84]. In this case, by using a core material of sufficiently high permeability, the number of turns can be significantly reduced. Figure 2.30(a) shows the circuit schematic of an unbalanced-to-unbalanced 1:4 transmission-line transformer, where terminal 4 is connected to input terminal 1. As a result, for $V = V_1 = V_2$, the output voltage is twice the input voltage, and the transformer has a 1:2 voltage step-up ratio. Because the ratio of input voltage to input current is one-fourth the load voltage to load current, the transformer is fully matched for maximum power transfer when $R_L = 4R_S$, and the transmission-line characteristic impedance Z_0 is equal to the geometric mean of the source and load impedances,

$$Z_0 = \sqrt{R_S R_L} \tag{2.110}$$

where R_S is the source resistance and R_L is the load resistance. Figure 2.30(b) shows the impedance transformer acting as a phase inverter, where the load resistance is included between terminals 1 and 4 to become a 1:4 balun. This technique is called the *bootstrap effect*, which does not have the same high-frequency response as the Guanella equal-delay approach because it adds a delayed voltage to a direct one [85]. The delay becomes excessive when the transmission line reaches a significant fraction of a wavelength.

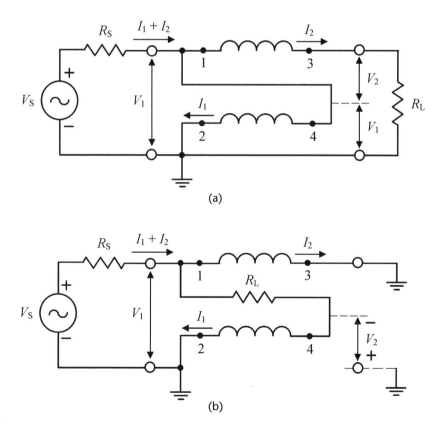

Figure 2.30 Schematic configurations of Ruthroff 1:4 impedance transformer.

Figure 2.31(a) shows the physical implementation of a 4:1 impedance Ruthroff transformer using a coaxial cable arranged inside the ferrite core. At lower frequencies, such a transformer can be considered an ordinary 2:1 voltage autotransformer. The insertion loss for a broadband 4:1 impedance transformer, as a function of the transmission-line electrical length θ, can be calculated from

$$\frac{P_L}{P_S} = \frac{4R_S R_L (1 + \cos\theta)^2}{\left[2R_S(1 + \cos\theta) + R_L \cos\theta\right]^2 + \left(\frac{R_S R_L + Z_0^2}{Z_0}\right)^2 \sin^2\theta} \tag{2.111}$$

where P_S is the maximum available power from the source with internal resistance R_S and P_L is the power delivered to load R_L [83, 86]. For a matched transformer when $R_L = 4R_S$ and $Z_0 = 2R_S$, (2.111) reduces to

$$\frac{P_L}{P_S} = \frac{4(1 + \cos\theta)^2}{(1 + 3\cos\theta)^2 + 4\sin^2\theta} \tag{2.112}$$

To improve the performance at higher frequencies, it is necessary to add an additional phase-compensating line of the same length, as shown in Figure 2.31(b), resulting in a Guanella ferrite-based 4:1 impedance transformer. In this case, a ferrite core is necessary only for the upper line because the outer conductor of the lower line is grounded at both ends, and no current is flowing through it. A current I driven

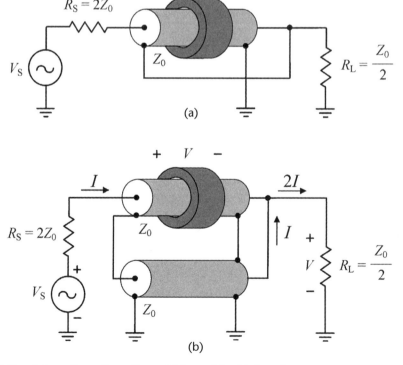

(a)

(b)

Figure 2.31   Schematic configurations of 4:1 coaxial cable transformer.

into the inner conductor of the upper line produces a current I that flows in the outer conductor of the upper line, resulting in a current $2I$ flowing into load R_L. Because the voltage $2V$ from the transformer input is divided in two equal parts between the coaxial line and the load, such a transformer provides an impedance transformation from $R_S = 2Z_0$ into $R_L = Z_0/2$, where Z_0 is the characteristic impedance of each coaxial line. The bandwidth extension for the Ruthroff transformers can also be achieved by using the transmission lines with a step function and exponential changes in their characteristic impedances [87, 88]. To adopt this transmission-line transformer for microwave planar applications, the coaxial line can be replaced by a pair of stacked strip conductors or coupled microstrip lines [89, 90].

2.8   Directional Couplers

The four-port networks are used for directional power coupling when, for a given input signal at port 1, the output signals are delivered to ports 2 and 3, and no power is delivered to port 4 (ideal case), as shown in Figure 2.32. The scattering S-matrix of a reciprocal four-port network matched at all its ports is given by

$$[S] = \begin{bmatrix} 0 & S_{12} & S_{13} & S_{14} \\ S_{12} & 0 & S_{23} & S_{24} \\ S_{13} & S_{23} & 0 & S_{34} \\ S_{14} & S_{24} & S_{34} & 0 \end{bmatrix} \tag{2.113}$$

where $S_{ij} = S_{ji}$ for the symmetric scattering S-matrix when all components are passive and reciprocal [91]. In this case, the power supplied to input port 1 is coupled to the coupled port, port 3, with a coupling factor $|S_{13}|^2$, whereas the remainder of the input power is delivered to the through port, port 2, with a coupling factor $|S_{12}|^2$.

For a lossless four-port network, the unitary condition of the fully matched S-matrix given by (2.113) results in

$$|S_{12}|^2 + |S_{13}|^2 = |S_{12}|^2 + |S_{24}|^2 = |S_{13}|^2 + |S_{34}|^2 = |S_{24}|^2 + |S_{34}|^2 = 1$$

which implies a full isolation between ports 2 and 3 and ports 1 and 4, respectively, when

$$S_{14} = S_{41} = S_{23} = S_{32} = 0$$

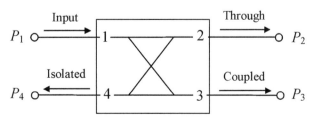

Figure 2.32   Schematic diagram of directional coupler.

and

$$|S_{13}| = |S_{24}| \qquad\qquad |S_{12}| = |S_{34}|$$

The scattering S-matrix of such a directional coupler, matched at all its ports with two decoupled two-port networks, reduces to

$$[S] = \begin{bmatrix} 0 & S_{12} & S_{13} & 0 \\ S_{12} & 0 & 0 & S_{24} \\ S_{13} & 0 & 0 & S_{34} \\ 0 & S_{24} & S_{34} & 0 \end{bmatrix} \qquad (2.114)$$

The directional coupler can be classified according to the phase shift ϕ between its two output ports 2 and 3 as the in-phase coupler with $\phi = 0$, quadrature coupler with $\phi = 90°$ or $\pi/2$, and out-of-phase coupler with $\phi = 180°$ or π. The following important quantities are used to characterize the directional coupler [91]:

- The power-split ratio or power division ratio K^2, which is calculated as the ratio of powers at the output ports when all ports are nominally (reflectionless) terminated:

$$K^2 = \frac{P_2}{P_3}$$

- The insertion loss C_{12}, which is calculated as the ratio of powers at input port 1 relative to output port 2:

$$C_{12} = 10\log_{10}\frac{P_1}{P_2} = -20\log_{10}|S_{12}|$$

- The coupling C_{13}, which is calculated as the ratio of powers at input port 1 relative to output port 3:

$$C_{13} = 10\log_{10}\frac{P_1}{P_3} = -20\log_{10}|S_{13}|$$

- The directivity C_{34}, which is calculated as the ratio of powers at output port 3 relative to isolated port 4:

$$C_{34} = 10\log_{10}\frac{P_3}{P_4} = 20\log_{10}\frac{|S_{13}|}{|S_{14}|}$$

- The isolation C_{14} and C_{23}, which are calculated as the ratios of powers at input port 1 relative to isolated port 4 and between the two output ports (output port 2 is considered an input port), respectively:

$$C_{14} = 10\log_{10}\frac{P_1}{P_4} = -20\log_{10}\left|S_{14}\right|$$

$$C_{23} = 10\log_{10}\frac{P_2}{P_3} = -20\log_{10}\left|S_{23}\right|$$

- The voltage standing wave ratio at each port or $VSWR_i$, where $i = 1, 2, 3, 4$, which is calculated as

$$VSWR_i = \frac{1+\left|S_{ii}\right|}{1-\left|S_{ii}\right|}$$

In an ideal case, the directional coupler would have an $VSWR_i = 1$ at each port, an insertion loss $C_{12} = 3$ dB, a coupling $C_{13} = 3$ dB, an infinite isolation, and a directivity $C_{14} = C_{23} = C_{34} = \infty$.

The first directional couplers were composed of either a two-wire balanced line coupled to a second balanced line along a distance of a quarter-wavelength, or a pair of rods a quarter-wavelength long between the ground planes [92]. Although the propagation of waves on systems of parallel conductors was investigated many decades ago in connection with the problem of crosstalk between open wire lines or cable pairs in order to eliminate the natural coupling rather than use it, the first exact design theory for transverse electromagnetic (TEM) transmission-line couplers was introduced by Oliver [93]. In terms of the even and odd electric-field modes describing a system of coupled conductors, it can be stated that the coupling is backward with a coupled wave on the secondary line propagating in the direction opposite to the direction of the wave on the primary line; the directivity will be perfect with VSWR equal to unity if $Z_0^2 = Z_{0e}Z_{0o}$ at all cross sections along the directional coupler; and the midband voltage coupling coefficient C of the directional coupler is defined as

$$C = \frac{Z_{0e} - Z_{0o}}{Z_{0e} + Z_{0o}} \tag{2.115}$$

where $C = 0$ for zero coupling and $C = 1$ for completely superposed transmission lines.

A coupled-line directional coupler, the stripline single-section topology of which is shown in Figure 2.33(a), can be used for broadband power dividing or combining. Its electrical properties are described using a concept of two types of excitations for the coupled lines in TEM approximation. In this case, for the even mode, the currents flowing in the strip conductors are equal in amplitude and flow in the same direction. The electric field has even symmetry about the center line, and no current flows between the two strip conductors. For the odd mode, the currents flowing in the strip conductors are equal in amplitude, but flow in opposite directions. The electric field lines have an odd symmetry about the center line, and a voltage null exists between these two strip conductors. An arbitrary excitation of the coupled lines can always be treated as a superposition of appropriate amplitudes of even and odd modes. Therefore, the characteristic impedance for even excitation mode Z_{0e}

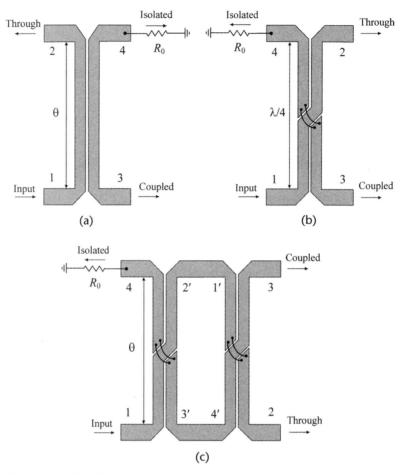

Figure 2.33 Coupled-line directional couplers.

and the characteristic impedance for the odd excitation mode Z_{0o} characterize the coupled lines. When the two coupled equal-striplines are used in a standard system with characteristic impedance $\sqrt{Z_{0e}Z_{0o}}$, then

$$Z_{0e} = Z_0\sqrt{\frac{1+C}{1-C}} \tag{2.116}$$

$$Z_{0o} = Z_0\sqrt{\frac{1-C}{1+C}} \tag{2.117}$$

An analysis in terms of the scattering S-parameters gives $S_{11} = S_{14} = 0$ for any electrical lengths of the coupled lines and output port 4 is isolated from the matched input port, port 1. Changing the coupling between the lines and their widths can change the characteristic impedances Z_{0e} and Z_{0o}. In this case,

$$S_{12} = \frac{\sqrt{1-C^2}}{\sqrt{1-C^2}\cos\theta + j\sin\theta} \tag{2.118}$$

$$S_{13} = \frac{jC\sin\theta}{\sqrt{1-C^2}\cos\theta + j\sin\theta} \tag{2.119}$$

where θ is the electrical length of the coupled-line section.

The voltage-split ratio K is defined as the ratio between voltages at port 2 and port 3 as

$$K = \left|\frac{S_{12}}{S_{13}}\right| = \frac{\sqrt{1-C^2}}{C\sin\theta} \tag{2.120}$$

where K can be controlled by changing the coupling coefficient C and electrical length θ.

For a quarter-wavelength-long coupler when $\theta = 90°$, (2.118) and (2.119) reduce to

$$S_{12} = -j\sqrt{1-C^2} \qquad\qquad S_{13} = C \tag{2.121}$$

from which it follows that an equal voltage split between output ports 2 and 3 can be provided with $C = 1/\sqrt{2}$.

If it is necessary to provide output ports 2 and 3 at one side, it is best to use the construction of a microstrip directional coupler with crossed bondwires, as shown in Figure 2.33(b). The strip crossover for a stripline directional coupler can be easily achieved with the three-layer sandwich. The microstrip 3-dB directional coupler fabricated on alumina substrate for idealized zero strip thickness should have the calculated strip spacing of less than 10 μm. Such a narrow value easily explains the great interest in the construction of directional couplers with larger spacing.

One of the effective solutions is to use a tandem connection for the two identical directional couplers, which alleviates the physical problem of tight coupling, since two individual couplers need only 8.34-dB coupling to achieve a 3-dB directional coupler [94, 95]. The tandem coupler shown in Figure 2.33(c) has the electrical properties of the individual coupler when output ports 1, 4 and 2, 3 are isolated in pairs, and the phase difference between output ports 2 and 3 is 90°.

From an analysis of the signal propagation from input port 1 to output ports 2 and 3 of the tandem coupler, when the signal from input port 1 propagates to output port 2 through traces 1-2′-1′-2 and 1-3′-4′-2, and the signal flowing through traces 1-2′-1′-3 and 1-3′-4′-3 is delivered to output port 3, the ratio of the scattering parameters S_{12}^T and S_{13}^T of a tandem coupler can be expressed through the corresponding scattering parameters S_{12} and S_{13} of the individual coupler as

$$\frac{S_{12}^T}{S_{13}^T} = \frac{S_{12}^2 + S_{13}^2}{2S_{12}S_{13}} = -j\frac{1 - C^2\left(1 + \sin^2\theta\right)}{2C\sqrt{1-C^2}\sin\theta} \tag{2.122}$$

As a result, the signal at output port 2 overtakes the signal at output port 3 by 90°. In this case, for a 3-dB tandem coupler with $\theta = 90°$, the magnitude of (2.122) must be equal to unity. Consequently, the required voltage coupling coefficient is calculated as

$$C = 0.5\sqrt{2 - \sqrt{2}} = 0.3827$$

or

$$C_{12} = C_{13} = 8.34 \text{ dB}$$

As an example, a tandem 8.34-dB directional coupler has the dimensions of $W/h = 0.77$ and $S/h = 0.18$ for alumina substrate with $\varepsilon_r = 9.6$, where W is the strip width, S is the strip spacing, and h is the substrate thickness [96].

Another way to increase the coupling between the two edge-coupled microstrip lines is to use several parallel narrow microstrip lines interconnected with each other by the bondwires, as shown in Figure 2.34. For a Lange coupler like that shown in Figure 2.34(a), four coupled microstrip lines are used, achieving 3-dB coupling over an octave or more bandwidth [97]. In this case, the signal flowing to input port 1 is distributed between output ports 2 and 3 with a phase difference of 90°. However, this structure is quite complicated for practical implementation when, for alumina substrate with $\varepsilon_r = 9.6$, the dimensions of a 3-dB Lange coupler are $W/h = 0.107$ and $S/h = 0.071$, where W is the width of each strip and S is the spacing between adjacent strips.

Figure 2.34(b) shows the unfolded Lange coupler with four strips of equal length, which offers the same electrical performance but is easier for circuit modeling [98]. The even-mode characteristic impedance Z_{e4} and odd-mode characteristic impedance Z_{o4} of the Lange coupler with $Z_0^2 = Z_{e4}Z_{o4}$ in terms of the characteristic impedances of a two-conductor line (which is identical to any pair of adjacent lines in the coupler) can be obtained by

$$Z_{e4} = \frac{Z_{0o} + Z_{0e}}{3Z_{0o} + Z_{0e}} Z_{0e} \tag{2.123}$$

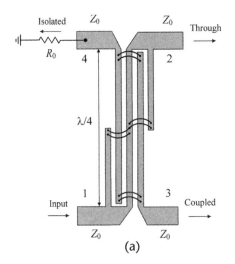
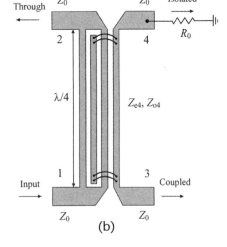

Figure 2.34   Lange directional couplers.

$$Z_{o4} = \frac{Z_{0e} + Z_{0o}}{3Z_{0e} + Z_{0o}} Z_{0o} \tag{2.124}$$

where Z_{0e} and Z_{0o} are the even- and odd-mode characteristic impedances of the two-conductor pair [99].

The midband voltage coupling coefficient C is given by

$$C = \frac{Z_{e4} - Z_{o4}}{Z_{e4} + Z_{o4}} = \frac{3\left(Z_{0e}^2 - Z_{0o}^2\right)}{3\left(Z_{0e}^2 + Z_{0o}^2\right) + 2Z_{0e}Z_{0o}} \tag{2.125}$$

The even- and odd-mode characteristic impedances Z_{0e} and Z_{0o}, as functions of the characteristic impedance Z_0 and coupling coefficient C, are determined by

$$Z_{0e} = Z_0 \sqrt{\frac{1+C}{1-C}} \frac{4C - 3 + \sqrt{9 - 8C^2}}{2C} \tag{2.126}$$

$$Z_{0o} = Z_0 \sqrt{\frac{1-C}{1+C}} \frac{4C + 3 - \sqrt{9 - 8C^2}}{2C} \tag{2.127}$$

For alumina substrate with $\varepsilon_r = 9.6$, the dimensions of such a 3-dB unfolded Lange coupler are $W/h = 0.112$ and $S/h = 0.08$, where W is the width of each strip and S is the spacing between the strips.

The design theory for TEM transmission-line couplers is based on an assumption of the same phase velocities of the even and odd propagation mode. However, this is not the case for coupled microstrip lines, since they have unequal even- and odd-mode phase velocities. In this case, the odd mode has more fringing electric field in the air region rather than the even mode with the electrical field concentrated mostly in the substrate underneath the microstrip lines. As a result, the effective dielectric permittivity in the latter case is higher, thus indicating a smaller phase velocity for

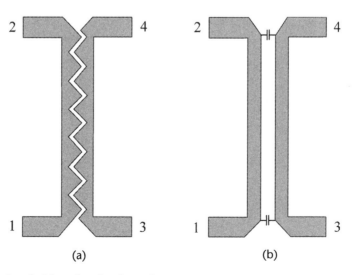

Figure 2.35   Coupled-line directional couplers.

the even mode. Consequently, phase velocity compensation techniques must be applied to improve coupler directivity, which decreases with increasing frequency.

Figure 2.35(a) shows the topology of a typical wiggly-line coupler (with a saw-tooth shape for the coupled lines), where wiggling the adjacent edges of the microstrip lines, which makes their physical lengths longer, slows the odd-mode wave without much affecting the even-mode wave [100]. High directivity can also be achieved by using a capacitive compensation. Figure 2.35(b) shows the capacitively compensated microstrip directional coupler, where the two identical lumped capacitors are connected between coupled lines at their edges. Physically, these edge capacitors affect the odd mode by equivalent extension of the transmission-line electrical lengths, with almost no effect for the even mode. For an ideal lossless operating condition at 12 GHz using standard alumina substrate, the compensated coupled-line microstrip directional coupler can improve directivity from 13.25 dB to infinity [101]. Capacitive compensation can be performed by gap coupling of the open-circuit stub formed in a subcoupled line [102]. In this case, the coupler directivity can be improved by 23 dB in a frequency range from 1 to 2.5 GHz, compared to the directivity of the conventional uncompensated microstrip coupler.

References

[1]   W. L. Everitt, "Output Networks for Radio-Frequency Power Amplifiers," *Proc. IRE*, Vol. 19, pp. 725–737, May 1931.

[2]   H. T. Friis, "Noise Figure of Radio Receivers," *Proc. IRE*, Vol. 32, pp. 419–422, July 1944.

[3]   S. Roberts, "Conjugate-Image Impedances," *Proc. IRE*, Vol. 34, pp. 198–204, Apr. 1946.

[4]   S. J. Haefner, "Amplifier-Gain Formulas and Measurements," *Proc. IRE*, Vol. 34, pp. 500–505, July 1946.

[5]   R. L. Pritchard, "High-Frequency Power Gain of Junction Transistors," *Proc. IRE*, Vol. 43, pp. 1075–1085, Sep. 1955.

[6]   A. R. Stern, "Stability and Power Gain of Tuned Power Amplifiers," *Proc. IRE*, Vol. 45, pp. 335–343, Mar. 1957.

[7]   L. S. Houselander, H. Y. Chow, and R. Spense, "Transistor Characterization by Effective Large-Signal Two-Port Parameters," *IEEE J. Solid-State Circuits*, Vol. SC-5, pp. 77–79, Apr. 1970.

[8]   B. T. Vincent, "Large Signal Operation of Microwave Transistors," *IEEE Trans. Microwave Theory Tech.*, Vol. MTT-49, pp. 865–866, Nov. 1965.

[9]   H. W. Bode, "A Method of Impedance Correction," *Bell Syst. Tech. J.*, Vol. 9, pp. 794–838, Oct. 1930.

[10] S. A. Schelkunoff, "The Impedance Concept and Its Application to Problems of Reflection, Refraction, Shielding and Power Absorption," *Bell Syst. Tech. J.*, Vol. 17, pp. 17–48, Jan. 1938.

[11] H. W. Bode, *Network Analysis and Feedback Amplifier Design*, New York: Van Nostrand, 1945.

[12] V. M. Bogachev and V. V. Nikiforov, *Transistor Power Amplifiers* (in Russian), Moskva: Energiya, 1978.

[13] A. Grebennikov, *RF and Microwave Power Amplifier Design*, New York: McGraw-Hill, 2014.

[14] A. Grebennikov, N. O. Sokal, and M. J. Franco, *Switchmode RF and Microwave Power Amplifiers*, New York: Academic Press, 2012.

[15] A. Tam, "Network Building Blocks Balance Power Amp Parameters," *Microwaves & RF*, Vol. 23, pp. 81–87, July 1984.

[16] P. H. Smith, *Electronic Applications of the Smith Chart*, New York: Noble Publishing, 2000.

[17] B. J. Thompson, "Oscillations in Tuned Radio-Frequency Amplifiers," *Proc. IRE*, Vol. 19, pp. 421–437, Mar. 1931.

[18] G. W. Fyler, "Parasites and Instability in Radio Transmitters," *Proc. IRE*, Vol. 23, pp. 985–1012, Sep. 1935.

[19] F. B. Llewellyn, "Some Fundamental Properties of Transmission Systems," *Proc. IRE*, Vol. 40, pp. 271–283, Mar. 1952.

[20] D. F. Page and A. R. Boothroyd, "Instability in Two-Port Active Networks," *IRE Trans. Circuit Theory*, Vol. CT-5, pp. 133–139, June 1958.

[21] J. M. Rollett, "Stability and Power Gain Invariants of Linear Two-Ports," *IRE Trans. Circuit Theory Appl.*, Vol. CT-9, pp. 29–32, Jan. 1962.

[22] J. G. Linvill and L. G. Schimpf, "The Design of Tetrode Transistor Amplifiers," *Bell Syst. Tech. J.*, Vol. 35, pp. 813–840, Apr. 1956.

[23] O. Muller and W. G. Figel, "Stability Problems in Transistor Power Amplifiers," *Proc. IEEE*, Vol. 55, pp. 1458–1466, Aug. 1967.

[24] O. Muller, "Internal Thermal Feedback in Fourpoles, Especially in Transistors," *Proc. IEEE*, Vol. 52, pp. 924–930, Aug. 1964.

[25] J. H. Morecroft and H. T. Friis, "The Vacuum Tubes as a Generator of Alternating-Current Power," *Trans. AIEE*, Vol. 38, pp. 1415–1444, Oct. 1919.

[26] D. C. Prince, "Vacuum Tubes as Power Oscillators, Part I," *Proc. IRE*, Vol. 11, pp. 275–313, June 1923.

[27] A. A. Oswald, "Power Amplifiers in Trans-Atlantic Radio Telephony," *Proc. IRE*, Vol. 13, pp. 313–324, June 1925.

[28] L. E. Barton, "High Audio Power from Relatively Small Tubes," *Proc. IRE*, Vol. 19, pp. 1131–1149, July 1931.

[29] F. E. Terman and J. H. Ferns, "The Calculation of Class C Amplifier and Harmonic Generator Performance of Screen-Grid and Similar Tubes," *Proc. IRE*, Vol. 22, pp. 359–373, Mar. 1934.

[30] A. I. Berg, *Theory and Design of Vacuum-Tube Generators* (in Russian), Moskva: GEI, 1932.

[31] P. H. Osborn, "A Study of Class B and C Amplifier Tank Circuits," *Proc. IRE*, Vol. 20, pp. 813–834, May 1932.

[32] C. E. Fay, "The Operation of Vacuum Tubes as Class B and Class C Amplifiers," *Proc. IRE*, Vol. 20, pp. 548–568, Mar. 1932.

[33] L. B. Hallman, "A Fourier Analysis of Radio-Frequency Power Amplifier Wave Forms," *Proc. IRE*, Vol. 20, pp. 1640–1659, Oct. 1932.

[34] W. L. Everitt, "Optimum Operating Conditions for Class C Amplifiers," *Proc. IRE*, Vol. 22, pp. 152–176, Feb. 1934.

[35] N. Chevaux and M. M. De Souza, "Comparative Analysis of VDMOS/LDMOS Power Transistors for RF Amplifiers," *IEEE Trans. Microwave Theory Tech.*, Vol. MTT-57, pp. 2643–2651, Nov. 2009.

[36] G. A. Holle and H. C. Reader, "Nonlinear MOSFET Model for the Design of RF Power Amplifiers," *IEE Proc. Circuits Devices Syst.*, Vol. 139, pp. 574–580, Oct. 1992.

[37] A. Grebennikov, *RF and Microwave Power Amplifier Design*, New York: McGraw-Hill, 2004.

[38] W. R. Curtice et al., "A New Dynamic Electro-Thermal Nonlinear Model for Silicon RF LDMOS FETs," *1999 IEEE MTT-S Int. Microwave Symp. Dig.*, Vol. 2, pp. 419–422.

[39] C. Fager et al., "Prediction of IMD in LDMOS Transistor Amplifiers Using a New Large-Signal Model," *IEEE Trans. Microwave Theory Tech.*, Vol. MTT-50, pp. 2834–2842, Dec. 2002.

[40] W. R. Curtice et al., "New LDMOS Model Delivers Powerful Transistor Library: The CMC Model," *High Frequency Electronics*, Vol. 3, pp. 18–25, Oct. 2004.

[41] Y. P. Tsividis, *Operation and Modeling of the MOS Transistor*, New York: McGraw-Hill, 1987.

[42] R. Sung, P. Bendix, and M. B. Das, "Extraction of High-Frequency Equivalent Circuit Parameters of Submicron Gate-Length MOSFET's," *IEEE Trans. Electron Devices*, Vol. ED-45, pp. 1769–1775, Aug. 1998.

[43] M. C. Ho et al., "A Physical Large Signal Si MOSFET Model for RF Circuit Design," *1997 IEEE MTT-S Int. Microwave Symp. Dig.*, pp. 391–394.

[44] B. J. Cheu and P. K. Ko, "Measurement and Modeling of Short-Channel MOS Transistor Gate Capacitances," *IEEE J. Solid-State Circuits*, Vol. SC-22, pp. 464–472, June 1987.

[45] J.-M. Dortu et al., "Accurate Large-Signal GaAs MESFET and HEMT Modeling for Power MMIC Amplifier Design," *Int. J. Microwave and Millimeter-Wave Computer-Aided Eng.*, Vol. 5, pp. 195–208, Sept. 1995.

[46] L.-S. Liu, J.-G. Ma, and G.-I. Ng, "Electrothermal Large-Signal Model for III-V FETs Including Frequency Dispersion and Charge Conservation," *IEEE Trans. Microwave Theory Tech.*, Vol. MTT-38, pp. 822–824, June 1990.

[47] C.-J. Wei, Y. Tkachenko, and D. Bartle, "An Accurate Large-Signal Model of GaAs MESFET Which Accounts for Charge Conservation, Dispersion, and Self-Heating," *IEEE Trans. Microwave Theory Tech.*, Vol. MTT-46, pp. 1638–1644, Nov. 1998.

[48] A. Jarndal and G. Kompa, "Large-Signal Model for AlGaN/GaN HEMTs Accurately Predicts Trapping- and Self-Heating-Induced Dispersion and Intermodulation Distortion," *IEEE Trans. Electron Devices*, Vol. ED-54, pp. 2830–2836, Nov. 2007.

[49] T. Kacprzak and A. Materka, "Compact DC Model of GaAs FET's for Large-Signal Computer Calculation," *IEEE J. Solid-State Circuits*, Vol. SC-18, pp. 211–213, Apr. 1983.

[50] I. Angelov, H. Zirath, and N. Rorsman, "A New Empirical Nonlinear Model for HEMT and MESFET Devices," *IEEE Trans. Microwave Theory Tech.*, Vol. MTT-40, pp. 2258–2266, Dec. 1992.

[51] O. S. A. Tang et al., "Design of High-Power, High-Efficiency 60-GHz MMIC's Using an Improved Nonlinear PHEMT Model," *IEEE J. Solid-State Circuits*, Vol. SC-32, pp. 1326–1333, Sep. 1997.

[52] I. Angelov et al., "On the Large-Signal Modelling of AlGaN/GaN HEMTs and SiC MESFETs," *13th Europ. GAAS Symp. Dig.*, pp. 309–312, 2005.

[53] A. Garcia-Osorio et al., "An Empirical *I-V* Nonlinear Model Suitable for GaN FET Class F PA Design," *Microwave and Optical Technology Lett.*, Vol. 53, pp. 1256–1259, June 2011.

[54] K. Joshin and T. Kikkawa, "High-Power and High-Efficiency GaN HEMT Amplifiers," *2008 IEEE Radio and Wireless Symp. Dig.*, pp. 65–68.

[55] T. Ishida, "GaN HEMT Technologies for Space and Radio Applications," *Microwave J.*, Vol. 54, pp. 56–66, Aug. 2011.

[56] U. K. Mishra et al., "GaN-Based RF Power Devices and Amplifiers," *Proc. IEEE*, Vol. 96, pp. 287–305, Feb. 2008.

[57] M. Berroth and R. Bosch, "Broad-Band Determination of the FET Small-Signal Equivalent Circuit," *IEEE Trans. Microwave Theory Tech.*, Vol. MTT-38, pp. 891–895, July 1990.

[58] Q. Fan, J. H. Leach, and H. Morkoc, "Small Signal Equivalent Circuit Modeling for AlGaN/GaN HFET: Hybrid Extraction Method for Determining Circuit Elements of AlGaN/GaN HFET," *Proc. IEEE*, Vol. 98, pp. 1140–1150, July 2010.

[59] G. Dambrine et al., "A New Method for Determining the FET Small-Signal Equivalent Circuit," *IEEE Trans. Microwave Theory Tech.*, Vol. MTT-36, pp. 1151–1159, July 1988.

[60] N. M. Rohringer and P. Kreuzgruber, "Parameter Extraction for Large-Signal Modeling of Bipolar Junction Transistors," *Int. J. Microwave and Millimeter-Wave Computer-Aided Eng.*, Vol. 5, pp. 161–272, Sep. 1995.

[61] J. P. Fraysee et al., "A Non-Quasi-Static Model of GaInP/AlGaAs HBT for Power Applications," *1997 IEEE MTT-S Int. Microwave Symp. Dig.*, Vol. 2, pp. 377–382.

[62] M. Reisch, *High-Frequency Bipolar Transistors*, Berlin: Springer, 2003.

[63] V. M. Bogachev and V. V. Nikiforov, *Transistor Power Amplifiers* (in Russian), Moskva: Energiya, 1978.

[64] P. M. Asbeck et al., "Heterojunction Bipolar Transistors for Microwave and Millimeter-Wave Integrated Circuits," *IEEE Trans. Microwave Theory Tech.*, Vol. MTT-35, pp. 1462–1468, Dec. 1987.

[65] Y.-S. Lin and J.-J. Jiang, "Temperature Dependence of Current Gain, Ideality Factor, and Offset Voltage of AlGaAs/GaAs and InGaP/GaAs HBTs," *IEEE Trans. Electron Devices*, Vol. ED-56, pp. 2945–2951, Dec. 2009.

[66] N. L. Wang et al., "28-V High-Linearity and Rugged InGaP/GaAs HBT," *2006 IEEE MTT-S Int. Microwave Symp. Dig.*, pp. 881–884.

[67] C. Steinberger et al., "250 W HVHBT Doherty with 57% WCDMA Efficiency Linearized to -55 dBc for 2c11 6.5 dB PAR," *IEEE J. Solid-State Circuits*, Vol. SC-43, pp. 2218–2228, Oct. 2008.

[68] D. Costa, W. U. Liu, and J. S. Harris, "Direct Extraction of the AlGaAs/GaAs Heterojunction Bipolar Transistor Small-Signal Equivalent Circuit," *IEEE Trans. Electron Devices*, Vol. ED-38, pp. 2018–2024, Sep. 1991.

[69] I. Angelov, K. Choumei, and A. Inoue, "An Empirical HBT Large-Signal Model for CAD," *Int. J. RF and Microwave Computer-Aided Eng.*, Vol. 13, pp. 518–533, Nov. 2003.

[70] Y. S. Noh, J. H. Park, and C. S. Park, "A Temperature and Supply Independent Bias Circuit and MMIC Power Amplifier Implementation for W-CDMA Applications," *IEICE Trans. Electron.*, Vol. E88-C, pp. 725–728, Apr. 2005.

[71] T. Sato and C. Grigorean, "Design Advantages of CDMA Power Amplifiers Built with MOSFET Technology," *Microwave J.*, Vol. 45, pp. 64–78, Oct. 2002.

[72] H. L. Krauss, C. W. Bostian, and F. H. Raab, *Solid State Radio Engineering*, New York: John Wiley & Sons, 1980.

[73] Z. I. Model, *Networks for Combining and Distribution of High Frequency Power Sources* (in Russian), Moskva: Sov. Radio, 1980.

[74] J. Sevick, *Transmission Line Transformers*, Norcross, GA: Noble Publishing, 2001.

[75] C. Trask, "Transmission Line Transformers: Theory, Design and Applications," *High Frequency Electronics*, Vol. 5, pp. 26–33, Jan. 2006.

[76] E. Rotholz, "Transmission-Line Transformers," *IEEE Trans. Microwave Theory Tech.*, Vol. MTT-29, pp. 148–154, Apr. 1981.

[77] J. Horn and G. Boeck, "Design and Modeling of Transmission Line Transformers," *Proc. 2003 IEEE SBMO/MTT-S Int. Microwave and Optoelectronics Conf.*, Vol. 1, pp. 421–424.

[78] G. Guanella, "New Method of Impedance Matching in Radio-Frequency Circuits," *The Brown Boveri Rev.*, Vol. 31, pp. 327–329, Sept. 1944.

[79] G. Guanella, "High-Frequency Matching Transformer," U.S. Patent 2,470,307, May 1949 (filed Apr. 1945).

[80] R. K. Blocksome, "Practical Wideband RF Power Transformers, Combiners, and Splitters," *Proc. RF Technology Expo 86*, pp. 207–227, 1986.

[81] J. L. B. Walker et al., *Classic Works in RF Engineering: Combiners, Couplers, Transformers, and Magnetic Materials*, Norwood, MA: Artech House, 2005.

[82] J. Sevick, "Magnetic Materials for Broadband Transmission Line Transformers," *High Frequency Electronics*, Vol. 4, pp. 46–52, Jan. 2005.

[83] C. L. Ruthroff, "Some Broad-Band Transformers," *Proc. IRE*, Vol. 47, pp. 1337–1342, Aug. 1959.

[84] C. L. Ruthroff, "Broadband Transformers," U.S. Patent 3,037,175, May 1962 (filed May 1958).

[85] J. Sevick, "A Simplified Analysis of the Broadband Transmission Line Transformer," *High Frequency Electronics*, Vol. 3, pp. 48–53, Feb. 2004.

[86] O. Pitzalis and T. P. M. Couse, "Practical Design Information for Broadband Transmission Line Transformer," *Proc. IEEE*, Vol. 56, pp. 738–739, Apr. 1968.

[87] R. T. Irish, "Method of Bandwidth Extension for the Ruthroff Transformer," *Electronics Lett.*, Vol. 15, pp. 790–791, Nov. 1979.

[88] S. C. Dutta Roy, "Optimum Design of an Exponential Line Transformer for Wide-Band Matching at Low Frequencies," *Proc. IEEE*, Vol. 67, pp. 1563–1564, Nov. 1979.

[89] M. Engels et al., "Design Methodology, Measurement and Application of MMIC Transmission Line Transformers," *1995 IEEE MTT-S Int. Microwave Symp. Dig.*, Vol. 3, pp. 1635–1638.

[90] S. P. Liu, "Planar Transmission Line Transformer Using Coupled Microstrip Lines," *1998 IEEE MTT-S Int. Microwave Symp. Dig.*, Vol. 2, pp. 789–792.

[91] D. M. Pozar, *Microwave Engineering*, New York: John Wiley & Sons, 2004.

[92] S. B. Cohn and R. Levy, "History of Microwave Passive Components with Particular Attention to Directional Couplers," *IEEE Trans. Microwave Theory Tech.*, Vol. MTT-32, pp. 1046–1054, Sep. 1984.

[93] B. M. Oliver, "Directional Electromagnetic Couplers," *Proc. IRE*, Vol. 42, pp. 1686–1692, Nov. 1954.

[94] G. D. Monteath, "Coupled Transmission Lines as Symmetrical Directional Couplers," *IEE Proc.*, Vol. 102, part B, pp. 383–392, May 1955.

[95] T. P. Shelton and J. A. Mosko, "Synthesis and Design of Wide-Band Equal-Ripple TEM Directional Couplers and Fixed Phase Shifters," *IEEE Trans. Microwave Theory Tech.*, Vol. MTT-14, pp. 462–473, Oct. 1966.

[96] O. A. Chelnokov, *Radio Transmitters* (in Russian), Moskva: Radio i Svyaz, 1982.

[97] J. Lange, "Interdigitated Stripline Quadrature Hybrid," *IEEE Trans. Microwave Theory Tech.*, Vol. MTT-17, pp. 1150–1151, Dec. 1969.

[98] R. Waugh and D. LaCombe, "Unfolding the Lange Coupler," *IEEE Trans. Microwave Theory Tech.*, Vol. MTT-20, pp. 777–779, Nov. 1972.

[99] W. P. Ou, "Design Equations for an Interdigitated Directional Coupler," *IEEE Trans. Microwave Theory Tech.*, Vol. MTT-23, pp. 253–255, Feb. 1975.

[100] A. Podell, "A High Directivity Microstrip Coupler Technique," *1970 G-MTT Int. Microwave Symp. Dig.*, pp. 33–36.

[101] M. Dydyk, "Accurate Design of Microstrip Directional Couplers with Capacitive Compensation," *1990 IEEE MTT-S Int. Microwave Symp. Dig.*, Vol. 1, pp. 581–584.

[102] C.-S. Kim, J.-S. Lim, D.-J. Kim, and D. Ahn, "A Design of Single and Multi-Section Microstrip Directional Coupler with the High Directivity," *2004 IEEE MTT-S Int. Microwave Symp. Dig.*, pp. 1895–1898.

CHAPTER 3

Overview of Broadband Power Amplifiers

In many telecommunication, radar, or testing systems, the transmitters operate in a very wide frequency range, for example, 1.5 to 30 MHz in high-frequency (HF) transceivers, 225 to 400 MHz in military frequency-agility systems, 470 to 860 MHz in ultrahigh-frequency (UHF) TV transmitters, or 2 to 8 GHz and 6 to 18 GHz in microwave applications. The power amplifier design based on a broadband concept provides some advantages when there is no need to tune resonant circuits, and it is possible to realize fast frequency agility or to transmit a wide multimode signal spectrum. However, many factors restrict the frequency bandwidth depending on the active device parameters. For example, it is quite easy to provide multiple-octave amplification from very low frequencies up to UHF band using the power MOSFET devices when lossy gain compensation is easily provided. This is possible due to the existence of some margin of power gain at lower frequencies for these devices, since its power gain value decreases with frequency by approximately 6 dB per octave. In addition, lossy gain-compensating networks can provide lower input reflection coefficients, smaller gain ripple, more predictable amplifier design, and can contribute to the amplifier stability factors that are superior to those of lossless matching networks. At higher frequencies when the device input impedance is significantly smaller and the influence of its internal feedback and parasitic parameters is substantially higher, it is necessary to use multisection matching networks with lumped and distributed elements.

Generally, the matching design procedure is based on the methods of circuit analysis, optimization, and synthesis. In the first method, the circuit parameters are calculated at one frequency chosen in advance (usually the center or high bandwidth frequency), and then power amplifier performance is analyzed across the entire frequency bandwidth. To synthesize the broadband matching/compensation network, it is necessary to choose the maximum attenuation level or reflection coefficient magnitude inside the operating frequency bandwidth and then to obtain the parameters of matching networks by using special tables and formulas to convert the lumped element into distributed ones. For push-pull power amplifiers, it is very convenient to use both lumped and distributed parameters when the lumped capacitors are connected in parallel to the microstrip lines due to the effect of a virtual ground.

3.1 Bode-Fano Criterion

Generally, the design for a broadband matching circuit should solve a problem with contradictory requirements when a wider matching bandwidth is required with minimum reflection coefficient; it is a matter of minimizing the number of matching network sections for a given wideband specification. The necessary requirements are determined by the Bode-Fano criterion, which gives (for certain canonical types of load impedances) a theoretical limit on the maximum reflection coefficient magnitude that can be obtained with an arbitrary matching network [1, 2].

For the lossless matching networks with a parallel RC load shown in Figure 3.1(a) and with a series LR load shown in Figure 3.1(b), the Bode-Fano criterion states that

$$\int_0^\infty \ln \frac{1}{|\Gamma(\omega)|}\, d\omega \le \frac{\pi}{\tau} \tag{3.1}$$

where $\Gamma(\omega)$ is the input reflection coefficient seen looking into the arbitrary lossless matching network and $\tau = RC = L/R$.

For the lossless matching networks with a series RC load shown in Figure 3.1(c) and with a parallel LR load shown in Figure 3.1(d), the Bode-Fano integral is written as

$$\int_0^\infty \omega^{-2} \ln \frac{1}{|\Gamma(\omega)|}\, d\omega \le \pi\tau \tag{3.2}$$

The mathematical relationships expressed by (3.1) and (3.2) reflect the flat responses of an ideal filter over the required frequency bandwidth, as shown in Figure 3.2 for two different cases. For the same load, both plots illustrate an important trade-off: The wider the matching network bandwidth, the worse the reflection coefficient magnitude. From (3.1) it follows that, when $|\Gamma(\omega)|$ is constant and equal to $|\Gamma|_{\max}$ over a frequency band of width $\Delta\omega$ and $|\Gamma(\omega)| = 1$ otherwise,

$$\int_0^\infty \ln \frac{1}{|\Gamma(\omega)|}\, d\omega = \int_{\omega_1}^{\omega_2} \ln \frac{1}{|\Gamma(\omega)|}\, d\omega = \Delta\omega \ln \frac{1}{|\Gamma|_{\max}} \le \frac{\pi}{\tau} \tag{3.3}$$

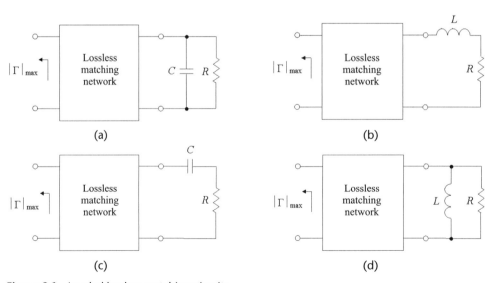

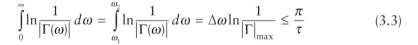

Figure 3.1 Loaded lossless matching circuits.

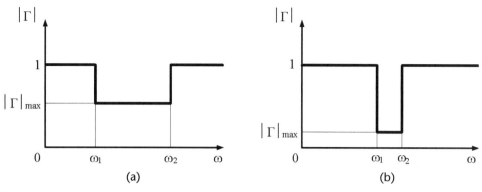

Figure 3.2   Ideal filter flat responses.

As a result,

$$|\Gamma|_{max} = \exp\left(\frac{-\pi}{\Delta\omega\tau}\right) \qquad (3.4)$$

where $\Delta\omega = \omega_2 - \omega_1$.

Similarly, for the lossless network with a series RC load and with a parallel LR load,

$$|\Gamma|_{max} = \exp\left(\frac{-\pi\omega_0^2\tau}{\Delta\omega}\right) \qquad (3.5)$$

where $\omega_0 = \sqrt{\omega_1\omega_2}$ is the center bandwidth frequency. Note that the theoretical bandwidth limits can be realized only with an infinite number of matching network sections. The frequency bandwidth with a maximum reflection coefficient magnitude is determined by a loaded quality factor $Q_L = \omega_0\tau$ for the series RL or parallel RC circuit and by $Q_L = 1/(\omega_0\tau)$ for the parallel RL or series RC circuit, respectively. The Chebyshev matching transformer with a finite number of sections can be considered as a close approximation to the ideal passband network when the ripple of the Chebyshev response is made equal to $|\Gamma|_{max}$. By combining matching theory with the closed formulas for the element values of a Chebyshev lowpass filter, explicit formulas for optimum matching networks can be obtained in certain simple but common cases [3]. For example, analytic closed-form solutions for the design of optimum matching networks up through order $n = 4$ can be derived [4].

Generally, (3.4) and (3.5) can be rewritten in a simplified form:

$$|\Gamma|_{max} = \exp\left(-\pi\frac{Q_0}{Q_L}\right) \qquad (3.6)$$

where $Q_0 = \omega_0/\Delta\omega$.

3.2 Matching Networks with Lumped Elements

To design correctly the broadband matching circuits for transistor power amplifiers, it is necessary to transform and match the device complex impedances with the source and load impedances, which are usually resistive and equal to 50Ω. For high-power or low-supply voltage cases, the device impedances may be small enough, and an ideal transformer (IT) needs to be included along with a matching circuit, as shown in Figure 3.3. In this case, such an ideal transformer provides only a required transformation between the source resistance R_S and the input impedance of the matching circuit and does not have any effect on the circuit frequency characteristics.

To implement such an ideal transformer for an impedance-transforming circuit, it is useful to operate with the Norton transform. As a result, an ideal transformer with two capacitors C_1 and C_2, which is shown in Figure 3.4(a), can be equivalently replaced by three capacitors C_I, C_{II}, and C_{III} connected in the form of a π-transformer, as shown in Figure 3.4(b). Their values are determined as follows:

$$C_I = n_T(n_T - 1)C_1 \tag{3.7}$$

$$C_{II} = n_T C_1 \tag{3.8}$$

$$C_{III} = C_2 - (n_T - 1)C_1 \tag{3.9}$$

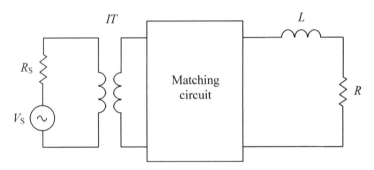

Figure 3.3 Matching circuit with ideal transformer.

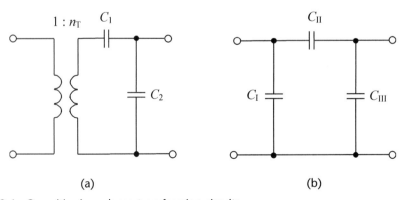

(a)   (b)

Figure 3.4 Capacitive impedance-transforming circuits.

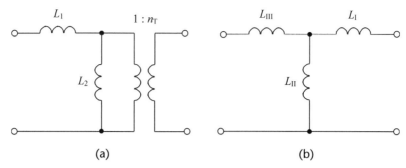

Figure 3.5 Inductive impedance-transforming circuits.

where n_T is the transformation coefficient. In this case, all of the parameters of these two-port networks are assumed identical at any frequency. However, such a replacement is possible only if the capacitance C_{III} obtained by (3.9) is positive and, consequently, physically realizable.

Similarly, an ideal transformer with two inductors L_1 and L_2, as shown in Figure 3.5(a), can be replaced by three inductors L_I, L_{II}, and L_{III} connected in the form of a T-transformer, as shown in Figure 3.5(b), with values determined by

$$L_I = n_T\left(n_T - 1\right)L_2 \tag{3.10}$$

$$L_{II} = n_T L_2 \tag{3.11}$$

$$L_{III} = L_1 - \left(n_T - 1\right)L_2 \tag{3.12}$$

Again, this replacement is possible only if the inductance L_{III} defined by (3.12) is positive and, consequently, physically realizable.

The broadband impedance-transforming circuits generally represent the transforming bandpass filters when the in-band matching requirements with specified ripple must be satisfied. In this case, the out-of-band mismatching can be very significant. One of the design methods of such matching circuits is based on the theory of transforming the lowpass filters of a ladder configuration of series inductors alternating with shunt capacitors, whose two-section equivalent representation is shown in Figure 3.6. For a large ratio of R_0/R_5, mismatching at zero frequency is

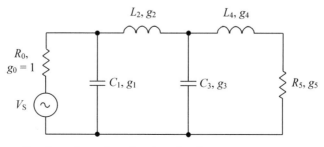

Figure 3.6 Two-section impedance-transforming circuit.

sufficiently high, and such a matching circuit can be treated as a bandpass imped-ance-transforming filter.

Table 3.1 gives the maximum passband ripples and coefficients g_1 and g_2 required to calculate the parameters of a two-section lowpass Chebyshev filter for different transformation ratios $r = R_0/R_S$ and frequency bandwidths $w = 2(f_2 - f_1)/(f_2 + f_1)$, where f_2 and f_1 are the high- and low-bandwidth frequencies, respectively [5]. The coefficients g_3 and g_4 are calculated as $g_3 = rg_2$ and $g_4 = g_1/r$, respectively, and the circuit elements can be obtained by

$$C_1 = \frac{g_1}{\omega_0 R_0} \qquad C_3 = \frac{g_3}{\omega_0 R_0} \qquad (3.13)$$

$$L_2 = \frac{g_2 R_0}{\omega_0} \qquad L_4 = \frac{g_4 R_0}{\omega_0} \qquad (3.14)$$

where $\omega_0 = \sqrt{\omega_1 \omega_2}$ is the center bandwidth frequency.

As an example, consider the design of a broadband input matching circuit in the form of a two-section lowpass transforming filter shown in Figure 3.6, with a center bandwidth frequency $f_0 = 3$ GHz, to match the source impedance $R_S = R_0 = 50\Omega$ with device input impedance $Z_{in} = R_{in} + j\omega_0 L_{in}$, where $R_{in} = R_S = 2\Omega$, $L_{in} = L_4 = 0.223$ nH, and $\omega_0 = 2\pi f_0$. The value of the series input device inductance $L_{in} = L_4$ is chosen to satisfy the requirements of Table 3.1 for $r = 25$ and $w = 0.4$ with maximum ripple of 0.156725 when $g_1 = 2.31517$ and $g_2 = 0.422868$. From (3.14), it follows that

$$L_4 = \frac{g_4 R_0}{\omega_0} = \frac{g_1 R_0}{\omega_0 r} = 0.223 \text{ nH}$$

Table 3.1 Two-Section Lowpass Chebyshev Filter Parameters

r	w	ripple, dB	g_1	g_2
5	0.1	0.000087	1.26113	0.709217
	0.2	0.001389	1.27034	0.704050
	0.3	0.007023	1.28561	0.695548
	0.4	0.022109	1.30687	0.638849
10	0.1	0.000220	1.60350	0.591627
	0.2	0.003516	1.62135	0.585091
	0.3	0.017754	1.65115	0.574412
	0.4	0.055746	1.69304	0.559894
25	0.1	0.000625	2.11734	0.462747
	0.2	0.009993	2.15623	0.454380
	0.3	0.050312	2.22189	0.440863
	0.4	0.156725	2.311517	0.422868
50	0.1	0.001303	2.57580	0.384325
	0.2	0.020801	2.64380	0.374422
	0.3	0.104210	2.75961	0.358638
	0.4	0.320490	2.92539	0.338129

As a result, the circuit parameters shown in Figure 3.7(a) are calculated from (3.13) and (3.14), resulting in the corresponding circuit frequency response shown in Figure 3.7(b) with the required passband from 2.6 to 3.4 GHz. The particular value of the inductance L_{in} is chosen for design convenience. If this value differs from the required value, then the maximum frequency bandwidth, the power ripple, or the number of ladder sections must be changed.

Another approach is based on the transformation from the lowpass impedance-transforming prototype filters, the simple L-, T-, and π-type equivalent circuits of which are shown in Figure 3.8, to the bandpass impedance-transforming filters. Table 3.2 gives the parameters for the lowpass impedance-transforming Chebyshev filter prototypes for different maximum in-band ripples and number of elements n [6]. This transformation can be obtained using the frequency substitution as

$$\omega \rightarrow \frac{\omega_0}{\Delta\omega}\left(\frac{\omega}{\omega_0} - \frac{\omega_0}{\omega}\right) \tag{3.15}$$

where $\omega_0 = \sqrt{\omega_1\omega_2}$ is the center bandwidth frequency, $\Delta\omega = \omega_2 - \omega_1$ is the passband, and ω_1 and ω_2 are the low and high edges of the passband, respectively.

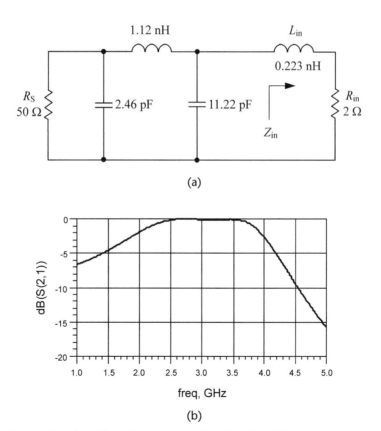

(a)

(b)

Figure 3.7 Two-section broadband lowpass matching circuit and its frequency response.

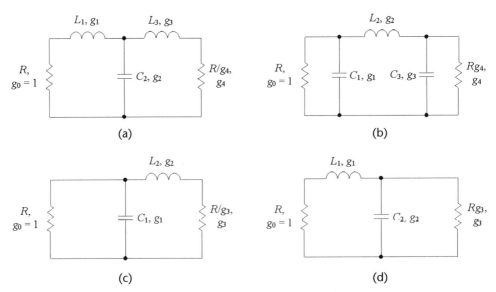

Figure 3.8 Lumped L-, π-, and T-type impedance-transforming circuits.

As a result, a series inductor L_k is transformed into a series LC circuit according to

$$\omega L_k = \frac{\omega_0}{\Delta\omega}\left(\frac{\omega}{\omega_0} - \frac{\omega_0}{\omega}\right)L_k = \omega L_k' - \frac{1}{\omega C_k'} \tag{3.16}$$

where

$$L_k' = \frac{L_k}{\Delta\omega} \qquad C_k' = \frac{\Delta\omega}{\omega_0^2 L_k} \tag{3.17}$$

Table 3.2 Parameters of Lowpass Chebyshev Filter Prototypes

ripple, dB	n	g_1	g_2	g_3	g_4
0.01	1	0.0960	1.0000		
	2	0.4488	0.4077	1.1007	
	3	0.6291	0.9702	0.6291	1.0000
0.1	1	0.3052	1.0000		
	2	0.8430	0.6220	1.3554	
	3	1.0315	1.1474	1.0315	1.0000
0.2	1	0.4342	1.0000		
	2	1.0378	0.6745	1.5386	
	3	1.2275	1.1525	1.2275	1.0000
0.5	1	0.6986	1.0000		
	2	1.4029	0.7071	1.9841	
	3	1.5963	1.0967	1.5963	1.0000

Similarly, a shunt capacitor C_k is transformed into a shunt LC-circuit as

$$\omega C_k = \frac{\omega_0}{\Delta\omega}\left(\frac{\omega}{\omega_0} - \frac{\omega_0}{\omega}\right)C_k = \omega C'_k - \frac{1}{\omega L'_k} \qquad (3.18)$$

where

$$C'_k = \frac{C_k}{\Delta\omega} \qquad L'_k = \frac{\Delta\omega}{\omega_0^2 C_k} \qquad (3.19)$$

The lowpass impedance-transforming prototype filter will be transformed to the bandpass impedance-transforming filter when all of its series elements are replaced by the series resonant circuits and all of its parallel elements are replaced by the parallel resonant circuits, each of which is tuned to the center bandwidth radian frequency ω_0. The bandpass filter elements can be calculated from

$$\Delta\omega C_k = \frac{g_k}{R} \qquad (3.20)$$

$$\Delta\omega L_k = g_k R \qquad (3.21)$$

where k is an element serial number for the lowpass prototype filter, and the g_k's are the appropriate coefficients given by Table 3.2.

Consider the design of a lowpass prototype filter for a given maximum ripple level in a frequency range up to $\omega_{2(2)}$ for a two-element filter and up to $\omega_{2(3)}$ for three elements, as shown in Figure 3.9(a). Then, for a selected arbitrary frequency ω_0, a series capacitance is added to each inductance and a parallel inductance is added to each capacitance on the assumption that all of these resonant circuits are tuned to the selected frequency ω_0. As a result, a new bandpass filter will be realized with the same ripple, as shown in Figure 3.9(b) for $n = 2$ and $n = 3$ with the passbands $\Delta\omega_{(2)}$ and $\Delta\omega_{(3)}$, respectively. Their elements are calculated according to (3.20) and (3.21).

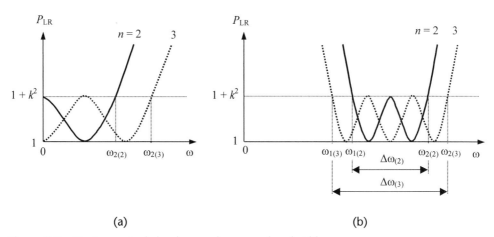

(a)   (b)

Figure 3.9 Maximum ripple level versus frequency bandwidth.

The maximum ripple level shown in Figure 3.9 determines the insertion loss IL (or power loss ratio P_{LR}) through the magnitude of the reflection coefficient Γ as

$$IL = 10 \ \log_{10} P_{LR} = -10 \ \log_{10}\left(1 - |\Gamma(\omega)|^2\right) \qquad (3.22)$$

Generally, the lowpass prototype filters and the bandpass filters obtained on their basis do not perform an impedance transformation. The input and output resistances are either equal for the symmetric T- or π-type filters shown in Figures 3.8(a) and (b) where $g_4 = 1$ or their ratio is too small for L-type filters, such as those shown in Figures 3.8(c) and (d), where $g_3 < 2$. Therefore, in this case, it is necessary to use an ideal transformer concept. This approach is based on using the existing data tables, from which the parameters of such impedance-transforming networks can be easily calculated for a given quality factor of the device input or output circuit. However, they can also be very easily verified or optimized by using a computer-aided design (CAD) optimization procedure that incorporates any comprehensive circuit simulator.

Consider the design example of a broadband interstage impedance-transforming filter with a center bandwidth frequency of 1 GHz to match the output driver-stage circuit with the input final-stage circuit of the power amplifier, as shown in Figure 3.10(a) [7]. In this case, it is convenient initially to convert the parallel connection of the device output resistance R_{out} and capacitance C_{out} into the corresponding series connection at the center bandwidth frequency ω_0, as shown in Figure 3.10(b).

For the three-element lowpass impedance-transforming prototype filter shown in Figure 3.8(a) with a maximum in-band ripple of 0.1 dB, we can obtain $g_1 = g_3 = 1.0315$, $g_2 = 1.1474$, and $g_4 = 1$ for $n = 3$ from Table 3.2. According to (3.21), the relative frequency bandwidth in this case is defined as

$$\frac{\Delta\omega}{\omega_0} = \frac{g_1 R_{in}}{\omega_0 L_{in}} = 16.5\%$$

based on a value of which the shunt capacitance C_2 can be calculated using (3.20), resulting in a capacitive reactance equal to 0.215Ω. The inductive reactance corresponding to a series inductance L_{in} is equal to 9.42Ω.

To convert the lowpass filter to its bandpass prototype, it is necessary to connect the capacitor in series to the input inductor and the inductor in parallel to the shunt capacitor and calculate with the same reactances to resonate at the center bandwidth frequency ω_0, as shown in Figure 3.10(c), where an ideal transformer IT with the transformation coefficient $n_T = \sqrt{9.8/1.5} = 2.556$ is included. Here, the reactances for each series element are equal to 9.42Ω, whereas the reactances for each parallel element are equal to 0.215Ω, respectively. Then, moving the corresponding elements with transformed parameters (each inductive and capacitive reactance is multiplied by n_T^2) to the left-hand side of IT in order to apply a Norton transform gives the circuit shown in Figure 3.10(d), where the required series elements with reactances of $9.42 n_T^2 \ \Omega$ are realized by the inductance L_{out}, converted device output capacitance C'_{out}, and additional elements L' and C'. Finally, by using a Norton transform (see Figure 3.4) with ideal transformer IT and two capacitors, the impedance matching bandpass filter shown in Figure 3.11(a) is obtained.

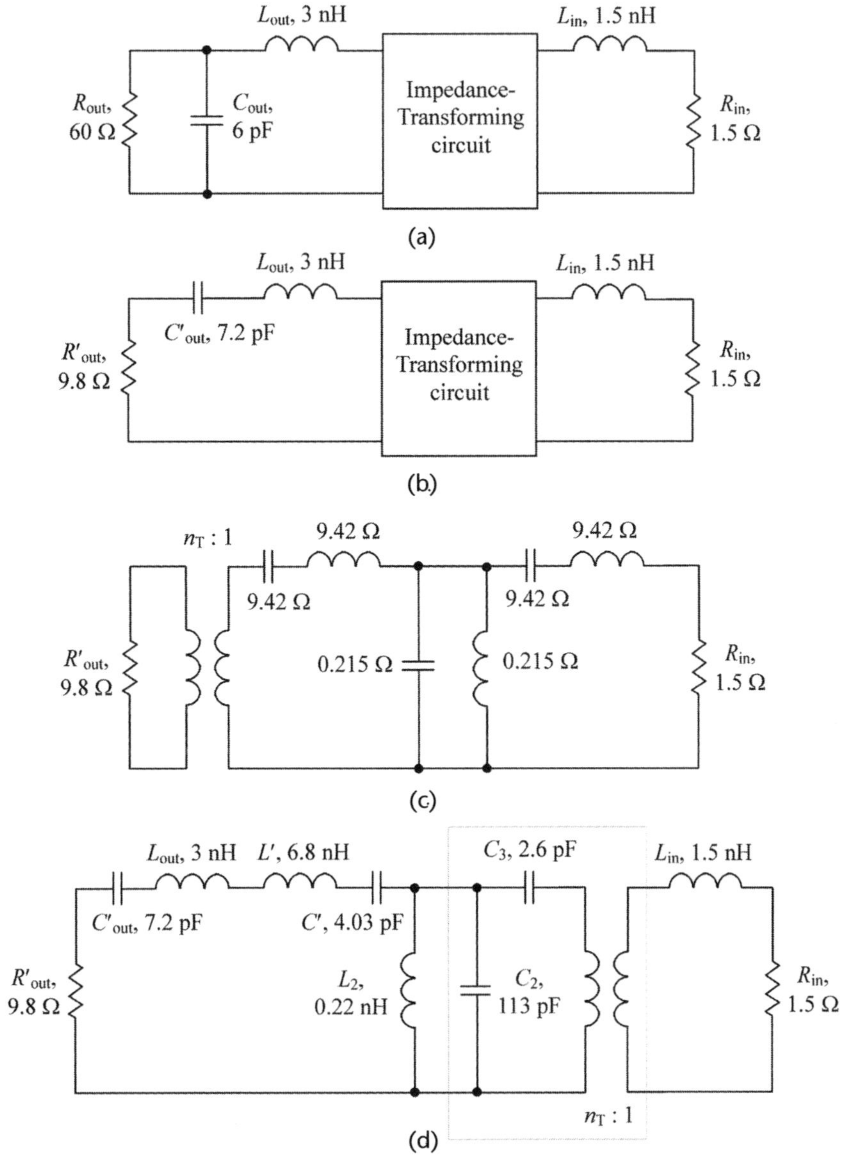

Figure 3.10 Impedance-transformer design procedure using a lowpass filter prototype.

The frequency response of the filter with minimum in-band ripple and significant out-of-band suppression is shown in Figure 3.11(b). In the case of serious difficulties with practical implementation of a very small inductance of 0.22 nH or a very large capacitance of 109 pF, it is possible to design a multisection lowpass impedance-transforming circuit.

Figure 3.12(a) shows the circuit schematic for a microwave broadband amplifier using a 1-μm GaAs FET packaged transistor, where the input multisection matching circuit is designed to provide the required gain taper and both input and output matching circuits are optimized to provide broadband impedance transformation [8]. As a result, a nominal power gain of 8 dB with a maximum deviation of ±0.07% in the frequency range of 7 to 14 GHz was achieved. In the first monolithic

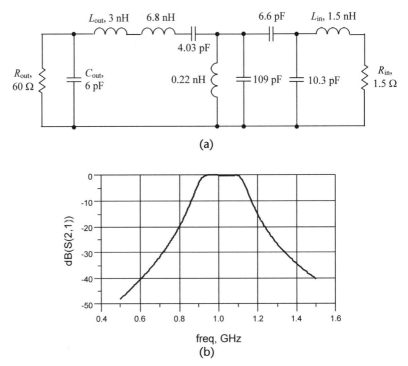

Figure 3.11   (a) Impedance-transforming bandpass filter and (b) its frequency response.

broadband GaAs FET amplifier, the input and output matching circuits were based on lumped elements fabricated together with the FET device on a semi-insulating high-resistivity gallium-arsenide substrate with a total size of 1.8×1.2 mm^2, providing a power gain of 4.5 ± 0.9 dB with an output power of 11 dBm at 1-dB gain compression from 7.0 to 11.7 GHz [9].

The circuit diagram of a two-stage pHEMT MMIC power amplifier for Ku-band applications is shown in Figure 3.12(b), where the lumped components were used in the input, interstage, and output matching circuits to minimize the overall chip size [10]. Here, the topology of each matching network represents a double-resonant circuit to form a broadband impedance transformer, which includes a shunt inductor in series with a bypass capacitor to provide a dc path, a series blocking capacitor, and a lowpass L-section transformer. In this case, for an 8.4-mm driver-stage pHEMT and a 16.8-mm power-stage pHEMT, a saturated output power of 38.1 dBm (6.5 W), a small-signal gain of 10.5 dB, and a peak PAE of 24.6% from 13.6 to 14.2 GHz were achieved with an MMIC chip size of as small as 3.64×2.35 mm^2. Based on the T-shape combining transformers with three individual inductors implemented in a 0.15-μm pHEMT technology, a broadband MMIC power amplifier combining two pHEMT devices with an overall 400-μm gate-width size achieved a saturated output power of 22 to 23.5 dBm and a power gain of over 10 dB from 17 to 35 GHz [11]. In a 90-nm standard CMOS process, a canonical doubly terminated third-order bandpass network was converted to the output matching topology, which provides both impedance transformation and differential-to-single-ended power combining [12]. The power amplifier achieved a 3-dB bandwidth from 5.2 to 13 GHz with a 25.2-dBm peak saturated output power and a peak PAE of 21.6%.

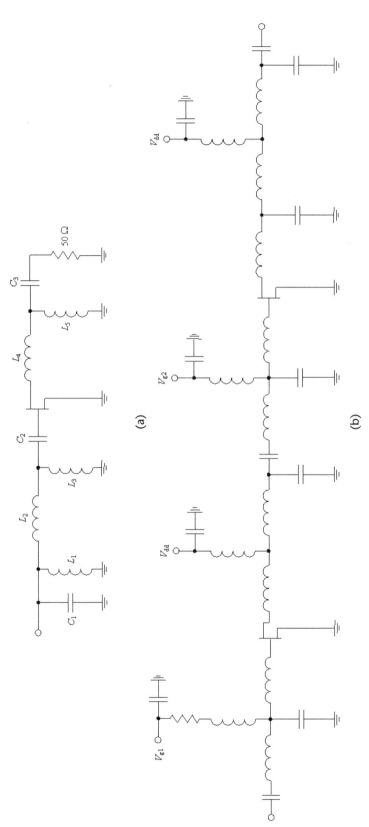

Figure 3.12   Schematics of broadband lumped-element microwave FET amplifiers.

3.3   Matching Networks with Mixed Lumped and Distributed Elements

The matching circuits, which incorporate mixed lumped and transmission-line elements, are widely used in both hybrid and monolithic design techniques. Such matching circuits are very convenient when designing push-pull power amplifiers with the effect of virtual grounding, where the shunt capacitors are connected between two series microstrip lines. According to the quasilinear transformation technique, the basic four-step design procedure consists of an appropriate choice of the lumped prototype schematic resulting in near-maximum gain across the required frequency bandwidth, its decomposition into subsections, their replacement by almost equivalent distributed circuits, and then the application of an optimization technique to minimize power variation over the operating frequency bandwidth [13].

A periodic lumped LC structure in the form of a lowpass ladder π-network is used as a basis for the lumped matching prototype. Then, the lumped prototype should be split up into individual π-type sections with equal capacitances by a consecutive step-by-step process and replaced by their equivalent distributed network counterparts. Finally, the complete mixed matching structure is optimized to

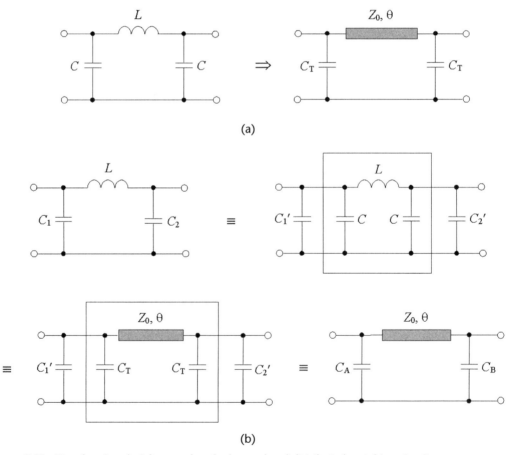

(a)

(b)

Figure 3.13 Transforming design procedure for lumped and distributed matching circuits.

improve the overall performance by employing a standard nonlinear optimization routine on the element values. Note that generally the lumped prototype structure can be decomposed into different subnetworks including L-type matching sections and individual capacitors or inductors.

For a single frequency equivalence between lumped and distributed elements, the lowpass lumped π-type ladder section can be made equivalent to a symmetrically loaded transmission line at a certain frequency, as shown in Figure 3.13(a). The transmission $ABCD$-matrices of these lumped and distributed ladder sections can be written, respectively, as

$$[ABCD]_L = \begin{bmatrix} 1 - \omega_0^2 LC & j\omega_0 L \\ j\omega_0 C \left(2 - \omega_0^2 LC\right) & 1 - \omega_0^2 LC \end{bmatrix} \tag{3.23}$$

$$[ABCD]_T = \begin{bmatrix} \cos\theta_0 - \omega_0 C_T Z_0 \sin\theta_0 & jZ_0 \sin\theta_0 \\ \dfrac{j}{Z_0}\left(2\omega_0 C_T Z_0 \cos\theta_0 + \sin\theta_0 - \omega_0^2 C_T^2 Z_0^2 \sin\theta_0\right) & \cos\theta_0 - \omega_0 C_T Z_0 \sin\theta_0 \end{bmatrix} \tag{3.24}$$

where θ_0 is the electrical length of a transmission line at the center bandwidth frequency ω_0.

Consequently, since these two circuits are equivalent, equal matrix elements $A_L = A_T$ and $B_L = B_T$ can be rewritten as

$$1 - \omega_0^2 LC = \cos\theta_0 - \omega_0 C_T Z_0 \sin\theta_0 \tag{3.25}$$

$$j\omega_0 L = jZ_0 \sin\theta_0 \tag{3.26}$$

After solving (3.25) and (3.26), the characteristic impedance Z_0 and shunt capacitance C_T can be explicitly calculated by

$$Z_0 = \frac{\omega_0 L}{\sin\theta_0} \tag{3.27}$$

$$C_T = \frac{\cos\theta_0 + \omega_0^2 LC - 1}{\omega_0^2 L} \tag{3.28}$$

To provide the design method using a single frequency equivalent technique, the following consecutive design steps can be performed:

1. Designate the lumped π-type C_1-L_1-C_2 section to be replaced.
2. From a chosen π-type C_1-L_1-C_2 section, form the symmetrical C-L-C ladder section with equal capacitances C, as shown in Figure 3.13(b). The choice

of capacitances is arbitrary but their values cannot exceed the minimum of (C_1, C_2).

3. Calculate the parameters of the symmetrical C_T-TL-C_T section using the parameters of the lumped equivalent π-section by setting the electrical length θ_0 of the transmission line according to (3.27) and (3.28). Here, it is assumed that the minimum of the capacitances C_1 and C_2 should be greater than or equal to C_T so that C_T can be readily embedded in the new C_T-TL-C_T section.

4. Finally, replace the π-type C_1-L_1-C_2 ladder section by the equivalent symmetrical C_T-TL-C_T section and combine adjacent shunt capacitances, as shown in Figure 3.13(b), where the loaded shunt capacitances C_A and C_B are given as $C_A = C_1' + C_T$ and $C_B = C_2' + C_T$.

Figure 3.14(a) shows the circuit schematic of a simulated broadband 28V LDMOSFET power amplifier. To provide an output power of about 15W with a power gain of more than 10 dB in the frequency range of 225 to 400 MHz, an LDMOSFET device with a gate geometry of 1.25 μm $\times$ 40 mm was chosen. In this case, the matching design technique is based on using multisection lowpass networks, with two π-type sections for the input matching circuit and one π-type section for the output matching circuit. The sections adjacent to the device input and output terminals incorporate the corresponding internal input gate-source and output drain-source device capacitances. Because a ratio between the device equivalent output resistance at the fundamental for several tens of watts of output power and the load resistance of 50Ω is not significant, it is sufficient to be limited to only one matching section for the output matching network.

Once a matching network structure is chosen, based on the requirements for the electrical performance and frequency bandwidth, the simplest and fastest way to optimize electrical performance using CAD simulators to satisfy certain criteria. For such a broadband power amplifier, these criteria can be the minimum output power ripple and input return loss with maximum power gain and efficiency. To minimize the overall dimensions of the power amplifier board, the shunt microstrip line in the drain circuit can be treated as an element of the output matching circuit, and its electrical length can be considered as a variable to be optimized. Applying a nonlinear broadband CAD optimization technique implemented in any high-level circuit simulator and setting the ranges of electrical length of the transmission lines between 0° and 90° and parallel capacitances from 0 to 100 pF, we can potentially obtain the parameters of the input and output matching circuits. The characteristic impedances of all transmission lines can be set to 50Ω for simplicity and convenience of the circuit implementation. However, to speed up this procedure, it is best to optimize circuit parameters separately for input and output matching circuits with the device equivalent input and output impedances: a series RC circuit for the device input and a parallel RC circuit for the device output. It is sufficient to use a fast linear optimization process, which will take only a few minutes to complete the matching circuit design. Then, the resulting optimized values are incorporated into the overall power amplifier circuit for each element and final optimization is performed using a large-signal active device model. In this case, the optimization process is finalized by choosing the nominal level of input power with optimizing elements in much narrower ranges of their values of about 10% to 20% for most

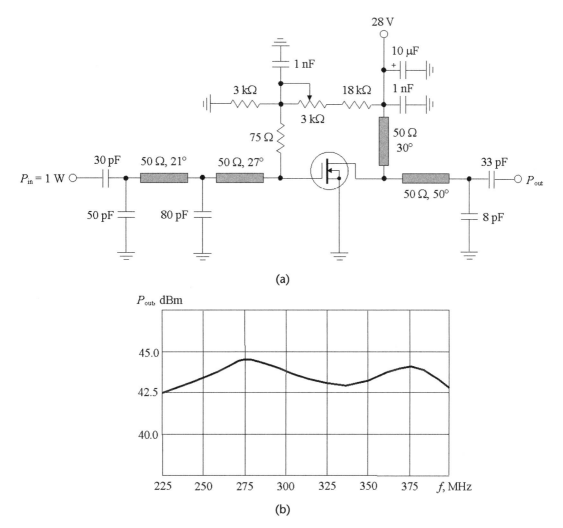

Figure 3.14   (a) Circuit schematic and (b) performance of broadband LDMOSFET power amplifier.

critical elements. Figure 3.14(b) shows the simulated broadband power amplifier performance, with an output power of 43.5±1.0 dBm and a power gain of 13.5±1.0 dB in a frequency bandwidth from 225 to 400 MHz.

3.4   Matching Networks with Transmission Lines

The lumped or mixed matching networks generally work well at sufficiently low frequencies (up to one or several gigahertz). However, the lumped elements such as inductors and capacitors are difficult to implement at microwave frequencies where they can be treated as distributed elements. In addition, the quality factors for inductors are sufficiently small that they contribute to additional losses.

Generally, the design of a practical distributed filter circuit is based on some approximate equivalence between lumped and distributed elements, which can be established by applying a Richards transformation [14]. This implies that the distributed circuits composed of equal-length open- and short-circuited transmission lines can be treated as lumped elements under the transformation

$$s = j\tan\frac{\pi\omega}{2\omega_0} \quad (3.29)$$

where $s = j\omega/\omega_c$ is the conventional normalized complex frequency variable, and ω_0 is the radian frequency for which the transmission lines are a quarter wavelength [15].

As a result, the one-port impedance of a short-circuited transmission line corresponds to the reactive impedance of a lumped inductor Z_L as

$$Z_L = sL = j\omega L = jL\tan\frac{\pi\omega}{2\omega_0} \quad (3.30)$$

Similarly, the one-port admittance of an open-circuited transmission line corresponds to the reactive admittance of a lumped capacitor Y_C as

$$Y_C = sC = j\omega C = jC\tan\frac{\pi\omega}{2\omega_0} \quad (3.31)$$

The results given by (3.30) and (3.31) show that an inductor can be replaced with a short-circuited stub of electrical length $\theta = \pi\omega/(2\omega_0)$ and characteristic impedance $Z_0 = L$, while a capacitor can be replaced with an open-circuited stub of electrical length $\theta = \pi\omega/(2\omega_0)$ and characteristic impedance $Z_0 = 1/C$ when a unity-filter characteristic impedance is assumed.

From (3.29), it follows that, for a lowpass filter prototype, the cutoff occurs when $\omega = \omega_c$, resulting in

$$\tan\frac{\pi\omega_c}{2\omega_0} = 1 \quad (3.32)$$

which gives a stub length of $\theta = 45°$ (or $\pi/4$) with $\omega_c = \omega_0/2$. Hence, the inductors and capacitors of a lumped-element filter can be replaced with the short-circuited and open-circuited stubs, as shown in Figure 3.15. Since the lengths of all stubs are

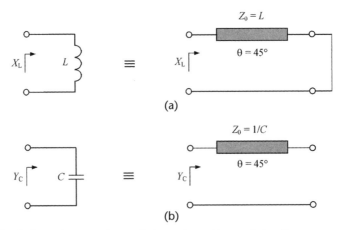

Figure 3.15 Equivalence between lumped elements and transmission lines.

the same and equal to $\lambda/8$ at the cutoff frequency ω_c, these lines are called the *commensurate lines*. At the frequency $\omega = \omega_0$, the transmission lines will be a quarter-wavelength long, resulting in an attenuation pole. However, at any frequency away from ω_c, the impedance of each stub will no longer match the original lumped-element impedances, and the filter response will differ from the desired filter prototype response. Note that the response will be periodic in frequency, repeating every $4\omega_c$.

Since the transmission line generally represents a four-port network, it is very convenient to use a matrix technique for filter design. In the case of a cascade of several networks, the rule is that the overall matrix of the new network is simply the matrix product of the matrices for the individual networks taken in the order of connection [16]. In terms of a Richards variable, an $ABCD$ matrix for a transmission line with characteristic impedance Z_0 can be written as

$$\begin{bmatrix} A & B \\ C & D \end{bmatrix} = \frac{1}{\sqrt{1-s^2}} \begin{bmatrix} 1 & sZ_0 \\ \dfrac{s}{Z_0} & 1 \end{bmatrix} \tag{3.33}$$

which represents a unit element that has a half-order transmission zero at $s = \pm 1$. The matrix of the unit element is the same as that of a transmission line of electrical length θ and characteristic impedance Z_0. Unit elements are usually introduced to separate the circuit elements in transmission-line filters, which are otherwise located at the same physical point.

The application of a Richards transformation provides a sequence of short-circuited and open-circuited stubs, which are then converted to a more practical circuit implementation. This can be done based on a series of equivalent circuits known as Kuroda identities, which allows these stubs to be physically separated, transforming the series stub into the shunt and changing impractical characteristic impedances into more realizable impedances [17]. The Kuroda identities use the unit elements, and these unit elements are thus commensurate with the stubs used to implement inductors and the capacitors of the prototype design. Connecting the unit element with the characteristic impedance Z_0 to the same load impedance Z_0 does not change the input impedance. The four Kuroda identities are illustrated in Figure 3.16, where the combinations of unit elements with the characteristic impedance Z_0 and electrical length $\theta = 45°$, the reactive elements, and the relationships between them are given.

To prove the equivalence, consider two circuits of identity at the first row in Figure 3.16 where the $ABCD$-matrix for the entire left-hand circuit can be written as

$$\begin{bmatrix} A & B \\ C & D \end{bmatrix}_L = \frac{1}{\sqrt{1-s^2}} \begin{bmatrix} 1 & 0 \\ sC & 1 \end{bmatrix} \begin{bmatrix} 1 & sZ_1 \\ \dfrac{s}{Z_1} & 1 \end{bmatrix}$$

$$= \frac{1}{\sqrt{1-s^2}} \begin{bmatrix} 1 & sZ_1 \\ s\left(C + \dfrac{1}{Z_1}\right) & 1 + s^2 Z_1 C \end{bmatrix} \tag{3.34}$$

where Z_1 is the characteristic impedance of the left-hand unit element.

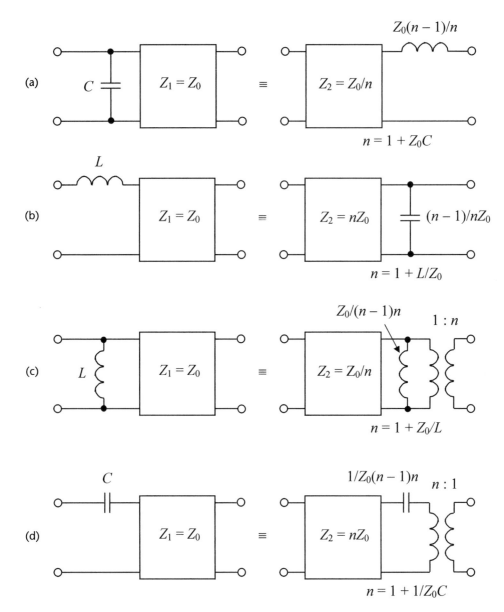

Figure 3.16 Four Kuroda identities.

Similarly, for the right-hand circuit,

$$\begin{bmatrix} A & B \\ C & D \end{bmatrix}_R = \frac{1}{\sqrt{1-s^2}} \begin{bmatrix} 1 & sZ_2 \\ \dfrac{s}{Z_2} & 1 \end{bmatrix} \begin{bmatrix} 1 & sL \\ 0 & 1 \end{bmatrix}$$

$$= \frac{1}{\sqrt{1-s^2}} \begin{bmatrix} 1 & s(Z_2 + L) \\ \dfrac{s}{Z_2} & 1 + \dfrac{s^2 L}{Z_2} \end{bmatrix} \tag{3.35}$$

where Z_2 is the characteristic impedance of the right-hand unit element.

The results in (3.34) and (3.35) are identical if

$$Z_1 = Z_2 + L \qquad \frac{1}{Z_1} + C = \frac{1}{Z_2} \qquad \frac{L}{Z_2} = Z_1 C$$

or

$$Z_2 = \frac{Z_1}{n} \qquad L = \frac{n-1}{n} Z_1 \qquad (3.36)$$

where $n = 1 + Z_1 C$.

As an example, consider the design of the broadband input transmission-line matching circuit based on a lumped two-section lowpass transforming filter shown in Figure 3.6, with a center bandwidth frequency $f_0 = 3$ GHz to match a 50Ω source impedance with the device input impedance $Z_{in} = R_{in} + j\omega_0 L_{in}$, where $R_{in} = 2\Omega$ and $L_{in} = 0.223$ nH. The value of the series input device inductance is chosen to satisfy Table 3.1 when, for $n = 4$, $r = 25$, $w = 0.4$, maximum ripple of 0.156725, and $g_1 = 2.31517$, from (3.14) it follows that

$$L_4 = \frac{g_4 R_0}{\omega_0} = \frac{g_1 R_0}{\omega_0 r} = 0.223 \text{ nH}$$

From Table 3.1, we obtain $g_2 = 0.422868$, which, from (3.13) and (3.14), gives the circuit parameters shown in Figure 3.17. The inductance value is chosen for design convenience. If this value differs from the required value, then the maximum frequency bandwidth, the power ripple, or the number of ladder sections must be changed.

Figure 3.18 shows the design transformation of a lumped lowpass transforming filter to a microstrip one using the Kuroda identities. The first step, which is shown in Figure 3.18(a), is to add a 50Ω unit element at the end of the circuit and convert a shunt capacitor to a series inductor using the second Kuroda identity, as shown in Figure 3.18(b). Then, adding another unit element and applying the first Kuroda identity, as shown in Figure 3.18(c), result in the circuit with two unit elements and

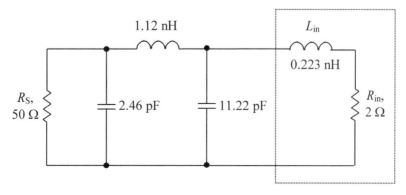

Figure 3.17 Two-section broadband matching circuit.

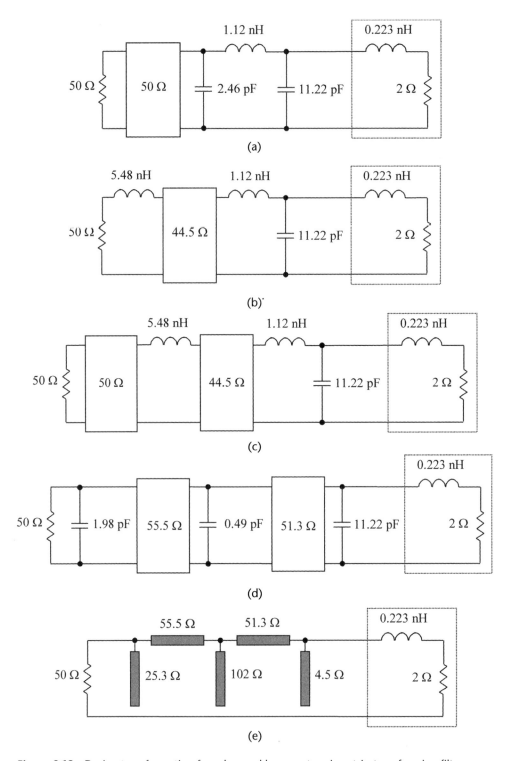

Figure 3.18 Design transformation from lumped lowpass to microstrip transforming filter.

three shunt capacitors shown in Figure 3.18(d). To keep the same physical dimensions during the calculation of the circuit parameters, the inductance should be taken in nanohenries, and the capacitance measured in nanofarads if the operating frequency is measured in gigahertz. Finally, a Richards transformation is used to convert the shunt capacitors to the corresponding transmission-line stubs. According to (3.31), the normalized characteristic impedance of a shunt stub is $1/C$, which is necessary to multiply by 50Ω.

Figure 3.18(e) shows the microstrip layout of the final lowpass transforming circuit, where the lengths of the shunt stubs are $\lambda/8$ at the cutoff frequency f_c, as well as the lengths of each unit element representing the series stubs. If the normalized frequency bandwidth and center bandwidth frequency are chosen to be $w = 0.4$ and $f_0 = 3\,\text{GHz}$, respectively, the cutoff frequency becomes equal to

$$f_c = f_0\left(1 + \frac{w}{2}\right) = 3.6\ \text{GHz}$$

In practical design of a microwave bipolar or GaAs FET amplifier, it is necessary to take into account that the intrinsic device generally exhibits a small-signal gain roll-off with increasing frequency at approximately 6 dB per octave [18, 19]. Therefore, to maintain a constant gain across the design frequency band, the matching network must be designed for maximum gain at the highest frequency of interest [20]. In this case, reflective mismatching conditions are provided to compensate for the increase in the intrinsic gain of a FET when the frequency is decreased. As a result, it is necessary to selectively mismatch the input of the transistor by employing the gain-tapered input matching circuit so that the overall gain of the amplifier will be flat [21]. Alternatively, the gain tapering could be done in the output network with input flat matching conditions. In a simplified practical implementation when two impedance-transforming L-sections with series microstrip lines and shunt microstrip stubs in the input matching circuit and a single impedance-transforming T-section with two series microstrip lines and one shunt microstrip stub in the output matching network are used, a flat gain of about 6 dB was achieved across the octave band of 4 to 8 GHz for a single-cell GaAs FET amplifier with the device transconductance $g_m = 55$ mS [22].

Figure 3.19 shows the matching circuit design steps and the circuit schematic of a broadband microwave GaN HEMT power amplifier [23]. In this case, the first step to design the octave-band power amplifier intended to operate across a frequency bandwidth of 2 to 4 GHz was to find the optimum source and load impedances that maximize the performance of the device in terms of efficiency in the required bandwidth. In view of a GaN HEMT Cree CGH60015DE device, since the optimum impedances were relatively close to each other across the band and the acceptable level of degradation in PAE was estimated to be less than 8%, the task was simplified to provide the optimum impedances seen by the device input and output at the center bandwidth frequency across the entire bandwidth. The bandpass matching network shown in Figure 3.19(a) was derived from a lowpass prototype matching circuit, assuming that the transistor output can be approximated by an ideal current source with a parallel RC network, where R_0 is the source resistance corresponding to device equivalent output resistance at the fundamental (or load-line

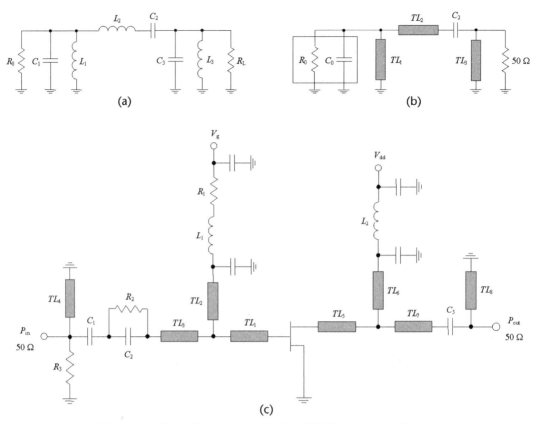

Figure 3.19 Schematics of broadband microwave GaN HEMT power amplifier.

resistance) and the capacitance C_0 is the total drain-source capacitance. To scale the obtained terminating resistor R_L upward to 50Ω, a Norton transformation of an ideal transformer ($n_T = 1.173$) with two series-shunt capacitors to an arrangement of three capacitors, as shown earlier in Figure 3.4, was used. Then, based on the transforms between lumped and distributed elements, the two resonant parallel LC circuits were approximated by the corresponding grounded shunt quarterwave transmission lines TL_1 and TL_3 with the characteristic impedance of each line equal to the reactance of the inductor or capacitor multiplied by $\pi/4$, whereas the lumped π-network with a series inductor and two shunt capacitors was approximated by the series transmission line TL_2, as shown in Figure 3.19(b). A similar approach can be applied to the design of the input matching circuit, which also includes lossy elements for better input return loss and stability. The entire circuit schematic of the designed broadband GaN HEMT power amplifier is shown in Figure 3.19(c), where the two series tapered transmission lines TL_1 and TL_5 are added at the input and output of the device. As a result, an output power of 41±1 dBm with a power gain of 10±1 dB and a drain efficiency of 52% to 72% was achieved across the frequency bandwidth of 1.9 to 4.3 GHz.

An alternative impedance matching technique is based on the multisection matching transformers consisting of the stepped transmission-line sections with different characteristic impedances and electrical lengths [24]. These transformers, in contrast to the continuously tapered transmission-line transformers, are significantly shorter and provide broader performance. The schematic structure of a stepped transmission-line transformer, which consists of a cascaded connection of n uniform sections of equal quarterwave lengths $l = \lambda_0/4$, where λ_0 is the wavelength corresponding to the center bandwidth frequency, represents an antimetric structure, for which the ratio between the characteristic impedances of its transmission-line sections can be written in the general form as

$$Z_i Z_{n+1-i} = Z_S Z_L \tag{3.37}$$

where $i = 1, 2, ..., n$ and n is the number of sections, Z_S is the source impedance, and Z_L is the load impedance [25]. In Figure 3.20, as a practical example, the minimum possible VSWR is plotted for a five-step transmission-line impedance transformer with a total characteristic-impedance variation of 8:1, which was designed for a maximum VSWR of 1.021 in an octave frequency bandwidth and where each section is of a quarterwave electrical length [26].

The main drawback of the stepped quarterwave transformers is their significant total length of $L = n\lambda_0/4$. However, it is possible to reduce the overall transformer length by applying other profiles of its structure. The stepped transformers using n cascaded uniform transmission-line sections of various lengths with alternating impedances are shorter by 1.5 to 2 times. In this case, the number of sections n is always an even number and the section impedances can be equal to the source and load impedances to be matched. For example, the input and output matching circuits of a microwave GaAs MESFET power amplifier, which was designed to operate in a frequency bandwidth of 4 to 8 GHz, were composed of the stepped microstrip lines where all the high-impedance sections were made 50Ω and all low-impedance sections were made 10Ω [22].

The total length of a stepped transmission-line transformer can be further reduced by using the structure representing the cascade connection of n transmission-line sections of the same length $l < \lambda_0/4$ with

$$\begin{aligned} Z_1 &< Z_3 < \cdots < Z_{n-1} \\ Z_2 &< Z_4 < \cdots < Z_n \end{aligned} \tag{3.38}$$

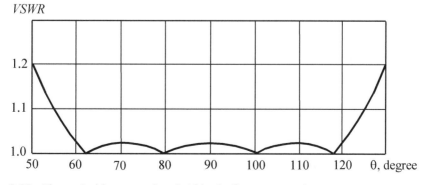

Figure 3.20 Theoretical frequency bandwidth of a five-step transformer.

where n is an even number and $Z_1 > Z_n$ when $Z_S < Z_L$ [27]. An example of the stepped transmission-line transformer to match a source impedance of 25Ω with a load impedance of 50Ω is shown in Figure 3.21(a), where the electrical length of each section is equal to $\lambda_0/12$. In this case, the total transformer length is shorter by three times compared to the basic structure with the quarterwave sections, and an octave passband from 2 to 4 GHz for the lossless ideal transmission-line sections is provided with an input return loss of better than 25 dB, as shown in Figure 3.21(b). However, it requires the use of a high impedance ratio for its sections reaching 30 to 50 when the source and load impedances differ significantly.

To reduce a high impedance ratio of the stepped transformers with a short total length, their generalized structure representing cascaded even n sections of different lengths l_i and impedances Z_i can be used. The optimum Chebyshev characteristics for this structure can be provided with the ratios between the lengths and characteristic impedances of its sections according to

$$l_i = l_{n+1-i}$$
$$Z_i Z_{n+1-i} = Z_S Z_L$$
(3.39)

where $i = 1, 2, ..., n/2$, and

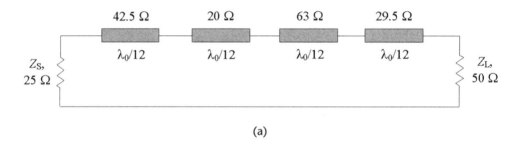

(a)

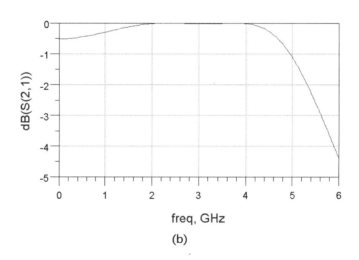

freq, GHz

(b)

Figure 3.21 Stepped transmission-line transformer with equal-length sections.

$$Z_{n-1} > Z_{n-3} > \cdots > Z_1 > Z_n > Z_{n-2} > \cdots > Z_2 \qquad (3.40)$$

where the impedances of both even and odd sections decrease in the direction from higher impedance Z_L to lower impedance Z_S and the impedance of any odd section is always larger than that of any even section [28]. The lengths of even sections decrease in the direction from the transmission line of a smaller impedance, whereas the lengths of odd sections increase in the same direction.

3.5   Lossy Matching Circuits

In many practical cases, to provide broadband matching with a minimum gain flatness and input reflection coefficient, it is sufficient to use the resistive shunt element at the transistor input. An additional matching improvement with reference to upper frequencies can be achieved by employing inductive reactive elements in series to the resistor. The resistive nature of this type of network may also improve amplifier stability and distortion. To provide a broadband performance for microwave GaAs MESFET power amplifiers, a resistively loaded shunt network, where the resistor is connected in series with a short-circuited quarterwave microstrip line to decrease the loaded quality factor without greatly reducing the maximum available gain, was used in the load network to provide a flat gain over 8 to 12 GHz, or in the input matching circuit to cover a frequency bandwidth of 2 to 6.2 GHz [29, 30]. For ultrabroadband high-gain multistage amplifiers, using a simple lossy compensation shunt circuit with a resistor in series with an inductor placed at the input and output of each transistor in parallel with the second-order LC circuits allows for a gain of 12 ± 1.5 dB with a VSWR of less than 2.5 from 150 MHz to 16 GHz to be achieved for a three-stage GaAs MESFET amplifier [31]. A 14-dB gain was obtained over the 3-dB bandwidth from 700 kHz to 6 GHz for a two-stage microstrip GaAs MESFET power amplifier, where a flat gain performance was achieved by using a shunt lossy gain-compensation circuit with a resistor in series with a short-circuited microstrip line placed at the input and output of the first-stage transistor in parallel to the input and interstage LC matching circuits [32].

A bandstop/bandpass diplexing RLC network is more useful than a simple lossy RL gain-compensation circuit because it provides an exact match at one frequency and an arbitrary amount of attenuation at any other frequency. Diplexing networks can be used in either input or output networks of the amplifier depending on noise figure, power output, and other amplifier constraints. Figure 3.22(a) shows the resonant diplexer LC network for lossy gain compensation, where the series L_sC_s and parallel L_pC_p resonant circuits are tuned to high bandwidth frequency and $R_L = R_0$ [33]. Here, the series capacitance C_s and shunt inductance L_p are obtained as $C_s = BW/\omega_h R_L$ and $L_p = BW(R_L/\omega_h)$, where BW is the normalized frequency bandwidth and $\omega_h = 2\pi f_h$ is the high-bandwidth radian frequency. The distributed form of a lossy gain-compensation network with additional input lowpass matching section is shown in Figure 3.22(b), where $Z_p = 4\omega_h L_p/\pi$, $Z_s = \omega_h L_s/\tan\theta_s$, and θ_s is the electrical length of the series transmission line.

Figure 3.23(a) shows the basic block of a microwave lossy match GaAs MESFET amplifier, where an input matching circuit and an open-circuit shunt stub

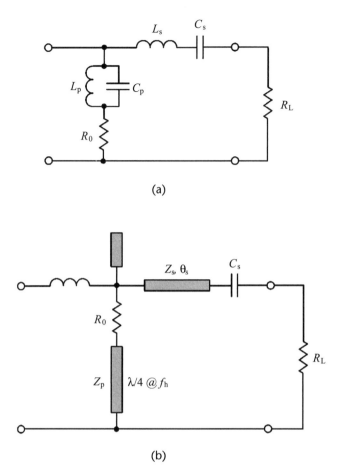

(a)

(b)

Figure 3.22 Circuit schematics of lossy gain-compensation circuits.

cascaded with a series transmission line at the device drain terminal are included to provide the amplifier desired frequency response [34]. For frequencies up to 1 GHz, the reactive elements of the transistor equivalent model have relatively little influence on the gain magnitude and reflection coefficients. As a result, the transistor described by S-parameters can be represented by its low-frequency model and the amplifier circuit can be significantly reduced to a simple network, where $S_{12} = 0$ and both S_{11} and S_{22} have negligible imaginary components. Then, the amplifier gain can be derived as

$$\text{Gain} = |S_{21}|^2 = \left[\frac{g_m Z_0}{2}\left(1 + S_{11}\right)\left(1 + S_{22}\right)\right]^2 \qquad (3.41)$$

which clearly expresses the trade-offs between the gain and the reflection coefficients, where g_m is the device transconductance and Z_0 is the characteristic impedance [34, 35]. The schematic of a multistage lossy match amplifier can be divided into three basic circuit functions: input matching, amplification, and interstage matching. Figure 3.23(b) shows the lossy match two-stage GaAs MESFET amplifier

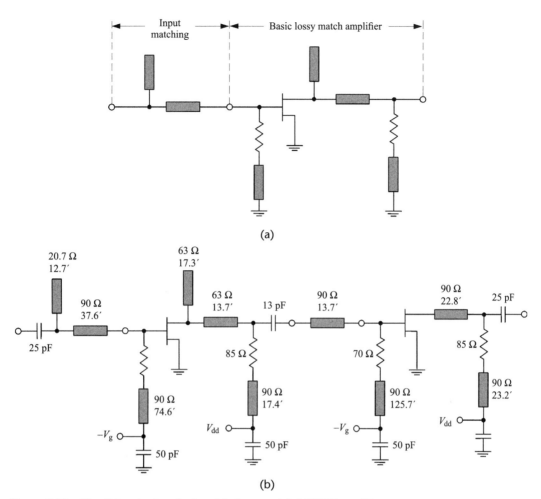

Figure 3.23 Circuit topologies of microstrip lossy match MESFET amplifiers.

with optimum values of the gate and drain shunt resistances to achieve flat gain performance over the frequency bandwidth of 2 to 8 GHz.

For a broadband lossy match silicon MOSFET high-power amplifier, it is sufficient to use a simple gain-compensation network with a resistor connected in series with a lumped inductor when operating frequencies are low enough compared to the device transition frequency f_T [36]. In this case, it is very important to optimize the elements of a lossy matching circuit to achieve minimum gain flatness over maximum frequency bandwidth. Let us consider the small-signal silicon MOSFET equivalent circuit, which is shown in Figure 3.24. When the load resistor R_L is connected between the drain and source terminals, an analytical expression for the input device impedance Z_{in} can be obtained as

$$Z_{in} = R_g + \left(R_{gs} + \frac{1}{j\omega C_{gs}} \right) \bigg/ \left[1 + \frac{C_{gd}}{C_{gs}} \frac{\left(1 + j\omega\tau_g\right)\left(1 + j\omega C_{ds}R_{L0}\right) + g_m R_{L0}}{1 + j\omega R_{L0}\left(C_{ds} + C_{gd}\right)} \right] \qquad (3.42)$$

where $R_{L0} = (R_L + R_d)/[1 + (R_L + R_d)/R_{ds}]$ and $\tau_g = R_{gs}C_{gs}$.

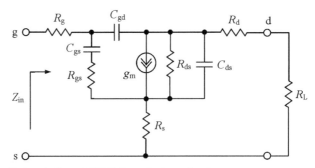

Figure 3.24 Small-signal silicon MOSFET equivalent circuit.

The modified circuit shown in Figure 3.25(a) describes adequately the frequency behavior of the input impedance of Figure 3.24. In (3.42), the series source resistance R_s and transit time τ are not taken into account due to their sufficiently small values for high-power MOSFETs in a frequency range of $f \leq 0.3f_T$, where $f_T = g_m/2\pi C_{gs}$. When $\omega \tau_g \leq 0.3$ and the device output capacitive impedance is inductively compensated, the input equivalent circuit simplifies significantly and can represent a capacitor and a resistor connected in series, as shown in Figure 3.25(b), where

$$R_{in} \cong R_g + R_{gs} \tag{3.43}$$

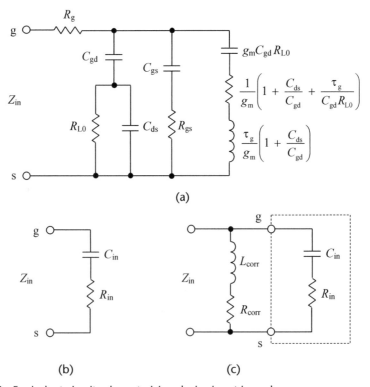

Figure 3.25 Equivalent circuits characterizing device input impedance.

$$C_{in} \cong C_{gs} + C_{gd}\left[1 + g_m\frac{R_L + R_d}{1 + R_L/(R_{ds} + R_d)}\right] \qquad (3.44)$$

To provide a constant real part of the input impedance Z_{in} in a frequency range up to $0.1f_T$, it is enough to use a simple lossy compensation circuit consisting of an inductor L_{corr} and a resistor R_{corr} connected in series, as shown in Figure 3.25(c).

The total input impedance of both a lossy match gain-compensation circuit and device input circuit is written as

$$Z_{in} = \frac{R_{corr} - \omega^2 C_{in}R_{in}L_{corr} + j\omega\left(L_{corr} + C_{in}R_{in}R_{corr}\right)}{1 - \omega^2 L_{corr}C_{in} + j\omega C_{in}\left(R_{corr} + R_{in}\right)} \qquad (3.45)$$

Under the condition $R = R_{corr} = R_{in}$, from (3.45) it follows that the reactive part of the input impedance Z_{in} becomes zero, that is, $\text{Im}Z_{in} = X_{in} = 0$, when

$$L_{corr} = C_{in}R^2 \qquad (3.46)$$

which leads to a pure active input impedance Z_{in} obtained as

$$Z_{in} = R = R_{in} \qquad (3.47)$$

At microwave frequencies, the short-circuited transmission line can be included instead of an inductor L_{corr} with the same input inductive reactance. When the frequency increases, the voltage amplitude applied to the input capacitance C_{in} decreases. This leads to the appropriate decrease of the operating power gain G_P at higher bandwidth frequencies. Due to the small values of R_{in} for high-power MOS-FETs, the value of G_P may not be high enough. Therefore, it is necessary to provide an additional impedance matching with lossless matching circuits to match with the source impedance of 50Ω or high output impedance of the active device of the previous power amplifier stage.

Figure 3.26(a) shows the circuit schematic of a broadband LDMOSFET power amplifier with device geometry of $1.25~\mu m \times 40$ mm. The optimized input three-element lossy matching circuit allows a very broadband operation to be provided with minimum power gain flatness, and a 1:2 output transformer contributes to increases in the output power level. The 20-pF capacitor connected in parallel with a 27Ω resistor provides an additional increase of power gain at higher bandwidth frequencies. The simulation results are shown in Figure 3.26(b), where an output power of 22W to 25W with a power gain of 13.7 ± 0.3 dB in a frequency range from 5 to 300 MHz can be achieved (curve 1). In this case, the input return loss is greater than 8 dB up to 225 MHz (curve 2). However, when a 50Ω load is directly connected to the device drain terminal through the blocking capacitor, this results in output power levels in the range of 6W to 7W.

For solid-state SSB and AM communication transmitters, linear amplification must be provided across the entire frequency range of 2 to 30 MHz, which was first covered by using bipolar technology based on a push-pull amplifier implementation

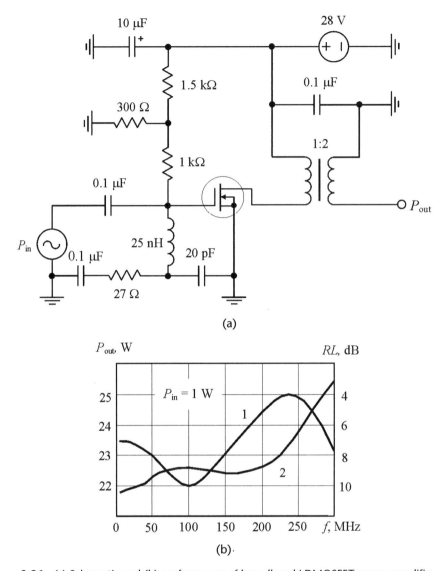

Figure 3.26 (a) Schematic and (b) performance of broadband LDMOSFET power amplifier.

with broadband toroidal transmission-line impedance transformers and combiners and interstage RLC gain-compensation networks. In this case, the driver stages are operated in a Class A mode for increased power gain, while the final stages are biased in Class AB with optimized quiescent currents for better linearity. As a result, the overall four-stage bipolar power amplifier achieved a PAE of more than 31% for two-tone 60W peak envelope power (PEP) signal over the entire frequency range of 2 to 30 MHz, with IM_3 equal to -30 dBc or better at output powers of 5W to 60W [37].

The circuit schematic of the input, interstage, and output networks intended to be implemented in microwave broadband power amplifiers are shown in Figure 3.27 [38]. A constant-resistance input network [Figure 3.27(a)] provides an input device impedance that is pure resistive and equal to $Z_{in} = R_{in}$ when $L_1 = C_{gs}R_{in}^2$, $C_1 =$

L_g/R_{in}^2, and $R_1 = R_{in}$, thus making wideband transformation of the input resistance to the source resistance much easier. In the output network shown in Figure 3.27(b), a value of the drain inductance L_d is properly chosen to compensate for the capacitive device output reactance at the center bandwidth frequency. Then, a resonant frequency of the parallel L_2C_2 circuit is set to be equal to the same center bandwidth frequency. In this case, for lower frequencies where the device output impedance Z_d is capacitive, the reactance of the parallel resonant circuit is inductive. On the other hand, for higher frequencies where the impedance Z_d is inductive, reactance of the parallel resonant circuit is capacitive. As a result, wideband reactance compensation is realized when the reactive part of the overall output impedance becomes very small over a wide frequency bandwidth. For microwave applications, such a parallel resonant circuit is fabricated by using a quarterwave short-circuit stub. The interstage network [see the circuit schematic in Figure 3.27(c)] comprises the input and output networks described above and a quarterwave microstrip transformer with the characteristic impedance of $Z_0 = \sqrt{R_{in}L_d/R_{out}C_{out}}$.

Figure 3.28 shows the circuit schematic diagram of a two-stage lossy match MESFET power amplifier [38]. By using a 1.05-mm device in the driver stage and two 1.35-mm devices in the final stage, a saturated output power of 27.7±2.7 dBm, a linear power gain of 8.3±2.8 dB, and a drain efficiency of 15.3±8.3% were measured in a frequency range of 4 to 25 GHz. The input and interstage constant-resistance networks are represented by the series connection of a resistor and a high-impedance microstrip line each. Two such networks connected in parallel provide pure resistive input impedance, where l_4 and l_5 are the series microstrip lines, and R_1 and R_2 are the series resistors. The short-circuited microstrip lines (l_7 and l_8 in the interstage network, l_{19} and l_{21} in the output network) with quarterwave electrical lengths at

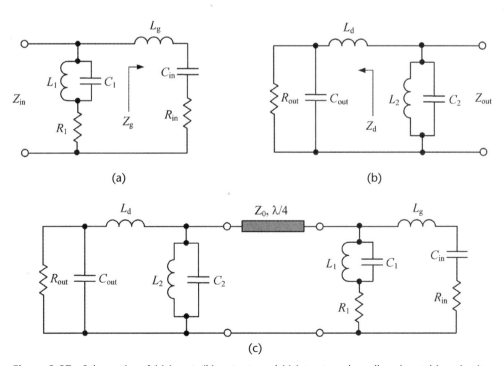

(a)

(b)

(c)

Figure 3.27 Schematics of (a) input, (b) output, and (c) interstage broadband matching circuits.

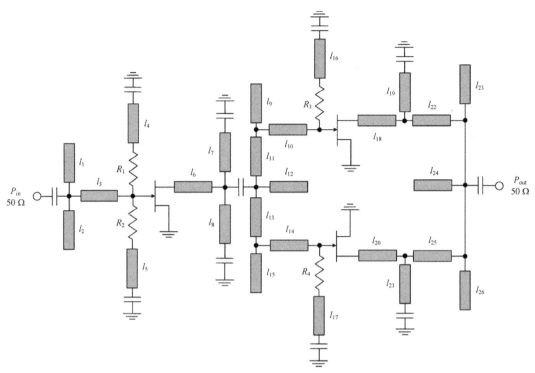

Figure 3.28 Microstrip two-stage lossy match MESFET power amplifier.

the center bandwidth frequency serve as the parallel resonant circuits connected at the device output terminals. Microstrip lines l_{10} and l_{14} in the interstage network represent the quarterwave impedance transformers, which provide matching between the output impedance of the driver-stage device and the input impedance of the second-stage devices connected in parallel. The input and output matching circuits are realized in the form of T-transformers, where the series microstrip lines and parallel open-circuit microstrip stubs replace the series inductors and shunt capacitors, respectively. To further increase an output power, the number of amplifying stages with lossy input and interstage matching circuits connected in parallel can be increased. As a result, by optimizing the output matching and combining circuits, for a three-stage MMIC 0.25-μm pHEMT power amplifier with a distributed amplifier used as a driver stage and four 1200-μm transistors in the output stage, an output power of 2.4±1.1W with a small-signal gain of 24±3.5 dB over the frequency range of 6 to 18 GHz was measured [39].

The output matching network can also represent the series connection of a π-type lowpass matching circuit and a lossy gain-compensation network, as shown in Figure 3.29 [35]. For both bipolar and MESFET broadband amplifiers, the π-type network comprises a device output capacitance and an open-circuit microstrip stub, which are connected to each side of a series lumped inductor, respectively. The output lossy gain-compensation network is connected between the π-type matching circuit and the load. This configuration is usually used for very broadband medium-power amplifiers. For example, the two-stage cascade of an L-band broadband bipolar amplifier [Figure 3.29(a)] was designed for a minimum input reflection coefficient with a VSWR of 1.78, a maximum gain variation of ±1.2 dB, and an approximately 16.5-dB power gain over a frequency range of 1 to 2 GHz. The two-stage cascade of a microwave MESFET amplifier [Figure 3.29(b)] was designed for maximum flat gain in a frequency range of 4 to 6 GHz when the power gain varies within 15.4±0.5 dB. To provide minimum loss at high bandwidth frequency, the short-circuited microstrip line in the output lossy gain-compensation circuit was chosen to be a quarter-wavelength long for each amplifier. For a two-stage 1-μm MESFET amplifier with the first stage designed for minimum noise figure and the second stage designed for maximum flat gain at higher bandwidth frequencies, a power gain of 9.5±1 dB over the frequency range of 6.5 to 12 GHz was achieved [40].

Figure 3.30(a) shows the circuit schematic of a broadband GaN HEMT microwave power amplifier implemented in the form of a flip-chip integrated circuit with a device geometry of 0.7 μm × 1 mm, transition frequency f_T = 18 GHz, and maximum frequency f_{max} = 35 GHz [41]. The optimized input three-element lossy LCR matching circuit provides a power gain up to 11.5 dB and a low input reflection of less than −10 dB over a frequency range of 3 to 9 GHz. Because the impedance at the input of a lossy match gain-compensation circuit is only about 10Ω, this necessitates an additional 50Ω to 10Ω broadband impedance

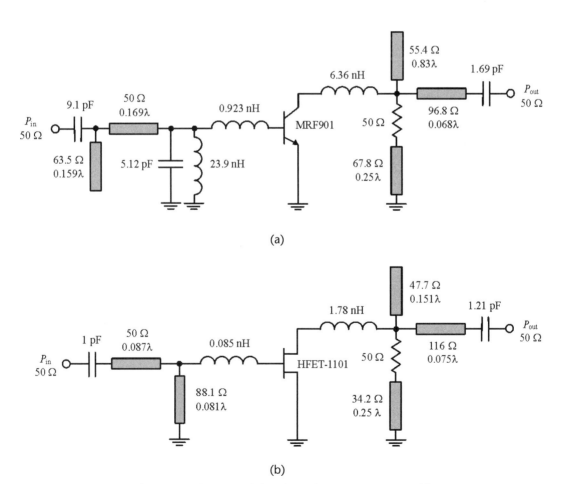

(a)

(b)

Figure 3.29   Broadband microstrip lossy match bipolar and MESFET power amplifiers.

transformation (Tr1), which was realized using a few sections of quarterwave coplanar transmission lines with decreasing characteristic impedances. The output network incorporates a lowpass LC circuit to compensate for the output device capacitance such that the intrinsic device sees an approximately real load within the entire frequency bandwidth. Since the optimum load for this 1-mm device with a supply voltage of 20V is about 50Ω, no output impedance transformation is needed. The output power was measured at about 1.6W with a PAE of 14% to 24% across the frequency bandwidth of 4 to 8 GHz. By combining four such GaN

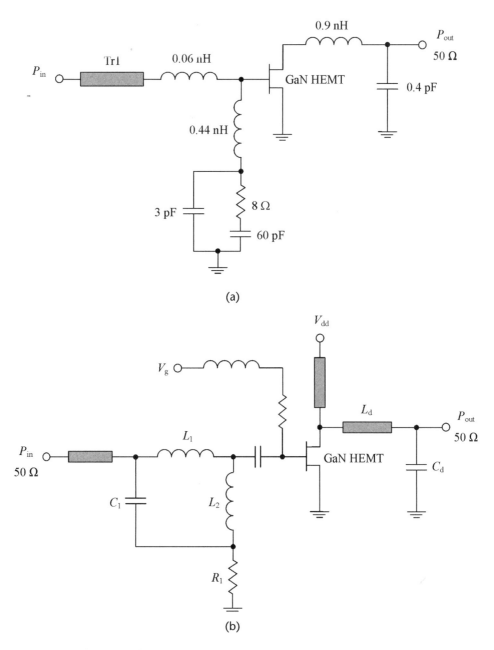

(a)

(b)

Figure 3.30 Schematics of microwave broadband GaN HEMT power amplifiers.

HEMT power amplifiers connected in parallel, the highest output power of 8W with a PAE of about 20% was obtained at 9.5 GHz, and the lowest output power of 4.5W was measured at 4.5 GHz, with a small-signal gain of 7 dB across the frequency bandwidth of 3 to 10 GHz [42].

To provide multidecade bandwidth with very good input return loss, a compact bridged-T all-pass input RLC matching network can be used, as shown in Figure 3.30(b), where the resistor R_1 was chosen to be 50Ω [43]. In this case, a GaN HEMT periphery of 2.2 mm was chosen to obtain an output power in the range of 10W. A simple two-element matching circuit consisting of a series microstrip line and a shunt capacitor was used at the output to provide optimum load impedance at the upper band edge. The power amplifier was packaged in a ceramic SO8 package including GaN on a SiC device operating at 28V and GaAs integrated passive matching circuitry. As a result, an output power of 8W and a power gain of 12 dB were measured over a frequency bandwidth from 50 MHz to 2 GHz with a drain efficiency of 36.7% to 65.4%.

3.6   Push-Pull and Balanced Power Amplifiers

Generally, if it is necessary to increase the overall output power of a power amplifier, several active devices can be used in parallel or push-pull configurations. In a parallel configuration, the active devices are not isolated from each other, which requires very good circuit symmetry, and the output impedance becomes too small in the case of high output power. The latter drawback can be eliminated in a push-pull configuration, which provides increased values for the input and output impedances. In this case, for the same output power level, the input impedance Z_{in} and output impedance Z_{out} are approximately four times as high as that of a parallel connection of the active devices since a push-pull arrangement is essentially a series connection. At the same time, the loaded quality factors of the input and output matching circuits remain unchanged because both the real and reactive parts of these impedances are increased by a factor of four. Very good circuit symmetry can be provided using balanced active devices with common emitters (or sources) in a single package. The basic concept of a push-pull operation can be analyzed by using the corresponding circuit schematic shown in Figure 3.31 [44].

3.6.1   Basic Push-Pull Configuration

It is most convenient to consider an ideal Class B operation, which means that each transistor conducts exactly half a cycle (equal to 180°) with zero quiescent current. Let us also assume that the number of turns of both primary and secondary windings of the output transformer T_2 is equal ($n_1 = n_2$) and the collector current of each transistor can be represented in the following half-sinusoidal form:

For the First Transistor:

$$i_{c1} = \begin{cases} +I_c \sin \omega t & 0 \leq \omega t < \pi \\ 0 & \pi \leq \omega t < 2\pi \end{cases} \tag{3.48}$$

For the Second Transistor:

$$i_{c2} = \begin{cases} 0 & 0 \leq \omega t < \pi \\ -I_c \sin \omega t & \pi \leq \omega t < 2\pi \end{cases} \qquad (3.49)$$

where I_c is the output current amplitude.

Being transformed through the output transformer T_2 with the appropriate phase conditions, the total current flowing through the load R_L is obtained as

$$i_R(\omega t) = i_{c1}(\omega t) - i_{c2}(\omega t) = I_c \sin \omega t \qquad (3.50)$$

The current flowing into the center tap of the primary windings of the output transformer T_2 is the sum of the collector currents, resulting in

$$i_{cc}(\omega t) = i_{c1}(\omega t) + i_{c2}(\omega t) = I_c |\sin \omega t| \qquad (3.51)$$

Ideally, the even-order harmonics being in phase are canceled and should not appear at the load. In practice, a level of the second-harmonic component of 30 to 40 dB below the fundamental is allowable. However, it is necessary to connect a bypass capacitor to the center tap of the primary winding to exclude power losses due to even-order harmonics. The maximum theoretical collector efficiency that can

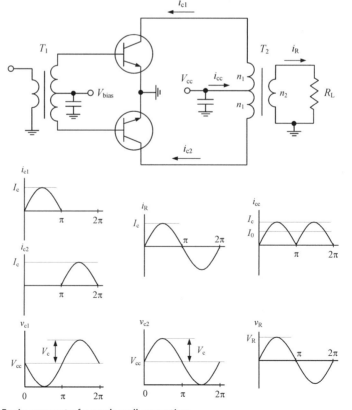

Figure 3.31   Basic concept of a push-pull operation.

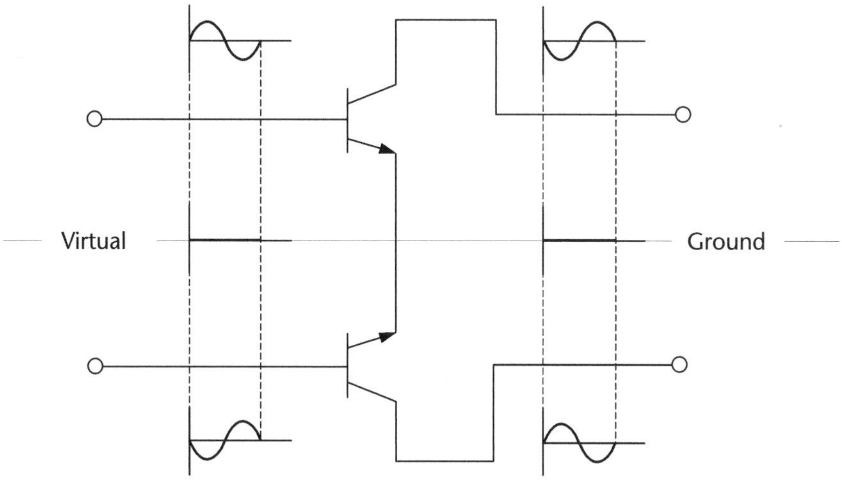

Figure 3.32 Basic concept of balanced transistor.

be achieved in push-pull Class B operation is equal to 78.5%, similar to a single-ended Class B mode.

In a balanced circuit, identical sides carry 90° quadrature or 180° out-of-phase signals of equal amplitude. In the latter case, if perfect balance is maintained on both sides of the circuit, the difference between signal amplitudes becomes equal to zero at each midpoint of the circuit, as shown in Figure 3.32. This effect is called the *virtual grounding*, and this midpoint line is referred to as the *virtual ground*. The virtual ground, which is actually inside the balanced transistor package having two identical transistor chips, reduces a common-mode inductance and results in better stability and usually higher power gain [45].

When using a balanced transistor, new possibilities for both internal and external impedance matching procedures emerge. For instance, for the push-pull operating mode of two single-ended transistors, it is necessary to provide reliable grounding for input and output matching circuits for each device, as shown in Figure 3.33(a). Using balanced transistors significantly simplifies the matching circuit topologies, with the series inductors and shunt capacitors connected between amplifying paths, as shown in Figure 3.33(b), where the dc-blocking capacitors are not needed [46]. Such an approach can provide additional design flexibility when, for example, a two-stage monolithic push-pull X-band GaAs MESFET power amplifier can be optimized for either small-signal, high-gain operation, or for large-signal power saturated operation by changing the lengths of the bondwires that form the shunt inductance at the drain circuits of each stage [47].

3.6.2   Baluns

For push-pull operation of a power amplifier with a balanced transistor, it is also necessary to provide the unbalanced-to-balanced transformation referenced to the ground both at the input and at the output of the power amplifier. The most suitable approach to solve this problem in the best possible manner at high frequencies and microwaves is to use the transmission-line baluns (balanced-to-unbalanced

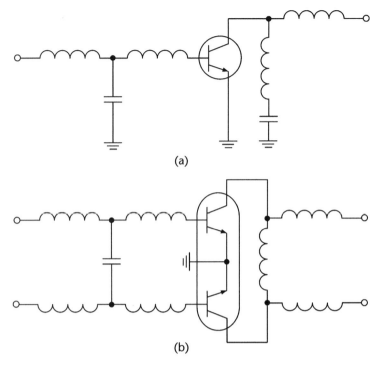

Figure 3.33 Matching technique for single-ended and balanced transistors.

transmission-line transformers). The first transmission-line balun for coupling a single coaxial line having a quarter wavelength at the center bandwidth frequency to a push-pull coaxial line (or a pair of coaxial lines), which maintains perfect balance over a wide frequency range, was introduced and described by Lindenblad in 1939 [48, 49].

Figure 3.34(a) shows the basic structure and equivalent circuit of a simple coaxial balun, where port A is the unbalanced port and port B is the balanced port. To be a perfect balun, when a balanced load is connected to port terminals B, the shield return current $I_2 - I_3$ would equal I_1, which would ideally represent the delayed input current, and the output terminal voltages would be equal and opposite with respect to ground. In this case, if the characteristic impedance Z_0 of the coaxial transmission line is equal to the input impedance at the unbalanced end of the transformer, the total impedance from both outputs at the balanced end of the transformer will be equal to the input impedance. Hence, such a transmission-line transformer can be used as a 1:1 balun. The equivalent circuit for this coaxial balun demonstrates the basic drawback of this balun, when its inner conductor is shielded from ground having practically an infinite impedance to ground, whereas the outer shield does have a finite impedance to ground when a balun is placed above a printed circuit board. The presence of the lower ground plane creates a shunt short-circuit stub with characteristic impedance Z_1 across one of the loads and this converts the highpass balun structure into a bandpass one. As a result, this stub has a dramatic effect on balun performance, with the bandwidth being reduced to about an octave based on phase imbalance. One of the solutions is simply to raise the transmission line above

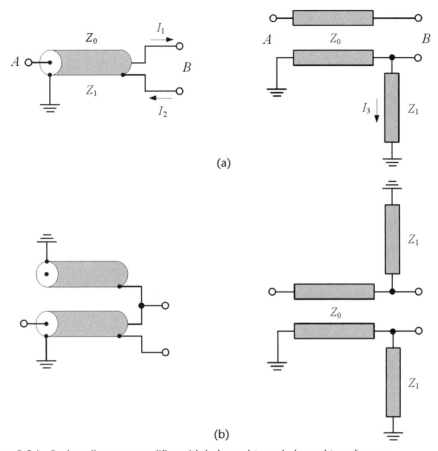

(a)

(b)

Figure 3.34   Push-pull power amplifier with balanced-to-unbalanced transformers.

the printed circuit board as high as possible and make both conductors symmetrical with respect to the lower ground plane. The other solution is to attach a compensating stub to the other load, as shown in Figure 3.34(b), which results in perfect amplitude and phase balance above the low-frequency cutoff region providing less than 1-dB insertion loss achieved from 5 MHz to 2.5 GHz [50].

Figure 3.35 shows the basic structure of a push-pull power amplifier with a balanced bipolar transistor including the input and output matching circuits. To extend the operating frequency range to lower frequencies, the outside of the coaxial line of the balun can be loaded with a low-loss ferrite core, which acts as a choke to force equal and opposite currents in the inner and outer conductor and isolate the 180° output from the input ground terminal by creating a high and lossy impedance for Z_1. In this case, the measured S-parameters of the back-to-back connected baluns showed an insertion loss of about 0.5 to 0.6 dB and a better than 20-dB return loss over 50 to 1000 MHz [51]. As a result, four broadband GaN HEMT power amplifier units were combined using such a low-loss coaxial balun that transforms an unbalanced 50Ω load into two 25Ω impedances that are 180° out of phase and each of the 25Ω ends is driven by a pair of power amplifier units connected in parallel. A similar balun is used at the input to create the 180° out-of-phase input to the two

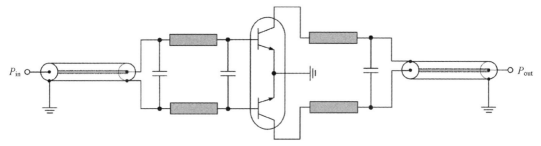

Figure 3.35 Push-pull bipolar power amplifier with input and output baluns.

pairs of power amplifier units, resulting in a greater than 100W output power and higher than 60% drain efficiency across the frequency bandwidth of 100 to 1000 MHz. The lower cutoff bandwidth frequency can be provided to cover down to 10 MHz by adding lower frequency ferrites, but it may affect performance at high bandwidth frequencies. Generally, because ferrite has a limited bandwidth, it is possible to use several ferrite cores to broaden the frequency bandwidth. For example, by using a low-frequency ferrite core covering 1 to 10 MHz, a medium-frequency ferrite core covering 10 to 200 MHz, and a high-frequency ferrite core covering high frequencies above 200 MHz, the balun can cover 1 MHz to 2.5 GHz with a loss of 0.25 dB at a low frequency and 1.3 dB at 2.5 GHz [52].

The miniaturized compact input unbalanced-to-balanced transformer shown in Figure 3.36 covers the frequency bandwidth up to an octave with well-defined rejection-mode impedances [53]. To avoid the parasitic capacitance between the outer conductor and the ground, the coaxial semirigid transformer T_1 is mounted atop microstrip shorted stub l_1 and soldered continuously along its length. The electrical length of this stub is usually chosen from the condition of $\theta \leq \pi/2$ on the high bandwidth frequency depending on the matching requirements. To maintain circuit symmetry on the balanced side of the transformer network, another semirigid coaxial section T_2 with unconnected center conductor is soldered continuously

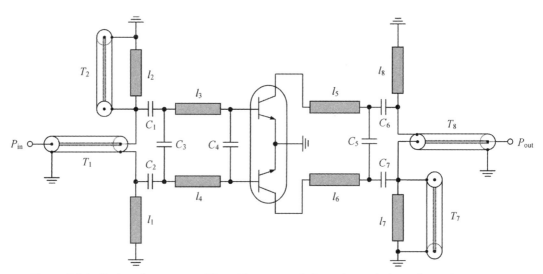

Figure 3.36 Push-pull power amplifier with compact balanced-to-unbalanced transformers.

along microstrip shorted stub l_2. The lengths of T_2 and l_2 are equal to the lengths of T_1 and l_1, respectively. Because the input short-circuited microstrip stubs provide inductive impedances, the two series capacitors C_1 and C_2 of the same value are used for matching purposes, thereby forming the first highpass matching section and providing dc blocking at the same time. The practical circuit realization of the output matching circuit and balanced-to-unbalanced transformer can be the same as for the input matching circuit.

Figure 3.37 shows the circuit schematic of a broadband microstrip balun with normalized parameters using a three-section Wilkinson divider for power splitting followed by Lange coupled-line directional couplers for phase shifting [54]. This planar balun structure can be easily fabricated on alumina substrate using a conventional monolithic process, resulting in good broadband amplitude and phase balance performance. To achieve the required tight coupling over a 3:1 bandwidth, interdigitated Lange couplers in an unfolded configuration to minimize bondwire connections were employed. The balun fabricated on a 10-mil alumina substrate achieved an amplitude imbalance of ±0.6 dB, average phase imbalance of 7° (with worst case of 11°), and maximum insertion loss of 1.2 dB from 6 to 20 GHz. With a single-section Wilkinson divider and additional short-length correction line on the noninverting arm, an amplitude response within ±1 dB, phase difference of 180±4°, and insertion loss on the order of 1 dB over 6 to 18 GHz were measured [55].

3.6.3   Balanced Power Amplifiers

The balanced amplifier technique using the quadrature 3-dB couplers for power dividing and combining represents an alternative approach to the push-pull operation. Figure 3.38(a) shows the basic circuit schematic of a balanced amplifier where two power amplifier units of the same performance are arranged between the input

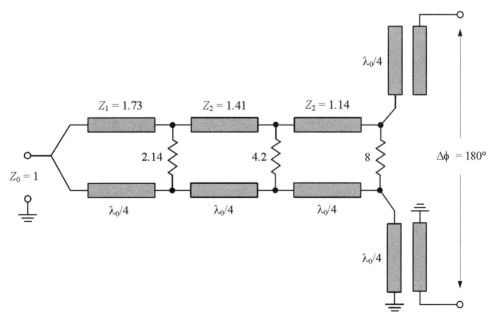

Figure 3.37   Circuit schematic of broadband planar balun.

splitter and output combiner, each having a 90° phase difference between coupled and through ports. The fourth port of each quadrature coupler must be terminated with a ballast resistor R_{bal}, which is equal to 50Ω for a 50Ω system impedance. The input signal is split into two equal-amplitude components by the first 90° hybrid coupler with 0° and 90° paths, then amplified, and finally recombined by the second 90° hybrid coupler. Due to proper phase shifting, both signals in the load of the isolated port of the combiner are canceled, and the load connected at the combiner output port sees the sum of these two signals. The theory of balanced amplifiers has been given by Kurokawa when the operating frequency bandwidth over 1.2 octaves can be obtained with one-section distributed quarterwave 3-dB directional couplers [56]. For a wide frequency range, the main advantages of the balanced design are the improved input and output impedance matching, gain flatness, intermodulation distortion, and potential design simultaneously for minimum noise figure and good input match. As an example, a four-stage balanced bipolar amplifier achieved a power gain of 20±0.5 dB and an input VSWR of less than 1.2 across the octave frequency bandwidth from 0.8 to 1.6 GHz [57]. By extending the wide operating frequency range to higher frequencies, an output power of around 23 dBm with gain variations close to 1 dB over 4.5 to 6.5 GHz and 8 to 12 GHz was achieved for the balanced microstrip GaAs MESFET amplifiers [58, 59].

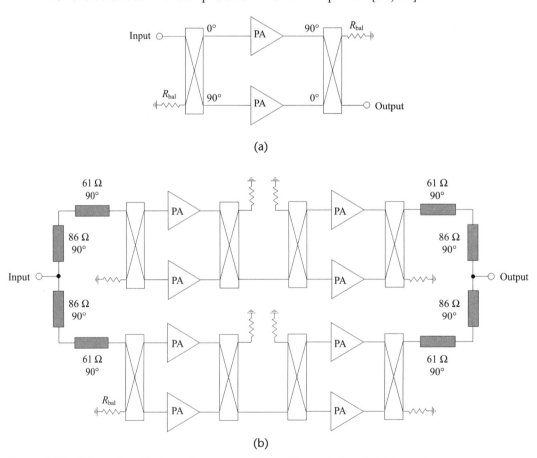

Figure 3.38 Schematics of balanced power amplifiers with quadrature hybrid couplers.

If the individual amplifiers with equal performance in the balanced pair are not perfectly matched at certain frequencies, then a signal in the 0° path of the coupler will be reflected from the corresponding amplifier and a signal in the 90° path of the coupler will be similarly reflected from the other amplifier. The reflected signals will again be phased with 90° and 0°, respectively, and the total reflected power as a sum of the in-phase reflected signals flows into the isolated port and dissipates on the ballast resistor R_{bal}. As a result, the input VSWR for a quadrature coupler does not depend on the equal load mismatch level. This gives a constant well-defined load to the driver stage, improving amplifier stability and driver power flatness across the operating frequency range. Generally, the stability factor of a balanced stage can be an order of magnitude higher than its single-ended equivalent, depending on the VSWR and isolation of the quadrature couplers. If one of the amplifiers fails or is turned off, the balanced configuration provides a gain reduction of −6 dB only. In addition, the balanced structure provides ideally the cancellation in the load of the third-order products such as $2f_1 + f_2$, $2f_2 + f_1$, $3f_1$, $3f_2$, ..., and attenuation by 3 dB of the second-order products such as $f_1 \pm f_2$, $2f_1$, $2f_2$, ... In a microstrip implementation for octave-band power amplifiers, one of the most popular couplers for power dividing and combining is a 3-dB Lange hybrid coupler.

Figure 3.38(b) shows the circuit schematic of a two-way balanced module consisting of two pairs of cascaded, balanced amplifier stages, where the respective output powers are combined using simple two-element power combiners, which are composed of two quarterwave transmission lines with different characteristic impedances [60]. Based on this architecture and GaAs MESFET devices with the gate periphery of $1 \times 1000~\mu m^2$, an output power of 1W across 7.25 to 12 GHz was achieved.

3.7   Practical Broadband RF and Microwave Power Amplifiers

Multisection matching networks based on the lowpass and highpass L-transformers for input and output matching circuits can provide a wide frequency bandwidth with minimum power gain ripple and significant harmonic suppression. Such a multisection matching circuit configuration using lumped elements was applied to the design of a 60W power amplifier operating in the frequency bandwidth of 140 to 180 MHz. The complete circuit schematic of the power amplifier is shown in Figure 3.39 [61]. To realize such technical requirements, an internally matched bipolar transistor for VHF applications, which provides a 100W output power level at a supply voltage of 28V, was used. According to the device data sheet, the input device impedance at the center bandwidth frequency $f_0 = \sqrt{140 \times 180} = 159$ MHz is equal to $Z_{in} = (0.9 + j1.8)\Omega$. Therefore, the input matching circuit was designed as a three-section network with two lowpass sections and one highpass section to minimize the circuit quality factor Q. In this case, the device input lead inductance of $1.8/(2\pi \times 0.159) = 1.8$ nH was considered as a series inductive element of the second lowpass section with a shunt capacitor of 540 pF. This power amplifier is operated in Class C mode due to the base bias circuit composed of the two inductors and a 15Ω resistor, which also provides low-frequency stability.

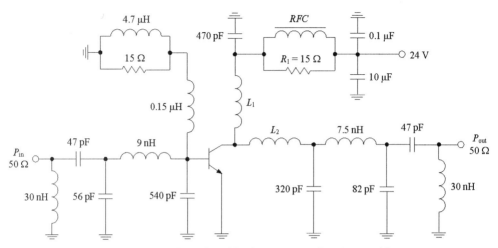

Figure 3.39 Circuit schematic of broadband high-power VHF bipolar amplifier.

A similar design philosophy was used to design the output matching circuit when the three-section network maintains a value of the quality factor close to unity or within the $Q = 1$ circle on a Smith chart. The output device impedance is practically resistive of 1.65Ω because the output device capacitive reactance is compensated by the device lead inductance. The series inductance L_2 of the first matching low-pass section adjacent to the collector terminal according to the Smith chart can be realized as a section of a 50Ω microstrip line with the electrical length of $0.011\lambda_0$, where λ_0 is the wavelength corresponding to the center bandwidth frequency f_0. The physical length of this microstrip line for 1/16-in. Teflon fiberglass with a dielectric permittivity of $\varepsilon_r = 2.55$ must be 0.51 in., whereas its width is equal to 0.4 in. The collector feed is provided through the combination of an inductor L_1, a resistor $R_1 = 15\Omega$, and an RF choke (RFC), which behaves as a high-impedance circuit at the operating frequencies but offers a very low resistance at dc. As a result, the designed broadband power amplifier achieved a power gain of at least 8 dB with a gain ripple of less than 3 dB, a collector efficiency of more than 50%, and an input VSWR below 3:1 [61]. As an alternative, the broadband input and output matching circuits can be composed of a single lowpass matching section followed by a 4:1 transmission-line transformer each. In this case, an output power of more than 25 W with a collector efficiency close to 70% was achieved across the frequency range of 118 to 136 MHz for an input power of 2 W using a 12.5 V bipolar device [62].

At microwave frequencies, an amplifier's bandwidth performance can also be improved by using an increased number of transmission-line transformer sections. For example, with the use of a multisection transformer with seven quarterwave transmission lines of different characteristic impedances, a power gain of 9±1 dB and a PAE of 37.5±7.5% over 5 to 10 GHz were achieved for a 15 W GaAs MESFET power amplifier [63]. The simplified schematic diagram of this microwave octave-band power amplifier is shown in Figure 3.40. To achieve minimum output power flatness, the number of sections of the output matching circuit is determined based on load-pull measurements. At the same time, the number of sections of the input matching circuit to compensate for the frequency-dependent power gain is chosen based on the small-signal S-parameter measurements. For a 5.25-mm GaAs MESFET device, the values of the input and output impedances at the fundamental derived from its large-signal model were assumed resistive and equal to $Z_{in} = 0.075\Omega$ and $Z_{out} = 1.32\Omega$, respectively. To achieve minimum gain flatness, the length of each microstrip section initially was chosen as a quarter wavelength at the highest frequency of 10 GHz. However, because the input and output device impedances are not purely resistive in practical implementation, the final optimized length of each microstrip section was reduced to be a quarter wavelength at around 15 GHz. The microstrip transformer sections $L_1 \ldots L_6$ and $L_{10} \ldots L_{14}$ were fabricated on alumina substrate with a dielectric permittivity of $\varepsilon_r = 9.8$ and a thickness of 0.635 mm for L_1 and L_2, 0.2 mm for $L_3 \ldots L_6$ and $L_{10} \ldots L_{12}$, and 0.38 mm for $L_{13} \ldots L_{14}$. The microstrip section L_7 was realized on a high-dielectric substrate with $\varepsilon_r = 38$ and thickness of 0.18 mm, whereas the microstrip sections L_8 and L_9 were fabricated on a high-dielectric substrate with $\varepsilon_r = 89$ and thickness of 0.15 mm. The final power amplifier represents a balanced configuration of the two 5.25-mm GaAs MESFETs with hybrid quadrature couplers.

The broadband power amplifier, whose circuit schematic is shown in Figure 3.41, was intended for TV transponders with complex video and audio TV signal amplification in the frequency bandwidth of 470 to 790 MHz. The power amplifier was implemented on a laminate substrate with $\varepsilon_r = 4.7$ for a 1.5-mm thickness. The lengths of the microstrip lines are given in terms of their lengths on the high-bandwidth frequency, and both collector RF chokes represent the three-turn air-core inductors. The device input and output impedances measured at the base and collector terminals at 600 MHz are equal to $Z_{in} = (6 + j4)\Omega$ and $Z_{out} = (15 + j17.5)\Omega$, respectively, which allows the corresponding two-section input matching circuit and a single-section output matching circuit to be used. In a Class A operating mode, such a power amplifier using a balanced TPV-595A bipolar transistor achieved a linear output power of 7W with a power gain of about 12 dB for a quiescent collector current of 1.3A.

Figure 3.42 shows the schematic diagram of a two-octave high-power transistor amplifier covering both the civil and military airbands between 100 and 450 MHz [64]. The BLF548 device is a balanced n-channel enhancement-mode VDMOS transistor designed for use in broadband amplifiers with an output power of 150W and a power gain of more than 10 dB in a frequency range of up to 500 MHz. In a frequency bandwidth from 100 to 500 MHz, the real part of its input impedance

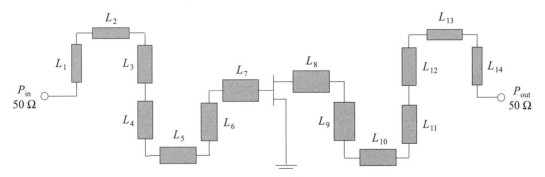

Figure 3.40 Microstrip broadband 15W GaAs MESFET power amplifier.

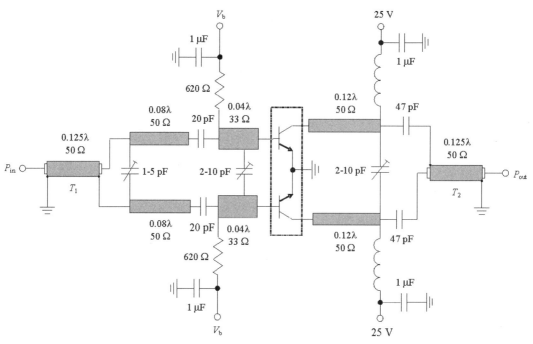

Figure 3.41 Circuit schematic of bipolar UHF power amplifier for TV applications.

$\mathrm{Re}Z_{\mathrm{in}}$ is almost constant and equal to 0.43Ω, whereas the imaginary part of the input impedance $\mathrm{Im}Z_{\mathrm{in}}$ changes its capacitive reactance of -4.1Ω at 100 MHz to the inductive reactance of 0.5Ω at 500 MHz. The required load impedance seen by the device output at the fundamental is inductive and equal to $Z_{\mathrm{L}} = (1.1 + j0.4)\Omega$ at the high bandwidth frequency of 500 MHz. Coaxial semirigid baluns are used to transform the unbalanced 50Ω source and load into two $180°$ out-of-phase 25Ω sections, respectively, followed by coaxial 4:1 transformers with the characteristic impedance of 10Ω for input matching and of 25Ω for output matching. This yields the lower impedance $R_{\mathrm{in}} = \sqrt{25 \times 10}/4 = 3.95\Omega$, which is then necessary to transform to the device input impedance of 0.43Ω, and the higher impedance $R_{\mathrm{out}} = \sqrt{25 \times 25}/4 = 6.25\Omega$, which is then necessary to transform to the load impedance of 2.8Ω seen by the device output at the center bandwidth frequency of 250 MHz. The final matching is provided by simple L-transformers with series microstrip lines and parallel variable capacitors. The microstrip lines were fabricated on a 30-mil substrate with a dielectric permittivity $\varepsilon_{\mathrm{r}} = 2.2$. In this case, the dimensions of each microstrip line with the characteristic impedance of 20Ω are as follows: L_1 and L_3 are 5×8 mm, L_2 and L_4 are 2.5×8 mm, L_5 and L_7 are 11.5×8 mm, and L_6 and L_8 are 4×8 mm. To compensate for the 6-dB/octave slope, conjugate matching is provided at 450 MHz, since at lower frequencies a mismatch gives the required decrease of a power gain to provide acceptable broadband power gain flatness. As a result, the gain variation of an output power of 150W is smaller than 1 dB with an input return loss better than 12 dB in a frequency range of 100 to 450 MHz.

Figure 3.43 shows the circuit schematic of a two-stage reactively matched GaN HEMT MMIC power amplifier, which operates as a driver amplifier for

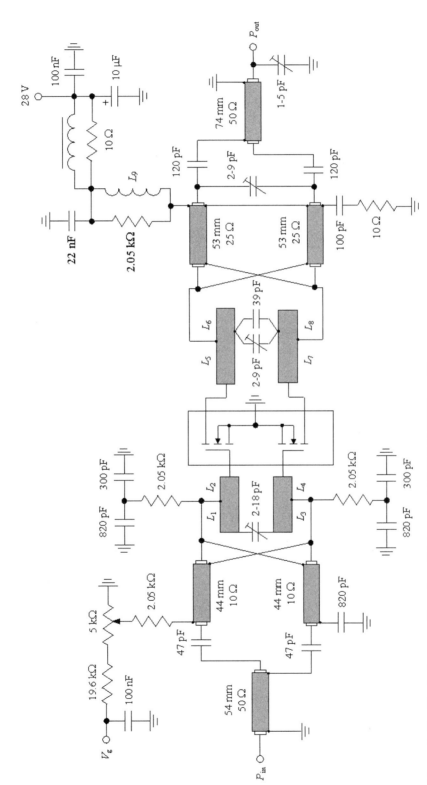

Figure 3.42 Circuit diagram of broadband high-power VHF-UHF MOSFET amplifier.

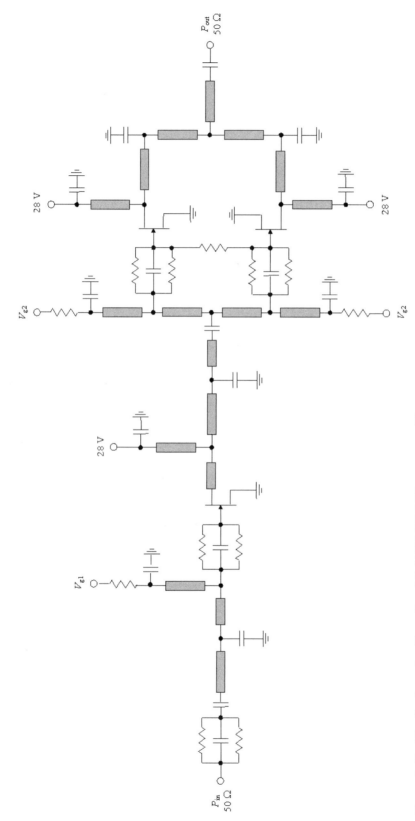

Figure 3.43   Circuit schematic of broadband GaN HEMT MMIC power amplifier.

ultrawideband high-power transmitting modules for multifunctional active electronically scanned antenna radar systems [65]. MMICs based on GaN HEMT technology can provide wider bandwidth, higher output power density, improved reliability at high junction temperature, better thermal properties, higher breakdown voltage, and higher operating efficiency compared to MMICs based on GaAs technology. For a 0.25-μm GaN HEMT technology using SiC substrate, the breakdown voltage of 120V allows operation with a supply voltage up to 40 V; and a maximum output power density of 5.6 W/mm for device gate periphery and capacitance sheet of 250 pF/mm^2 for MIM capacitor can be provided. In this case, the MMIC driver amplifier is based on three identical GaN HEMT cells, each with 8×100-μm gate periphery (one transistor in the first stage and two transistors in the second stage), to achieve the maximum output power of about 36 dBm with a parallel connection of two second-stage amplifying paths. The unconditional stability of the MMIC driver amplifier from 100 MHz to 6 GHz is provided by applying parallel RC networks at the gates of each transistor cell. The integrated resistors are also used in the gate bias circuits of each device cell to ensure stability without sacrificing gain or efficiency. The dc-feed paths, which consist of narrow microstrip lines to provide the corresponding inductive reactances and bypass MIM capacitors to provide isolation between the dc and RF paths, are constituent parts of the input, interstage, and output matching circuits, which are realized in the form of lowpass L- and T-transformers with the series microstrip lines and shunt MIM capacitors. The matching networks provide impedance transformation with low Q-factors enabling an increased frequency bandwidth.

References

[1]   H. W. Bode, *Network Analysis and Feedback Amplifier Design*, New York: Van Nostrand, 1945.

[2]   R. M. Fano, "Theoretical Limitations on the Broad-Band Matching of Arbitrary Impedances," *J. Franklin Institute*, Vol. 249, pp. 57–83, Jan. 1950; pp. 139–154, Feb. 1950.

[3]   R. Levy, "Explicit Formulas for Chebyshev Impedance-Matching Networks, Filters and Interstages," *Proc. IEE*, Vol. 111, pp. 1099–1106, June 1964.

[4]   D. E. Dawson, "Closed-Form Solutions for the Design of Optimum Matching Networks," *IEEE Trans. Microwave Theory Tech.*, Vol. MTT-57, pp. 121–129, Jan. 2009.

[5]   G. L. Matthaei, "Tables of Chebyshev Impedance-Transforming Networks of Lowpass Filter Form," *Proc. IEEE*, Vol. 52, pp. 939–963, Aug. 1964.

[6]   G. L. Matthaei, L. Young, and E. M. T. Jones, *Microwave Filters, Impedance-Matching Networks, and Coupling Structures*, Norwood, MA: Artech House, 1980.

[7]   A. Grebennikov, *RF and Microwave Power Amplifier Design*, New York: McGraw-Hill, 2004.

[8]   W. H. Ku et al., "Microwave Octave-Band GaAs FET Amplifiers," *1975 IEEE MTT-S Int. Microwave Symp. Dig.*, pp. 69–72.

[9]   R. S. Pengelly and J. A. Turner, "Monolithic Broadband GaAs F.E.T. Amplifiers," *Electronics Lett.*, Vol. 12, pp. 251–252, May 1976.

[10] C. H. Lin et al., "A Compact 6.5-W PHEMT MMIC Power Amplifier for Ku-Band Applications," *IEEE Microwave and Wireless Components Lett.*, Vol. 17, pp. 154–156, Feb. 2007.

[11] P. C. Huang et al., "A 17–35 GHz Broadband, High Efficiency PHEMT Power Amplifier Using Synthesized Transformer Matching Technique," *IEEE Trans. Microwave Theory Tech.*, Vol. MTT-60, pp. 112–119, Jan. 2012.

[12] H. Wang, C. Sideris, and A. Hajimiri, "A CMOS Broadband Power Amplifier with a Transformer-Based High-Order Output Matching Network," *IEEE J. Solid-State Circuits*, Vol. SC-45, pp. 2709–2722, Dec. 2010.

[13] B. S. Yarman and A. Aksen, "An Integrated Design Tool to Construct Lossless Matching Networks with Mixed Lumped and Distributed Elements," *IEEE Trans. Circuits and Systems—I: Fundamental Theory Appl.*, Vol. CAS-I-39, pp. 713–723, Sep. 1992.

[14] P. I. Richards, "Resistor-Transmission Line Circuits," *Proc. IRE*, Vol. 36, pp. 217–220, Feb. 1948.

[15] R. Saal and E. Ulbrich, "On the Design of Filters by Synthesis," *IRE Trans Circuit Theory*, Vol. CT-5, pp. 284–327, Dec. 1958.

[16] P. I. Richards, "Applications of Matrix Algebra to Filter Theory," *Proc. IRE*, Vol. 34, pp. 145–150, Mar. 1946.

[17] H. Ozaki and J. Ishii, "Synthesis of Transmission-Line Networks and the Design of UHF Filters," *IRE Trans Circuit Theory*, Vol. CT-2, pp. 325–336, Dec. 1955.

[18] H. F. Cooke, "Microwave Transistors: Theory and Design," *Proc. IEEE*, Vol. 59, pp. 1163–1181, Aug. 1971.

[19] R. S. Tucker, "Gain-Bandwidth Limitations of Microwave Transistor Amplifiers," *IEEE Trans. Microwave Theory Tech.*, Vol. MTT-21, pp. 322–327, May 1973.

[20] C. A. Liechti and R. L. Tillman, "Design and Performance of Microwave Amplifiers with GaAs Schottky-Gate Field-Effect Transistors," *IEEE Trans. Microwave Theory Tech.*, Vol. MTT-22, pp. 510–517, May 1974.

[21] W. H. Ku and W. C. Petersen, "Optimum Gain-Bandwidth Limitations of Transistor Amplifiers as Reactively Constrained Active Two-Port Networks," *IEEE Trans. Circuits and Systems*, Vol. CAS-22, pp. 523–533, June 1975.

[22] R. E. Neidert and H. A. Willing, "Wide-Band Gallium Arsenide Power MESFET Amplifiers," *IEEE Trans. Microwave Theory Tech.*, Vol. MTT-24, pp. 342–350, June 1976.

[23] P. Saad et al., "Design of a Highly Efficient 2-4-GHz Octave Bandwidth GaN-HEMT Power Amplifier," *IEEE Trans. Microwave Theory Tech.*, Vol. MTT-58, pp. 1677–1685, July 2010.

[24] H. Q. Tserng et al., "Microwave Power GaAs FET Amplifiers," *IEEE Trans. Microwave Theory Tech.*, Vol. MTT-24, pp. 936–943, Dec. 1976.

[25] H. J. Riblet, "General Synthesis of Quarter-Wave Impedance Transformers," *IRE Trans. Microwave Theory Tech.*, Vol. MTT-5, pp. 36–43, Apr. 1957.

[26] S. B. Cohn, "Optimum Design of Stepped Transmission-Line Transformers," *IRE Trans. Microwave Theory Tech.*, Vol. MTT-3, pp. 16–21, Apr. 1955.

[27] G. L. Matthaei, "Short-Step Chebyshev Impedance Transformers," *IEEE Trans. Microwave Theory Tech.*, Vol. MTT-14, pp. 372–383, Aug. 1966.

[28] V. P. Meschanov, I. A. Rasukova, and V. D. Tupikin, "Stepped Transformers on TEM-Transmission Lines," *IEEE Trans. Microwave Theory Tech.*, Vol. MTT-44, pp. 793–798, June 1996.

[29] C. A. Liechti and R. L. Tillman, "Design and Performance of Microwave Amplifiers with GaAs Schottky-Gate Field-Effect Transistors," *IEEE Trans. Microwave Theory Tech.*, Vol. MTT-22, pp. 510–517, May 1974.

[30] D. P. Hornbuckle and L. J. Kuhlman, Jr., "Broad-Band Medium-Power Amplification in the 2–12.4-GHz Range with GaAs MESFETs," *IEEE Trans. Microwave Theory Tech.*, Vol. MTT-24, pp. 338–342, June 1976.

[31] J. Obregon, R. Funck, and S. Barvet, "A 150MHz–16GHz FET Amplifier," *1981 IEEE Int. Solid-State Circuits Conf. Dig.*, pp. 66–67.

[32] K. Honjo and Y. Takayama, "GaAs FET Ultrabroad-Band Amplifiers for Gbit/s Data Rate Systems," *IEEE Trans. Microwave Theory Tech.*, Vol. MTT-29, pp. 629–633, July 1981.

[33] N. Riddle and R. J. Trew, "A Broad-Band Amplifier Output Network Design," *IEEE Microwave Theory Tech.*, Vol. MTT-30, pp. 192–196, Feb. 1982.

[34] K. B. Niclas, "On Design and Performance of Lossy Match GaAs MESFET Amplifiers," *IEEE Trans. Microwave Theory Tech.*, Vol. MTT-30, pp. 1900–1906, Nov. 1982.

[35] K. B. Niclas, "Multi-Octave Performance of Single-Ended Microwave Solid-State Amplifiers," *IEEE Trans. Microwave Theory Tech.*, Vol. MTT-32, pp. 896–908, Aug. 1984.

[36] V. V. Nikiforov, T. T. Kulish, and I. V. Shevnin, "Broadband HF-VHF MOSFET Power Amplifier Design (in Russian)," *Poluprovodnikovaya Elektronika v Tekhnike Svyazi*, Vol. 23, pp. 27–36, 1983.

[37] O. Pitzalis, Jr., R. E. Horn, and R. J. Baranello, "Broadband 60-W HF Linear Amplifier," *IEEE J. Solid-State Circuits*, Vol. SC-6, pp. 93–103, June 1971.

[38] Y. Ito et al., "A 4 to 25 GHz 0.5 W Monolithic Lossy Match Amplifier," *1994 IEEE MTT-S Int. Microwave Symp. Dig.*, pp. 257–260.

[39] A. R. Barnes, M. T. Moore, and M. B. Allenson, "A 6-18 GHz Broadband High Power MMIC for EW Applications," *1997 IEEE MTT-S Int. Microwave Symp. Dig.*, pp. 1429–1432.

[40] R. S. Pengelly, "Broadband Lumped-Element X Band GaAs F.E.T. Amplifier," *Electronics Lett.*, Vol. 11, pp. 58–60, Feb. 1975.

[41] Y. F. Wu et al., "3-9-GHz GaN-Based Microwave Power Amplifiers with L-C-R Broad-Band Matching," *IEEE Microwave and Guided Wave Lett.*, Vol. 9, pp. 314–316, Aug. 1999.

[42] J. J. Xu et al., "A 3–10-GHz GaN-Based Flip-Chip Integrated Broad-Band Power Amplifier," *IEEE Trans. Microwave Theory Tech.*, Vol. MTT-48, pp. 2573–2578, Dec. 2000.

[43] K. Krishnamurthy et al., "RLC Matched GaN HEMT Power Amplifier with 2 GHz Bandwidth," *2008 IEEE Compound Semiconductor Integrated Circuits Symp. Dig.*, pp. 1–4.

[44] H. L. Krauss, C. W. Bostian, and F. H. Raab, *Solid State Radio Engineering*, New York: John Wiley & Sons, 1980.

[45] L. B. Max, "Balanced Transistors: A New Option for RF Design," *Microwaves*, Vol. 16, pp. 42–46, June 1977.

[46] J. Johnson, "A Look Inside Those Integrated Two-Chip Amps," *Microwaves*, Vol. 19, pp. 54–59, Feb. 1980.

[47] V. Sokolov and R. E. Williams, "Development of GaAs Monolithic Power Amplifiers in X-Band," *IEEE Trans. Electron Devices*, Vol. ED-27, pp. 1164–1171, June 1980.

[48] N. E. Lindenblad, "Television Transmitting Antenna for Empire State Building," *RCA Rev.*, Vol. 3, pp. 387–408, Apr. 1939.

[49] N. E. Lindenblad, "Junction Between Single and Push-Pull Lines," U.S. Patent 2,231,839, Feb. 1941 (filed May 1939).

[50] A. Riddle, "Ferrite and Wire Baluns with Under 1dB Loss to 2.5 GHz," *1998 IEEE MTT-S Int. Microwave Symp. Dig.*, pp. 617–620.

[51] K. Krishnamurthy et al., "100 W GaN HEMT Power Amplifier Module with >60% Efficiency over 100–1000 MHz Bandwidth," *2010 IEEE MTT-S Int. Microwave Symp. Dig.*, pp. 940–943.

[52] A. K. Ezzedine and H. C. Huang, "10W Ultra-Broadband Power Amplifier," *2008 IEEE MTT-S Int. Microwave Symp. Dig.*, pp. 643–646.

[53] L. B. Max, "Apply Wideband Techniques to Balanced Amplifiers," *Microwaves*, Vol. 19, pp. 83–88, Apr. 1980.

[54] J. Rogers and R. Bhatia, "A 6 to 20 GHz Planar Balun Using a Wilkinson Divider and Lange Couplers," *1991 IEEE MTT-S Int. Microwave Symp. Dig.*, pp. 865–868.

[55] B. J. Minnis and M. Healy, "New Broadband Balun Structures for Monolithic Microwave Integrated Circuits," *1991 IEEE MTT-S Int. Microwave Symp. Dig.*, pp. 425–428.

[56] K. Kurokawa, "Design Theory of Balanced Transistor Amplifiers," *Bell Syst. Tech. J.*, Vol. 44, pp. 1675–1798, Oct. 1965.

[57] R. S. Engelbrecht and K. Kurokawa, "A Wideband Low Noise L-Band Balanced Transistor Amplifier," *Proc. IEEE*, Vol. 53, pp. 237–247, Mar. 1965.

[58] R. E. Neidert and H. A. Willing, "Wide-Band Gallium Arsenide Power MESFET Amplifiers," *IEEE Trans. Microwave Theory Tech.*, Vol. MTT-24, pp. 342–350, June 1976.

[59] K. B. Niklas et al., "A 12–18 GHz Medium-Power GaAs MESFET Amplifier," *IEEE J. Solid-State Circuits*, Vol. SC-13, pp. 520–527, Aug. 1978.

[60] K. B. Niklas et al., "Application of the Two-Way Balanced Amplifier Concept to Wide-Band Power Amplification Using GaAs MESFETs," *IEEE Trans. Microwave Theory Tech.*, Vol. MTT-28, pp. 172–179, Mar. 1980.

[61] A. Tam, "Network Building Blocks Balance Power Amp Parameters," *Microwaves & RF*, Vol. 23, pp. 81–87, July 1984.

[62] B. Becciolini, "Impedance Matching Networks Applied to RF Power Transistors," *Application Note AN721*, Freescale Semiconductor, 2005.

[63] Y. Ito et al., "A 5–10 GHz 15-W GaAs MESFET Amplifier with Flat Gain and Power Responses," *IEEE Microwave and Guided Wave Lett.*, Vol. 5, pp. 454–456, Dec. 1995.

[64] "100–450 MHz 250 W Power Amplifier with the BLF548 MOSFET," Application Note AN98021, Philips Semiconductors, Mar. 1998.

[65] U. Schmid et al., "Ultra-Wideband GaN MMIC Chip Set and High Power Amplifier Module for Multi-Function Defense AESA Applications," *IEEE Trans. Microwave Theory Tech.*, Vol. MTT-61, pp. 3043–3051, Aug. 2013.

CHAPTER 4

Distributed Amplification Concept and Its Design Methodology

The concept of a distributed amplifier [1–21] can be traced back to the patent specification entitled "Improvements In and Relating to Thermionic Valve Circuits" filed by Percival in 1935 [22]. The term *distributed amplifier* first appeared in 1948 [23]. Reports on the implementation and application of distributed amplifiers, primarily for pulse amplification, began to appear in the literature after 1945. A lucid account of the principle of distributed amplification was given in [23] covering frequency response, techniques for performance optimization, dissipation effects, and noise characteristics, each supported by elegant yet forthcoming analyses. After the publication of the Ginzton's paper [23], activities and discussions on the distributed amplifier immediately picked up momentum, as is evident from the number of articles appearing in scientific publications, with a wide range of applications including television, distributed amplifiers, testing of networks and lines, digital communication, oscillography, and nuclear research.

Distributed amplifiers employ a topology in which the gain stages are connected such that their capacitances are isolated, yet the output currents still combine in an additive fashion. Series-inductive elements are used to separate capacitances at the input and output of adjacent gain stages. The resulting topology, given by the interlaying series inductors and shunt capacitances, forms a lumped-component artificial transmission line. The additive nature of the gain dictates a relatively low gain; however, the distributed nature of the capacitance allows the amplifier to achieve very wide bandwidths (Figure 4.1).

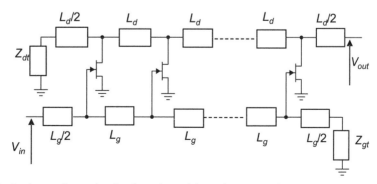

Figure 4.1 Basic topology of a distributed amplifier, where L_g and L_d denote gate and drain line, respectively; V_{in} is the input signal feeding; and V_{out} is output voltage [24].

4.1   Concept of Distributed Amplification

For gain and bandwidth products having an amplifier stage limited by intrinsic parameters to the active device employed, expanding the bandwidth will give rise to a reduction in the gain [1–10]. As the gain is made close to unity, it becomes inefficient to cascade amplifier stages. On the other hand, combining the outputs from a number of active devices in parallel will increase the output power but will produce no improvement in the gain bandwidth product [24]. The solution is to find an arrangement in such a way that the output currents from a number of devices are superimposed constructively while the effects of shunt capacitances are not accumulated, and this is the base of distributed amplification [24].

A transistor's input and output capacitance as part of the lumped elements of an artificial transmission line is formed with the series inductance that connects adjacent drains and gates. A schematic of a distributed amplifier is shown in Figure 4.2. The signal is coupled from the gate line to the drain line through transistors. The transmission lines can be of either the artificial type (i.e., made up of discrete-element inductors) or transmissions lines (i.e., microstrip or coplanar). The distributed amplifier concept was successfully applied to monolithic GaAs MESFET amplifiers at microwave frequencies in the 1980s for larger gain-bandwidth products [25]. Ayasli et al. have published design formulas for the gain of traveling-wave amplifier based on an approach that approximates gate and drain lines as continuous structures [25]. Similarly, Beyer et al. developed a closed-form expression for the gain that depends on the circuit propagation constants and the gate circuit cutoff frequency [1]. Niclas et al. have also developed a method based on the use of the admittance matrix employing the Y-parameters of the transistor model in an amplifier with either artificial or real transmission lines [26]. This method allows the use of much more sophisticated models for transistors developed from its measured S-parameters [27]. McKay et al. also proposed a formulation based on a normalized transmission using matrix formulation [27].

The operation of the distributed amplifier can be explained referring to Figure 4.2, where a RF signal applied to the input port of the gate line travels down the line to the termination where it is absorbed. The traveling signal is picked up by the gates of the individual transistor and transferred to the drain line via their transconductances g_m. If the phase velocities on the gate and the drain lines are

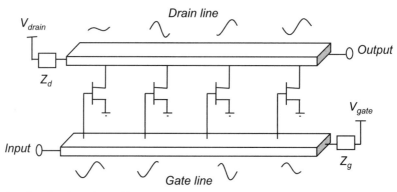

Figure 4.2   The distributed amplification concept.

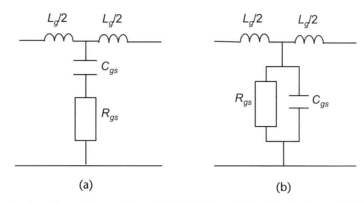

Figure 4.3   Lossless elementary section of distributed amplifier: (a) gate line and (b) drain line.

identical, the signals on the drain line add in the forward direction [1]. The phase velocities between gate and drain lines can be synchronized simply by setting the gate- and drain-line cutoff frequencies to be identical. Any signal that travels backward and is not entirely canceled by the out-of-phase additions will be absorbed by the drain-line termination [1].

The concept of distributed amplification is based on combining the input and output capacitances of the actives device with inductors in such a way that two artificial transmission lines are obtained. The input and output capacitance of each device becomes the capacitance per unit section for these lines (refer to Figure 4.3) and the lines are coupled by the g_m of the active device. As a result, it is possible to obtain amplification over a wider bandwidth than with conventional amplifiers [24]. Designers have concentrated mainly on increasing the gain-bandwidth product and the gain flatness, as well as on output power capabilities. In distributed amplifiers, the transmission structures employed are often analyzed as a cascade of two ports, as we will see in the following section. For the amplifiers employing transmission lines, the voltage developed along the output line tends to increase as the cutoff frequency is approached if the magnitudes of the current injected by active devices along the line remain constant [24]. This is the result of the factor of $(1 - \omega^2/\omega_c^2)^{-1/2}$ in $Z_{o\pi}$, the midshunt image impedance distributed amplifier.

4.2   Image Impedance Method

The image parameter method can be applied to a distributed amplifier because it consists of a cascade of identical two-port networks forming an artificial transmission line. The analysis can be conveniently accomplished using the $ABCD$-parameters, because the overall $ABCD$-matrix is the product of those of the cascaded two-port networks [24], as shown in Figure 4.4.

When considering signal transmission and impedance matching in cascaded two-port networks, each two-port should operate with the appropriate impedance terminations so that the maximum power transfer takes place over the prescribed bandwidth. Such a condition can be met by terminating the two-port with a pair of impedances known as image impedances so that the impedance appears the

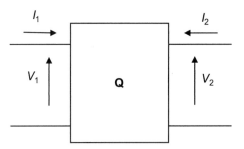

Figure 4.4 A two-port network.

same when one looks into either direction of each port as shown in Figure 4.5. The impedances, Z_{i1} and Z_{i2}, can be expressed as

$$Z_{i1} = \sqrt{Z_{sc1}Z_{oc1}} = \sqrt{\frac{B}{D}\frac{A}{C}} \tag{4.1}$$

$$Z_{i2} = \sqrt{Z_{sc2}Z_{oc2}} = \sqrt{\frac{B}{A}\frac{D}{C}} \tag{4.2}$$

where Z_{sc1} and Z_{oc1} are the impedances appearing at port 1, with port 2 short circuited and open circuited, respectively, and likewise for Z_{i2}. If the network is symmetrical, Z_{i1} and Z_{i2} become identical, and the characteristic impedance is denoted as Z_0.

Figure 4.6 shows the case of an infinite number of identical networks connected so that each junction is connected together. Due to the way the infinite chain of networks is connected in Figure 4.6, the impedances seen looking left and right at each junction are always equal; hence, there is never any reflection of a wave passing through a junction. Thus, from the wave point of view, the networks are perfectly matched [24]. The image impedance Z_i for a reciprocal symmetric two-port is defined as the impedance looking into port 1 or 2 of the two-port network when the other terminal is also terminated in Z_i [24]. To achieve an impedance match over

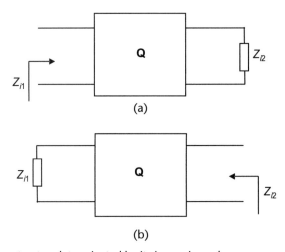

Figure 4.5 A two-port network terminated by its image impedance.

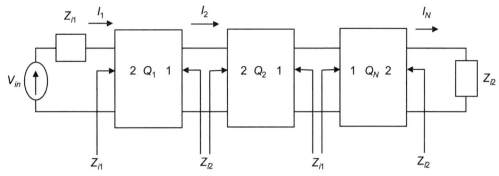

Figure 4.6 Artificial transmission line.

a broad range, the load and source impedance must be transformed into the image impedance. Otherwise, the gain response will not be flat as a function of frequency.

Having obtained expressions of the image impedance for a reciprocal two-port network, we now apply them to a number of elementary filter sections often found in distributed amplifiers. A simple filter can be constructed from two circuit elements as shown in Figure 4.7, known as an L-section or half section. The image impedances at port 1 and 2 are referred to as Z_{i1} and Z_{i2}, respectively, because they are also the characteristic impedances of the T-network and π-network formed by cascading two identical L-sections in a back-to-back fashion. From (4.1) and (4.2), we obtain

$$Z_{OT} = \sqrt{\frac{L}{C}\left(1 - \frac{\omega^2}{\omega_c^2}\right)} \tag{4.3}$$

$$Z_{O\pi} = \sqrt{\frac{L}{C}\left(1 - \frac{\omega^2}{\omega_c^2}\right)^{-1}} \tag{4.4}$$

where

$$\omega_c = 2/\sqrt{LC} \tag{4.5}$$

with ω_c being the cutoff frequency, which is the frequency where the image impedances go from real to imaginary.

If the desired characteristic impedance Z_0 of the transmission line is fixed, the cutoff frequency f_c can be expressed as

$$fc = \frac{1}{\pi \cdot C \cdot Z_0} \tag{4.6}$$

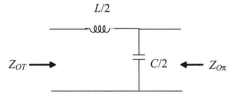

Figure 4.7 A lowpass L-section by means of constant-k image impedance.

This equation shows that the bandwidth of a distributed amplifier decreases as a value of the capacitance, and C increases (in reality, C refers to C_{gs} and C_{ds}). Because this capacitance is proportional to the dimension of the employed transistors and the gain of the distributed amplifier, there is a trade-off between gain and bandwidth in distributed amplifier design.

4.3   Theoretical Analysis of Distributed Amplification

Analysis of distributed amplifiers is facilitated by the assumption of lossless transmission networks that are realized from ladder networks based on constant-k low-pass filters and unilateral active devices [24]. A simplified equivalent circuit of the transistor[1] is shown in Figure 4.8, where R_{gs}, C_{gs}, R_{ds}, and C_{ds} are the gate-to-source and drain-to-source resistance and capacitance, respectively, and g_m is the device transconductance. The next subsections provide detailed analyses of distributed amplifiers using different approaches.

4.3.1   Analytical Approach to Two-Port Theory

As given in [1], the device is considered unilateral, that is, C_{gd} (the gate-to-drain capacitance) is neglected. The equivalent gate and drain transmission lines are shown in Figures 4.9(a) and (b). The lines are assumed to be terminated by their image impedances at both ends. With the unilateral device model employed, the two transmission lines are nonreciprocally coupled through the action of the transconductance g_m. From Figure 4.9(b), the current delivered to the load [1] is given as

$$I_0 = \frac{1}{2}g_m e^{-\gamma_d/2}\left[\sum_{k=1}^{n}V_k e^{-(n-k)\gamma_d}\right] \qquad (4.7)$$

where

V_k = voltage across C_{gs} of the kth transistor

$\gamma_d = \alpha_d + j\beta_d$ = the propagation factor of the drain line

α_d and β_d = attenuation and phase shift per section on the drain line, respectively

n = number of transistors in the amplifier.

The value of V_k can be expressed in terms of the voltage at the gate terminal of the kth field-effect transistor (FET) [1] as follows:

$$V_k = \frac{V_i e^{-(2k-1)\gamma_g/2 - j\tan^{-1}(\omega/\omega_g)}}{\left[1+\left(\dfrac{\omega}{\omega_g}\right)^{1/2}\left[1-\left(\dfrac{\omega}{\omega_c}\right)^2\right]\right]} \qquad (4.8)$$

---

[1] For simplicity, a basic transistor model excludes the feedback effect C_{gd}, due to the fact that the analysis becomes more complex. Furthermore, this method gives use freedom to isolate the gate and drain to form an effective artificial transmission line by means of a constant-k ladder network.

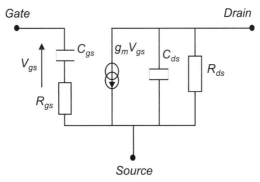

Figure 4.8   Simplified small-signal circuit model of a transistor.

where

V_i = voltage at the input terminal of the amplifier

$\gamma_g = \alpha_g + j\beta_g$ = propagation factor of the gate line

α_g and β_g = attenuation and phase shift per section on the gate line, respectively

$\omega_g = 1/(R_{gs}C_{gs})$ = gate circuit cutoff frequency

$\omega_c = 2\pi f_c$ = cutoff frequency of the lines ($f_c = 1/L_g C_{gs}$).

For constant-k type transmission lines, the phase velocity is a well-known function of the cutoff frequency ω_c of the line. By requiring gate and drain lines to have the same cutoff frequency, the phase velocities are constrained to be equal. Therefore, we have $\beta_g = \beta_d = \beta$. Then, output current I_0 [1] can be expressed as

$$I_0 = \frac{g_m V_i \sinh\left[\dfrac{n}{2}(\alpha_d - \alpha_g)\right] e^{-n(\alpha_d - \alpha_g)/2} e^{-jn\beta - j\tan^{-1}(\omega/\omega_g)}}{2\left[1 + \left(\dfrac{\omega}{\omega_g}\right)^2\right]^{1/2}\left[1 - \left(\dfrac{\omega}{\omega_g}\right)^2\right]\sinh\left[\dfrac{1}{2}(\alpha_d - \alpha_g)\right]}$$ (4.9)

The power delivered to the load and input power to the amplifier are given, respectively, by

$$P_0 = |I_0|^2 \,\Re\left[Z_{ID}\right]$$ (4.10)

$$P_i = \frac{|V_i|^2}{2|Z_{IG}|^2}\,\Re\left[Z_{IG}\right]$$ (4.11)

where Z_{ID} and Z_{IG} are the image impedances of the drain and gate lines, respectively. Therefore, the power gain of the amplifier is given as

$$G_P = \frac{g_m 2 Z_{0g} Z_{0d} \sinh^2\left[0.5n(\alpha_d - \alpha_g)\right] e^{-0.5n(\alpha_g + \alpha_d)}}{4\left[1 + \left(\dfrac{\omega}{\omega_g}\right)^2\right]\left[1 - \left(\dfrac{\omega}{\omega_g}\right)^2\right]\sinh^2\left[\dfrac{1}{2}(\alpha_d - \alpha_g)\right]}$$ (4.12)

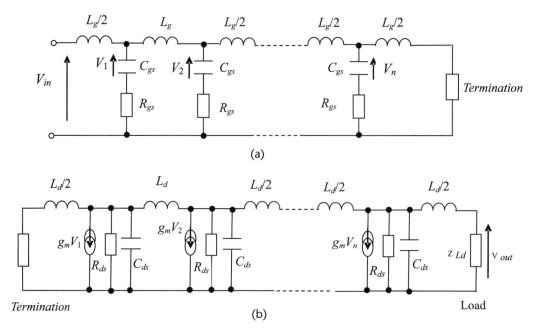

Figure 4.9 (a) Gate transmission line and (b) drain transmission line.

where $Z_{0g} = \sqrt{L_g/C_g}$ and $Z_{0d} = \sqrt{L_d/C_d}$ are the characteristic impedances of the gate and drain line, respectively.

The most commonly used definition of power transducer gain is the so-called transducer gain G_T defined as

$$G_T = \frac{P_{\text{load}}}{P_{\text{av}}} \tag{4.13}$$

where P_{load} is the power delivered to the load by the amplifier, and P_{av} is the power available from the source. The latter is the same as the power delivered to the amplifier input by the source under the condition that the amplifier input impedance is conjugate matched to the source impedance.

4.3.2 Analytical Approach to Admittance Theory

The elementary circuit of a lumped-element distributed amplifier can be represented by a four-port network as shown in [26]. Replacing the transistor by its two-port representation with the current source i_k leads to the equivalent circuit shown in Figure 4.10. The matrix equation that relates the voltage and current in Figure 4.10 takes the following form:

$$\begin{bmatrix} V_{Dk-1} \\ I_{Dk-1} \\ V_{Gk-1} \\ I_{Gk-1} \end{bmatrix} = A_k \begin{bmatrix} V_{Dk} \\ -I_{Dk} \\ V_{Gk} \\ -I_{Gk} \end{bmatrix} \tag{4.14}$$

where

$$A_k = A_{1k}A_{Fk}A2_k \tag{4.15}$$

Here, $[A_{1k}]$ is the matrix of the input link and $[A_{2k}]$ is that of the output link as shown in Figure 4.10, whereas $[A_{FK}]$ constitutes the transistor admittance matrix. Cascading n elementary circuits and terminating the idle ports with R_G and R_D (the gate and drain loads, respectively) yields the following matrix equation [26]:

$$\begin{bmatrix} V_{D0} \\ -R_D{}^{-1}V_{D0} \\ V_{G0} \\ I_{G0} \end{bmatrix} = D \begin{bmatrix} V_{Dn} \\ -I_{Dn} \\ V_{Gn} \\ -R_G{}^{-1}V_{Gn} \end{bmatrix} \tag{4.16}$$

where

$$D = \prod_n^{k=0} A_k \tag{4.17}$$

The insertion gain is expressed as the ratio of the signal power delivered to the load by the circuit to the signal power delivered directly to that load. The insertion gain can now be determined after some algebraic steps [26] as follows:

$$\text{Gain} = \left| 2Y_0 \frac{C_2}{C} \right| \tag{4.18}$$

with

$$Y_0 = 1/Z_0 \tag{4.19}$$

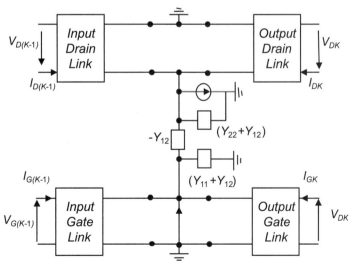

Figure 4.10 Equivalent four-port representation of a circuit with a distributed amplifier form [26].

$$C_1 = D(4,3) + R_G^{-1}D(4,4) + Y_0\left(D(3,3) + R_D^{-1}D(3,4)\right) \qquad (4.20)$$

$$C_2 = D(2,3) + R_G^{-1}D(2,4) + Y_0\left(D(1,3) + R_D^{-1}D(1,4)\right) \qquad (4.21)$$

$$C = C_1\left[D(2,1) + Y_0^{-1}D(2,2) + Y_0\left(D(1,1) + R_D^{-1}D(1,2)\right)\right]$$
$$-C_2\left[D(4,1) + Y_0^{-1}D(4,2) + Y_0\left(D(3,1) + R_D^{-1}D(3,2)\right)\right] \qquad (4.22)$$

This equation represents the exact solution for the gain of a distributed amplifier in its most general form. In the case of the distributed amplifier structure, the load and the input impedance are the same, typically (50Ω). The insertion and the transducer gain define the same quantity.

4.3.3   Analytical Approach to Wave Theory

The normalized transmission matrix approach was presented by McKay et al. [27]. This theory applies to a general class of distributed amplifiers with discrete sampling points on the gate line that couple to discrete excitation points on the drain line. Moussa et al. [28] extend this concept by considering the bilateral case obtained by including the gate-drain capacitance C_{gd} of the transistor. Using the scattering formalism, the wave quantities [27] as shown in Figure 4.11 are given by

$$b_n^{\pm} = \frac{V_{bn}}{\sqrt{Z_{0d}}} \pm i_{bn}\sqrt{Z_{0d}} \qquad (4.23)$$

$$a_n^{\pm} = \frac{V_{an}}{\sqrt{Z_{0g}}} \pm i_{an}\sqrt{Z_{0g}} \qquad (4.24)$$

where V_{an}, i_{an}, V_{bn}, and i_{bn} are the total voltages and currents at section n and the a and b denote the complex wave amplitudes on the gate and the drain line, respectively. The Z_{0g} and Z_{0d} terms represent the characteristic impedances of the gate and drain lines, respectively. It is the last equation that allows us to analyze the bilateral case [27].

The transfer matrix $[M]$, defined as $[M] = [G_{-1/2}][T]^N[G_{1/2}]$ [27], is given by

$$W_{out} = \left[G_{-1/2}\right]\left[T^N\right]\left[G_{1/2}\right]W_{in} \qquad (4.25)$$

where

$$W_i = \left[a_i^+ b_i^+ a_i^- b_i^-\right]^T \qquad (4.26)$$

and where T denotes the operator transpose and in and out are the input and the output vectors [28], respectively.

Additionally note the following:

$$\left[G_{1/2}\right] = \mathrm{diag}\left\{\exp\left(-\theta_g/2\right), \exp\left(-\theta_d/2\right), \exp\left(\theta_g/2\right), \exp\left(\theta_d/2\right)\right\} = \left[G_{-1/2}\right]^{-1} \quad (4.27)$$

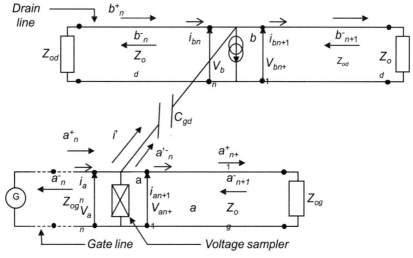

Figure 4.11 Elementary section of bilateral distributed amplifier. The variables b_n and a_n represent scattering waves [27].

Note that the propagation constants θ_g and θ_d of, respectively, the gate and the drain line, are complex:

$$[T] = [G][H] \tag{4.28}$$

where

$$[G] = \mathrm{diag}\left\{\exp\left(-\theta_g\right), \exp\left(-\theta_d\right), \exp\left(\theta_g\right), \exp\left(\theta_d\right)\right\} \tag{4.29}$$

$$[H] = \begin{bmatrix} 1 + jZ_{0g}C_{gd}\omega & -\dfrac{1}{2}jC_{gd}\omega\sqrt{Z_{0_g}Z_{0_d}} & \dfrac{1}{2}jZ_{0_d}C_{gd}\omega & -\dfrac{1}{2}jC_{gd}\omega\sqrt{Z_{0_g}Z_{0_d}} \\[2mm] H + \dfrac{1}{2}jC_{gd}\omega\sqrt{Z_{0_g}Z_{0_d}} & 1 - \dfrac{1}{2}jZ_{0_d}ZC_{gd}\omega & -H - \dfrac{1}{2}jC_{gd}\omega\sqrt{Z_{0_g}Z_{0_d}} & -\dfrac{1}{2}jZ_{0_d}C_{gd}\omega \\[2mm] -\dfrac{1}{2}jZ_{0_g}C_{gd}\omega & \dfrac{1}{2}jC_{gd}\omega\sqrt{Z_{0_g}Z_{0_d}} & 1 - jZ_{0g}C_{gd}\omega & \dfrac{1}{2}jC_{gd}\omega\sqrt{Z_{0_g}Z_{0_d}} \\[2mm] -H - \dfrac{1}{2}jC_{gd}\omega\sqrt{Z_{0_g}Z_{0_d}} & \dfrac{1}{2}jZ_{0_d}C_{gd}\omega & -H - \dfrac{1}{2}jC_{gd}\omega\sqrt{Z_{0_g}Z_{0_d}} & 1 + \dfrac{1}{2}jZ_{0_d}C_{gd}\omega \end{bmatrix}$$

$$\tag{4.30}$$

where

$$H = -\frac{1}{2}g_m D(\omega)\sqrt{Z_{0_g}Z_{0_d}} \tag{4.31}$$

ω is the pulsation, and

$$D(\omega) = \frac{1}{1 + jR_{gs}C_{gs}\omega} \tag{4.32}$$

Under the assumption of perfect matching at the input and output lines, the transmission coefficient S_{21}, which relates the incident gate at the input to the incident drain signal at the output, has the following form:

$$S_{21} = \frac{b_{out}^+}{a_{in}} \qquad (4.33)$$

where b_{out}^+ is the output wave of the last section on the drain line and a_{in} is the input wave of the first section of the gate line [28].

4.4 Gain/Power-Bandwidth Trade-Off

The transfer characteristics of an amplifier with lumped components as the coupling elements will reveal that the gain and the bandwidth cannot be simultaneously increased beyond a certain limit. As a result, these two quantities are often considered trade-offs in the design of an amplifier. To begin our understanding of the trade-offs, let's consider a simple bandpass amplifier consisting of an active device like that shown in Figure 4.12. The voltage transfer function [24] is given as

$$A_v(\omega) = \frac{-g_m R}{1 + jQ\left(\dfrac{\omega}{\omega_0} - \dfrac{\omega_0}{\omega}\right)} \qquad (4.34)$$

where $Q = \omega_0 RC$. The maximum gain occurs at midband and has a magnitude of $g_m R$. The bandwidth (BW; -3 dB is $\omega_0/2\pi Q$ or $1/2\pi RC$ [24]. Hence, the gain-bandwidth product of the amplifier is

$$A_{v0}BW = \frac{g_m}{2\pi C} \qquad (4.35)$$

It is clear from (4.35) that the gain-bandwidth product is proportional to g_m/C. The R that influences the maximum gain in the passband does not appear in the gain-bandwidth product. From a device point of view, the gain-bandwidth product can be explained (where L can be eliminated in Figure 4.12). The amplifier has a maximum gain of $-g_m/R$ at dc, and a bandwidth of $1/2\pi RC$, so (4.35) is still valid in this case. Hence, the gain-bandwidth product cannot be overcome by connecting in parallel configuration.

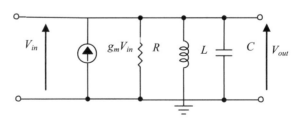

Figure 4.12 A simple bandpass amplifier schematic. Feedback of the amplifier is neglected.

In a distributed amplifier, the parasitic capacitance (shunted) of the device is well isolated from the device to form a lowpass filtering response with the transmission line inductance, where a wide bandwidth is promising. To explain the gain-bandwidth product in a distributed amplifier, we review Figure 4.9. The voltage gain of the distributed amplifier [24] can be shown to be

$$A = \frac{g_m \left(Z_{0g} Z_{0d}\right)^{1/2} \sinh\left[0.5n\left(\alpha_d - \alpha_g\right)\right] e^{-0.5n(\alpha_g + \alpha_d)}}{2\left[1 + \left(\dfrac{\omega}{\omega_g}\right)^2\right]^{1/2} \left[1 - \left(\dfrac{\omega}{\omega_g}\right)^2\right]^{1/2} \sinh\left[\dfrac{1}{2}\left(\alpha_d - \alpha_g\right)\right]} \tag{4.36}$$

From (4.36), we can show that the number of the devices that maximizes gain at any given frequency is

$$n_{opt} = \frac{\ln\left(\alpha_d/\alpha_g\right)}{\alpha_d - \alpha_g} \tag{4.37}$$

Podgorski and Wei [29] have shown that the optimum gate width of a traveling-wave amplifier has a relation similar to that of (4.37). Therefore, it is clear that in the presence of attenuation, the gain of a distributed amplifier cannot be increased indefinitely by adding devices [4]. Additional transistors not only decrease the excitation of the last device but also increase the overall attenuation on the drain line. The variables α_g and α_d are gate and drain-line attenuation [24], respectively, and are given by

$$\alpha_g = \frac{\left(\omega/\omega_c\right)^2 \left(\omega_c/\omega_g\right)}{\sqrt{1 - \left(1 - \left(\omega_c/\omega_g\right)^2\right)\left(\omega/\omega_c\right)}} \tag{4.38}$$

$$\alpha_d = \frac{\left(\omega_d/\omega_c\right)}{\sqrt{1 - \omega^2/\omega_c^2}} \tag{4.39}$$

where ω_g is the gate corner frequency, given by $1/R_{gs}C_{gs}$, and ω_d is the drain corner frequency, given by $1/R_{ds}C_{ds}$.

Reference [1] showed that by extending the analysis of [30], (4.38) and (4.39) can be rearranged as

$$\alpha_g = \frac{2aX_k^2}{n\sqrt{1 + \left[\dfrac{4a^2}{n^2} - 1\right]X_k^2}} \tag{4.40}$$

$$\alpha_d = \frac{2b}{n\sqrt{1 - X_k^2}} \tag{4.41}$$

where

$$a = \frac{n\omega_c}{2\omega_g}, \; b = \frac{n\omega_d}{2\omega_c}, \text{ and } X_k = \omega/\omega_c$$

At dc, (4.36), after substituting (4.40) and (4.41), can be rewritten as

$$A_0 = \frac{g_m \left(Z_{0g}Z_{0d}\right)^{1/2} \sinh(b)e^{-b}}{2\sinh(b/n)} \tag{4.42}$$

To explain the trade-off between the gain and bandwidth of a distributed amplifier, voltage gain from (4.36) can be used. Normalizing the voltage gain to dc operation is convenient to understand the gain and bandwidth trade-off. The terms a and b are used to simplify the normalized gain.

Normalized gain A/A_0 can be deduced from (4.36) and (4.42), and the expression shown as

$$A_N = A/A_0 = \frac{\sinh\left(\frac{b}{n}\right)e^b \sinh\left[b/\left(1-X_k^2\right)^{1/2} - aX_k^2/\left[1+\left(\frac{4a^2}{n^2}-1\right)X_k^2\right]^{1/2}\right]e^{-[b/(1-X_k^2)^{1/2}+aX_k^2/[1+(4a^2/n^2-1)X_k^2]^{1/2}}}{\sinh(b)\left[1+\frac{4a^2}{n^2}X_k^2\right]^{1/2}\left[1-X_k^2\right]^{1/2}\sinh\frac{1}{n}\left[b/\left(1-X_k^2\right)^{1/2}-aX_k^2/\left[1+\left(\frac{4a^2}{n^2}-1\right)X_k^2\right]^{1/2}\right]}$$

$$(4.43)$$

The normalized gains of (4.43) over frequency response X_k for various values of a and b are illustrated in Figure 4.13. In this figure $n = 4$ is selected to understand the gain-bandwidth trade-off for a few variations of a and b. Note that beyond $n = 4$, the gain response degraded for the same a and b. Nevertheless, lower a and b values led to a broadband frequency response, as shown in Figure 4.13. For example, the selection of $a = 1$ gives a poorer bandwidth response than that of $a = 0.6$. Therefore, it is important to select a device having suitable ω_g and ω_d for wideband operation.

Nonuniform distributed amplifier design applying drain-line impedance tapered eliminating drain-line reverse wave and maximizing output power combination at load termination [31]. Consider a nonuniform distributed amplifier (Figure 4.14) with a cutoff frequency ω_c, gate-line characteristic impedance Z_G, and maximum output power $P_{\max}$ consisting of n sections and having transistors with maximum power P_o and gate-source capacitance C_{gs}. Its output power-bandwidth product (PBW) [31] is defined by

$$\text{PBW} = P_{\max} \cdot f_c = n \cdot P_o \cdot f_c = \frac{n \cdot \left(V_{bk} - V_k\right)^2}{8 \cdot Z_D} \cdot \frac{1}{\pi \cdot Z_G \cdot C_{gs}} \tag{4.44}$$

where Z_D is the load presented to its drain terminal, and V_{bk} and V_k are the breakdown and knee voltages, respectively.

For the $(V_{bk} - V_k)/C_{gs}$ ratio, which is typically a fixed value, P_o depends on V_{bk} and V_k, which along with C_{gs}, increases with gate length. The P_o term is contributed

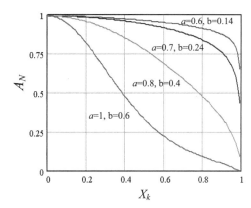

Figure 4.13 Normalized gains over frequency response for various values for a and b, where $n = 4$.

mainly by the power transistor. However, for moderate bandwidth operation (i.e., 2 GHz), a higher P_{max} could be achieved.

The increased single-stage gain permits a proportional P_{max} increase in those amplifiers limited by dynamic (linear) range of the input signal [31]. The maximum transistor output power P_o is given by $G_p \cdot P_{in(max)}$, which is proportional to the single-stage amplifier power gain G_p and to the maximum limited input strength $P_{in(max)}$. Therefore, connecting a few stages of nonidentical transistors with interstage tapered impedance can increase $P_{in(max)}$ to the power transistor, and having high-f_τ transistor (lower C_{gs}) can be coupled to the gate line [32], as shown in Figure 4.15.

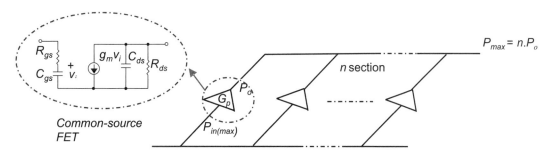

Figure 4.14 Maximum transistor output is given by $G_p \cdot P_{in(max)}$ for a nonuniform distributed amplifier.

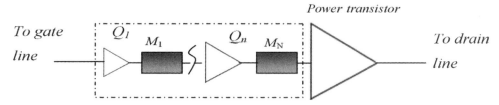

Figure 4.15 Connecting a few stages of nonidentical transistors with interstage tapered impedance (broadband matching networks, e.g., M_1, ... , M_2) can increase $P_{in(max)}$ to the power transistor, and having high-f_τ transistor (lowest C_{gs}) can be coupled to the gate-line input. Q_1, ... , Q_N are high-f_τ transistors.

As given in (4.44), n sections will increase the PBW products. However, adding more n sections will not improve the products because of the losses associated with active transistor resistances. Beyer et al. have shown that gain in a conventional distributed amplifier cannot be increased indefinitely by adding more sections [1]. The following discussion will determine an optimum number of sections to maximize output power at a given frequency associated with device input and output corner frequencies. The output power of a distributed amplifier is given by

$$P_{max} = \frac{g_m^2 V_{in}^2 e^{-n(\alpha_g + \alpha_d)}}{\left(1 + \omega^2/\omega_g^2\right)\left(1 - \omega^2/\omega_c^2\right)} \frac{\sinh^2\left[\frac{n}{2}\left(\alpha_g - \alpha_d\right)\right]}{\sinh^2\left[\frac{1}{2}\left(\alpha_g + \alpha_d\right)\right]} \qquad (4.45)$$

where α_g and α_d are gate and drain-line attenuation, given in (4.38) and (4.39), respectively [1]; ω_g is the gate corner frequency, given by $1/R_{gs}C_{gs}$; and ω_d is the drain corner frequency, given by $1/R_{ds}C_{ds}$. Note that α_g and α_d are the critical factors controlling the frequency response.

The plots of α_g and α_d for a GaN power transistor are shown in Figure 4.16. The information of ω_g and ω_d of the power transistor can be computed from the intrinsic elements. It is strongly evident from Figure 4.16 that the gate line is more sensitive to frequency response than drain-line attenuation, and the drain-line attenuation does not vanish at a low frequency limit. Therefore, the sensitivity of the distributed amplifier frequency response can be minimized as the signal move toward ω_c by coupling a high-f_τ transistor to the gate line that is having high ω_g. To improve drain-line attenuation, use of a compensation network is necessary to introduce to the output of the power transistor. Typically, a bigger device periphery causes the loading effect to become stronger. Attenuation compensation with an

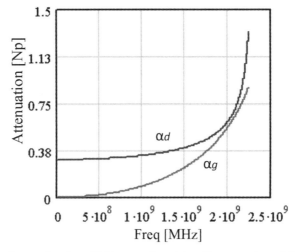

Figure 4.16 Plots of α_g and α_d for a GaN power transistor. The α_g and α_d of the transistor are computed from the intrinsic elements of the GaN device.

active load to reduce drain-line losses dominated by R_{ds} (drain-source resistance) [33] is a common technique, but the cost and area required are increased.

The number of sections that maximizes output power is given in [34] as

$$n_{\text{opt}} = \frac{\ln\left[\dfrac{3\alpha_d - \alpha_g}{3\alpha_g - \alpha_d}\right]}{\alpha_d - \alpha_g} \qquad (4.46)$$

Referring to [34], for a larger device periphery (i.e., GaN device having a gate width of 3.6 mm) and to deliver 30W from each device, n_{opt} is approximately 3 ... 4. To understand the trade-off relation between output power and bandwidth, (4.45) can be plotted for various ω_g and ω_d.

Additionally, from (4.44), reducing Z_D would benefit the products. The following section explains the optimum impedance that will be synthesized when Z_D is reduced. Consequently, wideband impedance transformation is required. Bear in mind that by simply reducing Z_D will not guarantee the multiple-source current source combining to a single load due to the fact that an optimum virtual impedance in two directions (i.e., $Z_{u(k)}$ and $Z_{r(k)}$) is not fulfilled, nevertheless, this can be achieved and detailed explanations will be given in Chapter 5.

4.5   Practical Design Methodology

Distributed amplifiers are known for their flat gain, linear phase, and low return loss over a wide frequency band [1–20]. The major challenges in the design of a high-power distributed amplifier are related to active input/output device losses [1–5]. In multilayer printed circuit boards (PCBs), all RF and dc lines are often laid out together on the same PCB using commercial layout tools. As a result, parasitic coupling between different parts of the circuit may occur, thus degrading the device's overall performance. This can also lead to stability problems and undesired oscillations. Typically, a PCB design is verified through measurements. Modifications are then made based on experimental data, which often lead to a long development time. Hence, it is very important to have an accurate simulation tool that can predict undesired coupling between the circuit elements before the fabrication of the PCB.

This section explains design methodologies for practical distributed amplifiers, including a few steps from the design goal/specifications process, device selection, analytical approach, and so forth. These steps are applied in the distributed amplifier development discussions throughout this text, and experimentally demonstrate good performance without any tuning at the board level [8, 9, 20, 21, 32–36], which is especially significant for high-power development (Chapter 8); the resulting device is referred to as a distributed power amplifier (DPA).

4.5.1   Design Goal/Specifications

Understanding the design goals/specifications is the first step when starting a distributed amplifier design. The basic design requirement is to achieve a high-output-power (e.g., P_{out} >10W) distributed amplifier that meets software defined radio (SDR) applications.[2] Knowing the basic goal is P_{out} >10W, breakdown voltage V_{bk} can be computed from

$$V_{bk} - V_k = \sqrt{8P_{out}R_L} \tag{4.47}$$

where V_k and R_L are knee voltage and load impedance of the distributed amplifier, respectively.

4.5.2   Device Selection

If V_k in (4.47) is neglected (or assumed zero for simplicity), then V_{bk} over P_{out} can be computed with condition $R_L = 50\Omega$ (to eliminate impedance transformation) and plotted in Figure 4.17. To achieve 10W or more, at 50Ω impedance, the device must have a V_{bk} of at least 70V (from Figure 4.17). Due to power requirements and a limited operating voltage for the 50Ω condition, LDMOS and HBT devices are not suitable. However, this can be achieved with $R_L < 50\Omega$ impedance, where an additional impedance transformation network is required. Due to high bandgap, a GaN HEMT device is selected as the power device. The device from CREE Inc. (part number CGH40010F) offers a V_{bk} of ~70V for the 50Ω condition, which was found to be suitable for this application. However, many GaN devices are available from other device manufacturers, for example, Nixtonic Inc., Fujitsu Inc., and NXP Inc.

4.5.3   Theoretical Analysis (Zero-Order Analysis)

Various distributed amplifier topologies have been suggested, and it is essential to understand the electrical performance of the topologies. Therefore, the circuit analysis begins with a simplified FET model. The intrinsic elements of the small-signal FET model (g_m, C_{gs}, C_{ds}, R_{gs}, R_{ds}, and so on) are extracted by means of device modeling (Figure 4.18). For simplicity, feedback element C_{gd} is neglected. It is convenient to investigate gate and drain lines independently to understand their performance over the frequency range. Basic theoretical approaches (superposition, Thevenin, distributed amplifier, Darlington theorem, and so on) are employed in this thesis. As an example, in Chapter 5, the generalized design equation for the virtual impedance of a distributed amplifier by means of current source properties and a simplified FET model are developed. This provides a good initial guess of the circuit elements with which to begin the next steps.

4.5.4   Analysis with VCCS and C_{gd} (First-Order Analysis)

As the next step, the voltage-controlled current source (VCCS) and C_{gd} are taken into consideration to understand the electrical performance of the proposed topologies (Figure 4.19). Due to the feedback effect, electrical performance would be degraded and ripple may be expected near cutoff frequency f_c. Therefore, it is necessary to

---

[2] Such an amplifier must meet the parameters of two-way radio requirements: output power, efficiency, stability, transient response, and so on, over the frequency band of interest.

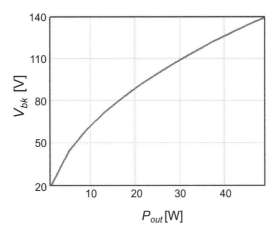

Figure 4.17   Breakdown voltage V_{bk} versus output power P_{out} for 50Ω condition.

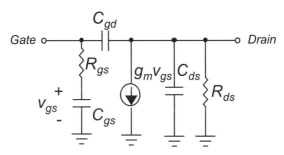

Figure 4.18   Simplified small-signal FET model.

identify the circuitry elements that could compensate for the electrical degradation of the distributed amplifier. Typically, gate and drain-line impedances (including the dummy termination impedances) are adjusted from the theoretical analysis in many distributed amplifier applications. Reference [36] showed that the implementation of a nonuniform gate line can be employed in a distributed amplifier design to compensate for degradation of electrical performance

4.5.5   DC Biasing Circuitry Design

DC biasing circuitry needs to be designed carefully because of the influence over the frequency response. Figure 4.20 shows a typical FET dc bias arrangement. A dc feeding network, L_d, together with C_d selection must provide high impedance over the frequency bandwidth. Recommended values for L_d and C_d for bandwidth operation from 10 to 2000 MHz are 180 nH and 33 pF, respectively[3],[4] [11]. As shown in [37], multiple chokes and capacitors offer broadband response up to 3 GHz (Figure 4.21). For a dc biasing network, L_g and C_g must provide a stable quiescent point.

---

[3] To obtain 180 nH, high-Q ceramic 0603HP series chip inductors were provided by Coilcraft, Inc., with Q up to ~150 at 1.7 GHz.
[4] To obtain 33 pF, a broadband high-Q capacitor 3060 from Murata, Inc., was used. Details can be obtained at www.murata.com.

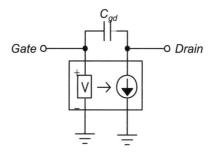

Figure 4.19 VCCS and C_{gd} for a single distributed amplifier. Other intrinsic elements are included in the VCCS section.

In some cases L_g is replaced with a resistor [11], which provides transient controlled and stable operation. Coupling capacitor C_b is necessary to block dc from flowing into the RF path. With proper selection, it offers a broadband response; for instance, 120 pF is adequate to satisfy up to ~2 GHz. A proper dc turn-on sequence is applied: First apply negative bias to the voltage (i.e., $V_{gate} < 0V$) and then apply the drain voltage ($V_{drain} > 0V$); during turn-off, V_{drain} must be switched to 0V followed by V_{gate}.

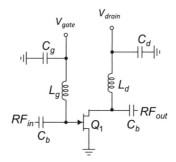

Figure 4.20 Basic FET dc bias arrangement.

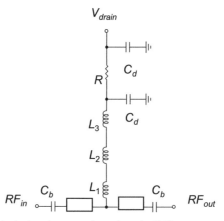

Figure 4.21 Broadband choke implementation given in [38].

4.5.6   Device Modeling of GaN HEMTs

The enhanced equivalent circuit of a high-power GaN HEMT device can include the delay network to describe the high-frequency delay effects, the source spreading resistance to describe the influence of the device channel to the increase in magnitude of S_{21} with frequency, and the electrothermal elements to estimate channel temperature rise due to power dissipation [39]. However, it is most important to evaluate properly the main nonlinear intrinsic elements of the device equivalent circuit; the extrinsic linear elements, however, whose effects are not so significant especially at lower frequencies, can be included within the distributed circuit parameters.

To characterize the intrinsic device, first consider the admittance Y-parameters derived from the intrinsic small-signal equivalent circuit as

$$Y_{11} = \frac{j\omega C_{gs}}{1 + j\omega C_{gs} R_{gs}} + j\omega C_{gd} \tag{4.48}$$

$$Y_{12} = -j\omega C_{gd} \tag{4.49}$$

$$Y_{21} = \frac{g_m}{1 + j\omega C_{gs} R_{gs}} + j\omega C_{gd} \tag{4.50}$$

$$Y_{22} = \frac{1}{R_{ds}} + j\omega \left(C_{ds} + C_{gd} \right) \tag{4.51}$$

By dividing these equations into their real and imaginary parts, the parameters of the device intrinsic small-signal equivalent circuit can be determined as [40]

$$C_{gd} = -\frac{\mathrm{Im}\, Y_{12}}{\omega} \tag{4.52}$$

$$C_{gs} = -\frac{\mathrm{Im}\, Y_{11} - \omega C_{gd}}{\omega} \left[1 + \left(\frac{\mathrm{Re}\, Y_{11}}{\mathrm{Im}\, Y_{11} - \omega C_{gd}} \right)^2 \right] \tag{4.53}$$

$$R_{gs} = \frac{\mathrm{Re}\, Y_{11}}{\left(\mathrm{Im}\, Y_{11} - \omega C_{gd} \right)^2 + \left(\mathrm{Re}\, Y_{11} \right)^2} \tag{4.54}$$

$$g_m = \sqrt{\left(\mathrm{Re}\, Y_{21}\right)^2 + \left(\mathrm{Im}\, Y_{21} + \omega C_{gd}\right)^2} \sqrt{1 + \left(\omega C_{gs} R_{gs}\right)^2} \tag{4.55}$$

$$C_{ds} = \frac{\mathrm{Im}\, Y_{22} - \omega C_{gd}}{\omega} \tag{4.56}$$

$$R_{ds} = \frac{1}{\mathrm{Re}\, Y_{22}} \tag{4.57}$$

where the intrinsic admittance Y-parameters can be defined from measurements after extraction of all extrinsic parasitic parameters [40].

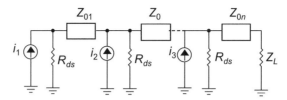

Figure 4.22 Circuit showing an analysis of drain line tapered with constant-k terminated with 50Ω, where Z_{0i}, with i = 1, 2 ... n, refers to a line impedance distributed amplifier; consists of shunt capacitance and inductance. Note that dummy termination is neglected to push all current to the load Z_L.

4.5.7 Loading Device into Distributed Output Network

Generally in a distributed amplifier, output capacitance C_{ds} becomes a shunt element in a constant-k ladder network (Figure 4.22), and this determines the cutoff frequency f_c of the line. Synthesizing the inductance required careful consideration to provide desired impedance(with minimum flatness) over the frequency bandwidth response. Loading properties of the line strictly rely on real part R_{ds}, and significant loading takes place with a larger device periphery. Throughout Chapters 5, 6, and 7, the distributed amplifier prototype boards discussed are using medium-power devices, for which the loading effect is not significant. However, in Chapter 8, due to the use of high-power devices, the loading effect of the device is taken into consideration.

Due to packaged device implementation in distributed amplifier development, keep in mind that to extract effective real and imaginary parts of the input and output packaged device, $Z_{in}(\omega) = R_{in}(\omega) + jX_{in}(\omega)$ and $Z_{opt}(\omega) = R_{opt}(\omega) + jX_{opt}(\omega)$, respectively. For example, Figure 4.23 shows a plot of S_{11} for a transistor in packaged and die form for a medium-power device [36]. It is important to note that the resonance frequency of a die is much higher than that of a packaged device. Figure 4.24(a) shows an example of the packaged properties of high-power GaN device. Table 4.1 shows the extraction values of $Z_{in}(\omega)$ and $Z_{opt}(\omega)$ for high-f_τ and power transistors.

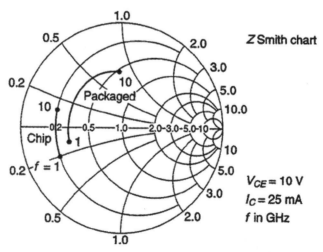

Figure 4.23 Plot of S_{11} for a common emitter medium-power device in packaged and die form [41].

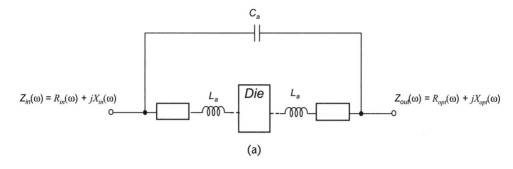

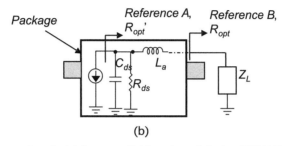

Figure 4.24 (a) Properties of a high-power GaN packaged device (CGH40010F) [42] and (b) illustration of optimum load impedance R'_{opt} (at reference A) and R_{opt} (at reference B).

As shown in Figure 4.24(b), the optimum load impedance R'_{opt} and R_{opt} (at both reference planes, i.e., A and B) are different, since package effects have to be taken into consideration. In practice, it is not possible to observe the optimum RF voltage or current swing at reference A, but the load pull impedance contours of the device can be extracted at reference B. As a result, R_{ds} will be replaced with $R_{opt}(\omega)$, and $X_{opt}(\omega)$ will be absorbed into the drain line to form desired ω_c, as shown in Figure 4.25. Therefore, the line inductance L_i is realized by means of (4.58).

4.5.8   Synthesizing Distributed Input/Output Networks

Figure 4.25 shows when a device is loaded by artificial transmission line (drain line). Therefore, with packaged device selection, the cutoff frequency f_c can be defined as

$$f_c = \frac{1}{\pi\sqrt{L_i C_{opt,i}}} \tag{4.58}$$

Table 4.1 Extraction Value of $Z_{in}(\omega)$ and $Z_{out}(\omega)$ for High-f_τ and Power Transistors*

	R_{in} (Ω)	C_{in} (pF)	R_{opt} (Ω)	C_{opt} (pF)
First high-f_τ transistor[a]	600	2.4	470	1.2
Second high-f_τ transistor[b]	378	3.3	220	2.8
Power transistor (CGH40010F)	102	5.9	40	3.7

* Note that R_{in} and C_{in} are in parallel form.

[a] The manufacturer part number is ATF51143, from Avago Inc. This is an enhancement mode pHEMT device, designed with a 6,400-μm gate width and 46 gate fingers. The packaged device information is given in [43].

[b] The manufacturer part number is ATF51143, from Avago Inc. This is an enhancement mode pHEMT device, designed with an 800-μm gate width and 16 gate fingers. The packaged device information is given in [44].

where C_{opt} is the element extracted from $X_{opt}(\omega)$, and L_i is line inductance. In most cases, $C_{opt} < C_{in}$, where phase synchronization between gate and drain lines is achieved using a capacitively coupled technique [13].

Once $Z_{in}(\omega)$ and $Z_{opt}(\omega)$ have been identified, we are ready to synthesize the gate/drain-line networks by means of image impedance characteristics. Bear in mind that selection of C_{opt} from $X_{opt}(\omega)$ requires careful analysis and is valid below a resonance frequency, and that $R_{opt}(\omega)$ only plays a role in high-power distributed amplifier design, and typically does not cause a loading effect for small- or medium-output-power devices.

4.5.9 Layout Design

PCB selection (i.e., thickness h and dielectric constant ε_r) is important; for instance, using standard routing to connect two lumped elements of size 3060, with the allowable minimum width of 30 mils. A PCB with a thicker h has the advantage of being able to minimize line inductance, assuming ε_r is identical. However, as a trade-off stray capacitance increases, although via-hole inductance is reduced [45]. As a guideline, it is strongly recommended to use an h of 0.762 mm and an ε_r of 3.66 (Rogers' properties) to design a distributed amplifier covering bandwidths up to 2 GHz.

As far as the PCB is concerned, the self-resonance frequency of the board should be as high as possible (as a rule of thumb). It is not easy to measure the frequency, but some details are given in [46]. For high current handling and to keep the frequency as high as possible, coupling effects can be investigated with the aid of a full-wave simulator (e.g., CST, HFSS, or others). Figure 4.26 shows a via-hole full-wave simulation that was designed to study optimum spacing for a selected diameter via hole [47]. The results revealed that the spacing must be at least two times the diameter in order to reduce the coupling at higher frequencies (up to 4 GHz).

4.5.10 Full-Wave Simulation/Layout Optimization

Typically, a PCB design goes through many prototypes to solve unintended coupling problems through measurements and modifications based on experience, often late in the development cycle. Hence, it is highly desirable to have an accurate simulation

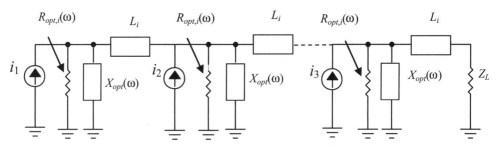

Figure 4.25 Circuit showing an analysis of a drain line tapered with constant-k terminated with 50Ω. Note that $i = 1, 2 \ldots n$, and dummy termination is neglected to push all current to the load R_L.

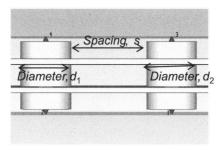

Figure 4.26 Differential via-hole study to understand the coupling effect for optimum grounding potential, especially at higher frequencies.

tool that can predict unwanted coupling in the layout without actually fabricating a PCB [48, 49]. A design methodology that uses full-wave electromagnetic (EM) simulation is adopted in this work to minimize design cycle time and avoid tedious optimization.

Irrelevant and insignificant details such as curved transmission line bends, small holes, or gaps could cause long simulation times due to the fine meshing requirements. Fine meshing produces more accurate results, but requires longer computational times due to the high density of meshing, whereas coarse meshing requires less computational time, but may compromise the accuracy [50, 51]. Therefore, there is always a trade-off between computational time and accuracy.

The two most common mesh types offered in available 3D EM simulators (such as CST) are hexahedral and tetrahedral meshing (Figure 4.27).

Normally for an electrically large simulation model, hexahedral mesh is preferred. Tetrahedral mesh is preferred for electrically small models. In this simulation, we use a tetrahedral mesh because the model is electrically small and covers a broadband frequency. The mesh type used is always related to the accuracy and the speed of the simulation. This is a trade-off that needs to be taken care of before we start the simulation. To achieve good accuracy, a curvature refinement option is

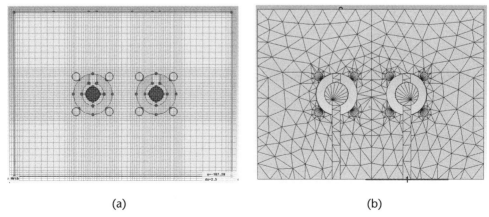

(a)   (b)

Figure 4.27 Common mesh types offered in available 3D EM tools: (a) hexahedral and (b) tetrahedral.

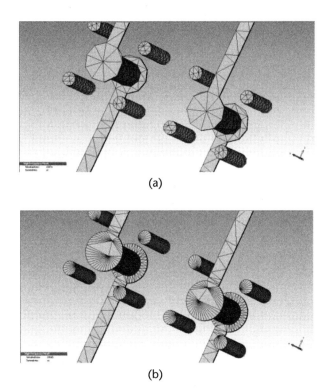

(a)

(b)

Figure 4.28   (a) Without and (b) with curvature refinement for the accuracy refinement options.

used at the via section for better meshing (Figure 4.28). However, this will increase the meshing size from 25,000 to 111,255 as shown in Figure 4.29.

Radiation boundaries are used in order to terminate the field so that no electromagnetic energy reflected back to the object that is being simulated. The minimum distance to the radiation boundary is a quarter wavelength ($\lambda/4$) at the lowest frequency of interest. CST has several types of solvers. The time domain solver is a powerful algorithm that is able to perform simulations for a broad frequency range. The frequency domain solver is able to solve an electromagnetic problem a single frequency at a time.

Modeling with a full-wave EM simulator including PCB layer stack-up, via holes, indium foil, grounded heat sink, and RF connector modeling [34, 45] are considered. For a GaN power transistor, a grounded heat sink is attached at the bottom layer via indium foil. The indium foil (part number IN52-48SN, 0.004-in. thickness, from Indium Corporation Inc., North Carolina, US); the thermal conductivity of the copper foil is 0.34 W/cm at 85°C. A cross section of the indium foil and its

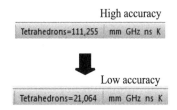

Figure 4.29   Trade-off between meshing size and accuracy.

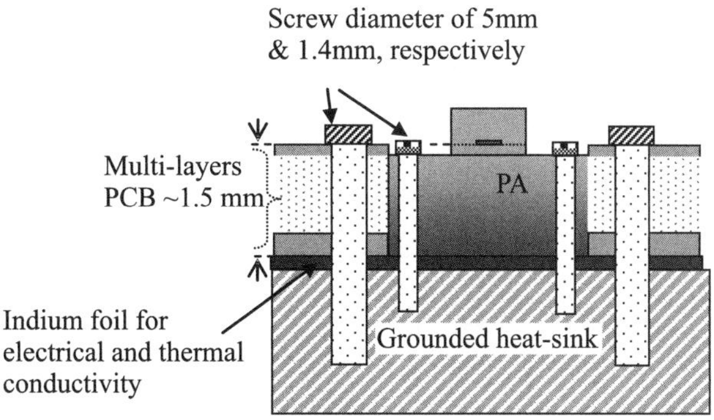

Figure 4.30   Cross-sectional diagram of a high-power GaN amplifier and PCB.

multiple screws is shown in Figure 4.30. The board has four bigger diameter (5-mm) screws mounted to hold the grounded chassis and two smaller diameter (1.4-mm) screws to mount the GaN device amplifier to the chassis. The bottom layer of the PCB has solder resist, and a bigger area of indium foil is filled in between the bottom layer and the grounded heat sink. Note that active and passive device models are not placed during EM simulation.

The EM simulation design flow begins by importing the ODB++ file (PCB Gerber file, which was created to bring some order to the transfer of board date to manufacturer) from Cadence software into the CST environment, defining port and boundary regions, and exporting the *S*-parameter to the ADS simulator until optimizing the layout in a harmonic balance (HB) simulation before releasing the Gerber file for fabrication. ADS cosimulation assisted by CST is performed in the ADS environment. This simulation allows for optimization of the geometry of the layout, while performing complete optimization in the HB simulation (assisted CST). Each active and passive component is terminated with a discrete lumped port in CST, and a total of 90 ports for the first DPA development were created. Therefore, the layout information (*.s90 file) is exported to ADS to evaluate complete power performance. All passive and active components modeled are connected to the appropriate ports in the ADS.

References

[1]   J. B. Beyer et al., "MESFET Distributed Amplifier Design Guidelines," *IEEE Trans. Microwave Theory and Techniques*, Vol. MTT-32, No. 3, pp. 268–275, Mar. 1984.

[2]   B. Kim, H. Q. Tserng, and H. D. Shih, "High Power Distributed Amplifier Using MBE Synthesized Material," *IEEE Microwave and Millimeter Monolithic Integrated Circuits Symposium*, pp. 35–37, 1985.

[3]   P. H. Ladbrooke, "Large Signal Criteria for the Design of GaAs FET Distributed Power Amplifier," *IEEE Trans. Electron Devices*, Vol. ED-32, No. 9, pp. 1745–1748, Sep. 1985.

[4]   B. Kim and H. Q. Tserng, "0.5W 2–21 GHz Monolithic GaAs Distributed Amplifier," *Electronics Letters*, Vol. 20, No. 7, pp. 288–289, Mar. 1984.

[5]   M. J. Schindler et al., "A 15 to 45 GHz Distributed Amplifier Using 3 FETs of Varying Periphery," *IEEE GaAs IC Symposium Technical Digest*, pp. 67–70, 1986.

[6]   L. Zhao et al., "A 6 Watt LDMOS Broadband High Efficiency Distributed Power Amplifier Fabricated Using LTCC Technology," *IEEE MTT-S Int. Microwave Symp. Dig.*, pp. 897–900, June 2002.

[7]   J. Gassmann et al., "Wideband High Efficiency GaN Power Amplifiers Utilizing a Non-Uniform Distributed Topology," *IEEE MTT-S Int. Microwave Symp. Dig.*, pp. 615–618, June 2007.

[8]   K. Narendra et al., "Vectorially Combined pHEMT/GaN Distributed Power Amplifier for SDR Applications," *IEEE Trans. Microwave Theory and Techniques*, Vol. 60, No. 10, pp. 3189–3200, Oct. 2012.

[9]   K. Narendra et al., "Vectorially Combined Distributed Power Amplifier with Load Pull Determination," *Elect. Letters*, Vol. 46, No. 16, pp. 1137–1138, Aug. 2010.

[10]   E.W. Strid and K. R. Gleason, "A DC-12 GHz Monolithic GaAsFET Distributed Amplifier," *IEEE Trans. Electron Devices*, Vol. ED-29, No. 7, pp. 1065–1071, July 1982.

[11]   K. Krishnamurthy et al., "Broadband GaAs MESFET and GaN HEMT Power Amplifiers," *IEEE Journal Solid State Circuits*, Vol. 35, No. 9, pp. 1285–1292, 2000.

[12]   H. Amasuga et al., "A High Power and High Breakdown Voltage Millimeter-Wave GaAs pHEMT with Low Nonlinear Drain Resistance," *IEEE MTT-S Int. Microwave Symp. Dig.*, pp. 821–824, June 2007.

[13]   Y. Ayasli et al., "Capacitively Coupled Traveling Wave Power Amplifier," *IEEE Trans. Microwave Theory Techniques*, Vol. MTT-32, No. 12, pp. 1704–1709, Dec. 1984.

[14]   A. Ayasli et al., "A Monolithic GaAs 1–13 GHz Traveling Wave Amplifier," *IEEE Trans. Microwave Theory Techniques*, Vol. 30, No. 12, pp. 976–981, July 1982.

[15]   K. B. Niclas, R. R. Pereira, and A. P. Chang, "On Power Distribution in Additive Amplifiers," *IEEE Trans. Microwave Theory and Techniques*, Vol. 38, No. 3, pp. 1692–1699, Nov. 1990.

[16]   J. L. B. Walker, "Some Observations on the Design and Performance of Distributed Amplifiers," *IEEE Trans. Microwave Theory and Techniques*, Vol. 38, pp. 164–1698, Jan. 1992.

[17]   R. Halladay, M. Jones, and S. Nelson, "2–20 GHz Monolithic Distributed Power Amplifiers," *IEEE Microwave and Millimeter-Wave Monolithic Integrated Circuits Symp.*, pp. 35–37, 1985.

[18]   S. N. Prasad, J. B. Beyer, and I. S. Chang, "Power-Bandwidth Considerations in the Design of MESFET Distributed Amplifiers," *IEEE Trans. Microwave Theory Techniques*, Vol. 36, pp. 1117–1123, July 1988.

[19]   S. N. Prasad, S. Reddy, and S. Moghe, "Cascaded Transistor Cell Distributed Amplifiers," *Microwave and Optical Technology Lett.*, Vol. 12, No. 3, pp. 163–167, June 1996.

[20]   K. Narendra et al., "Dual Fed Distributed Power Amplifier with Controlled Termination Adjustment," *Progress in Electromagnetic Research*, Vol. 139, pp. 761–777, May 2013.

[21]   K. Narendra et al., "Discrete Component Design of a Broadband Impedance Transforming Filter for Distributed Power Amplifier," *10th IEEE Microwave Mediterranean Symp.*, pp. 292–295, Aug. 2010.

[22]   W. S. Percival, "Improvements In and Relating to Thermionic Valve Circuits," British Patent 460,562, 1937.

[23]   E. L. Ginzton et al., "Distributed Amplification," *Proc. IRE*, pp. 956–969, 1948.

[24]   T. T. Y. Wong, *Fundamentals of Distributed Amplification*, Norwood, MA: Artech House, 1993.

[25]   Y. Ayasli et al., "2-to-20 GHz GaAs Traveling Wave Power Amplifier," *IEEE Trans. Microwave Theory Techniques*, Vol. 32, No. 3, pp. 290–295, Mar. 1984.

[26]   K. B. Niclas et al., On Theory Performance of SolidState Microwave Distributed Amplifiers, *IEEE Trans. Microwave Theory Techniques*, Vol. MTT-31, No. 6, pp. 447–456, June 1983.

[27]   T. McKay, J. Eisenberg, and R. E. Williams, "A High Performance 2–18.5 GHz Distributed Amplifier Theory and Experiment," *IEEE Trans. Microwave Theory and Techniques*, Vol. 12, No. 34, pp. 1559–1568, Dec. 986.

[28]   M. S. Moussa, M. Trabelsi, and R. Aksas, "Analysis of a Bilateral Distributed Amplifier Using Scattering Parameters," *Microwave Technology and Optical Lett.*, Vol. 36, No. 2, pp. 120–122, Jan. 2003.

[29]   A. W. Podgorski and L. Y. Wei, "Theory of Traveling-Wave Transistors," *IEEE Trans. Electron Devices*, Vol. ED-29, No. 12, pp. 1845–1853, Dec. 1982.

[30]   W. H. Horton, J. H. Jasberg, and J. D. Noe, "Distributed Amplifiers: Practical Considerations and Experimental Results," *Proc. IRE*, Vol. 38, pp. 748–753, July 1950.

[31]   B. M. Green et al., "High-Power Broadband AlGaN/GaN HEMT MMICs on SiC Substrates," *IEEE Trans. Microwaves Theory and Techniques*, Vol. 49, No. 12, pp. 2486–2493, Dec. 2001.

[32]   K. Narendra et al., "Cascaded Distributed Power Amplifier with Non-Identical Transistors and Inter-Stage Tapered Impedance," *40th European Microwave Conf.*, pp. 549–522, Sep. 2010.

[33]   S. Diebele and J. B. Beyer, "Attenuation Compensation in Distributed Amplifier Design," *IEEE Trans. Microwave Theory and Techniques*, Vol. 37, No. 9, pp.1425–1433, Sep. 1989.

[34]   K. Narendra et al., "Design Methodology of High Power Distributed Amplifier Employing Broadband Impedance Transformer," *IEEE Int. Conf. of Antenna, Systems and Propagation 2009*, Sep. 2009.

[35]   K. Narendra et al., "PHEMT Distributed Power Amplifier Adopting Broadband Impedance Transformer," *Microwave Journal*, Vol. 56, pp. 76–82, 2013.

[36]   K. Narendra et al., "High Efficiency Applying Drain Impedance Tapering for 600mW pHEMT Distributed Power Amplifier," *IEEE Int. Conf. on Microwave and Millimeter Wave Technology*, pp. 1769–1772, Apr. 2008.

[37]   L. Zhao et al., "A 6 Watt LDMOS Broadband High Efficiency Distributed Power Amplifier Fabricated Using LTCC Technology," *IEEE MTT-S Int. Microwave Symp. Dig.*, pp. 897–900, June 2002.

[38]   A. Sayed and G. Boeck, "Two-Stage Ultrawide Band 5-W Power Amplifier Using SiC MESFET," *IEEE Trans. Microwave Theory and Techniques*, Vol. 53, No. 7, pp. 2441–2449, July 2005.

[39]   I. Angelov et al., "On the Large-Signal Modeling of High Power AlGaN/GaN HEMTs," *IEEE MTT-S Int. Microwave Symp. Dig.*, pp. 1–3, June 2012.

[40]   M. Berroth and R. Bosch, "Broad-Band Determination of the FET Small-Signal Equivalent Circuit," *IEEE Trans. Microwave Theory Tech.*, Vol. 38, No. 7, pp. 891–895, July 1990.

[41]   G. Gonzalez, *Microwave Transistor Amplifiers—Analysis and Design*, Upper Saddle River, NJ: Prentice Hall, 1997.

[42]   Application Notes CGH40010P, www.cree.com.

[43]   Application Notes ATF511P8, www.avagotech.com.

[44]   Application Notes ATF54143, www.avagotech.com.

[45]   M. I. Montrose, *EMC and the Printed Circuit Board—Design, Theory, and Layout Made Simple*, New York: John Wiley & Sons, 1996.

[46]   E. Laermans et al., "Modeling Differential Via Holes," *IEEE Trans. Advanced Packaging*, Vol. 24, No. 3, pp. 357–363, Aug. 2001.

[47]   M. R. A. Gaffoor et al., "Simple and Efficient Full-Wave Modeling of Electromagnetic Coupling in Realistic RF Multilayer PCB Layouts," *IEEE Trans. Microwave Theory Tech.*, Vol. 50, No. 6, pp. 1445–1457, June 2002.

[48]   M. R. A. Gaffoor, "Simple and Efficient Full-Wave Analysis of Electromagnetic Coupling In Realistic RF Multilayer PCB Layouts," Ph.D. dissertation, Department of Electrical Engineering, University of Mississippi, Dec. 2000.

[49] G. Dambrine et al., "A New Method for Determining the FET Small-Signal Equivalent Circuit," *IEEE Trans. Microwave Theory Tech.*, Vol. 36, No. 7, pp. 1151–1159, July 1988.

[50] R. A. Amy, G. S. Aglietti, and G. Richardson, "Accuracy of Simplified Printed Circuit Board Finite Element Models," *Microelectronics Reliability*, Vol. 50, pp. 86–97, 2010.

[51] I. F. Kovacevic et al., "Full PEEC Modeling of EMI Filter Inductors in the Frequency Domain," *IEEE Trans. Magnetics*, Vol. 49, No. 10, pp. 5248–5257, Oct. 2013.

CHAPTER 5

Efficiency Analysis of Distributed Amplifiers

Distributed amplifiers are known for their flat gain, linear phase, and low return losses over wide bandwidths [1–26]. They represent an attractive candidate for software defined radio (SDR) applications [6, 27–30]. One of the major challenges in designing a distributed amplifier is to achieve high efficiency [26–32]. Several researchers have addressed the problems of low efficiency and output power of distributed amplifiers and suggested solutions to overcome these limitations [31–42]. To increase the efficiency of a distributed amplifier, the drain current from each transistor must be pushed to the load termination while mitigating the effect of the drain termination [6, 8, 9, 28, 29]; this is known as a nonuniform drain line or impedance tapering, in which backward current waves are canceled at each junction for characteristic impedance equal to Z_o/n, where Z_o is the characteristic impedance of the first section and n is the number of sections [41]. Work by Krishnamurthy et al. [42] discussed a distributed amplifier without an output synthesis transmission line; delay equalization is instead provided by impedance matched line sections between common-source (CS) and common-gate (CG) devices. A nonuniform distributed amplifier design methodology as a function of optimum power load of each device for maximizing power and efficiency has been demonstrated [36, 37].

5.1 Efficiency Limitations of Distributed Amplifiers

The efficiency of conventional distributed amplifiers has never demonstrated a PAE of more than 20% [31, 41]. This is primarily due to the current splitting on the drain line into two branches that form waves traveling toward the output load termination, and waves traveling toward the dummy termination. Each device transistor injects a current of $g_m v_{gs}$ into each drain of the transistor, where g_m and v_{gs} are transconductance and gate-source voltage, respectively. Because the drain of each transistor sees an impedance of Z_π in both directions, half of this current travels to the left and half to the right, as illustrated in Figure 5.1. The drain impedance seen in each direction Z_π is given as

$$Z_\pi = \sqrt{\frac{L_d\left(1 - \dfrac{f^2}{f_c^2}\right)^{-1}}{4C_d}} \qquad (5.1)$$

where L_d and C_d are inductive and capacitance elements, respectively, that form an artificial transmission line along the drain line, and f_c is the cutoff frequency of the transmission line.

Current flowing to the load termination I_d^R and dummy termination I_d^L can be deduced from the network shown in Figure 5.1. Each lowpass network, $L_d - C_d$ is contributing an image propagation factor θ_d along the drain line. Applying the superposition theorem [42] to the network shown in Figure 5.1, the current toward drain termination I_d^R can be derived as follows:

$$
\begin{aligned}
I_d^R &= \frac{1}{2}\Big[I_1 e^{-(n-1/2)\theta_d} + I_2 e^{-(n-3/2)\theta_d} + I_3 e^{-(n-5/2)\theta_d} + \cdots I_n e^{-(n-(n+2)/2)\theta_d} \Big] \\
&= \frac{1}{2} e^{-\frac{\theta_d}{2}} \sum_{K=1}^{n} I_K e^{-\theta_d(n-K)}
\end{aligned}
\tag{5.2}
$$

where n is the number of transistor sections. The network is assumed to be loss-less. Similarly, applying the superposition theorem [42], the current toward dummy termination I_d^L is derived as follows:

$$
I_d^L = \frac{1}{2}\Big[I_1 e^{-(1/2)\theta_d} + I_2 e^{-\theta_d} + I_3 e^{-2\theta_d} + \cdots I_n e^{-(n-1)\theta_d} \Big] = \frac{1}{2} e^{-\frac{\theta_d}{2}} \sum_{K=1}^{n} I_K e^{-\theta_d(K-1)}
\tag{5.3}
$$

For the real resistive termination for load and dummy termination, the ratio of power absorption between the load termination and dummy termination is given by

$$
P_{\text{ratio}} = \frac{P_{\text{load}}}{P_{\text{dummy}}} = \frac{\frac{1}{2}\big|I_d^R\big|^2 \Re\{Z_{OT}^R\}}{\frac{1}{2}\big|I_d^L\big|^2 \Re\{Z_{OT}^L\}} = \frac{\left| e^{-\frac{\theta_d}{2}} \sum_{K=1}^{n} I_K e^{-\theta_d(n-K)} \right|^2}{\left| e^{-\frac{\theta_d}{2}} \sum_{K=1}^{n} I_K e^{-\theta_d(K-1)} \right|^2}
\tag{5.4}
$$

where Z_{OT}^R and Z_{OT}^L are load termination and dummy drain termination, respectively. The notation $\Re$ refers to the real part and in the following section, Á refers to the imaginary part.

Equation (5.4) shows that the power ratio between load termination and dummy termination is equal if the both termination impedances have equal characteristics. It is clear that half of the fundamental RF energy is wasted in the dummy termination

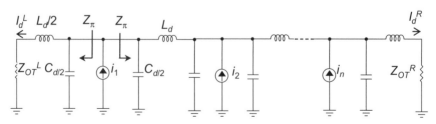

Figure 5.1 Impedance Z_π seen by each transistor in both directions, half right and half left.

from the overall fundamental RF energy generated by all transistor sections. The following section leads into a discussion of the technique to improve efficiency.

5.2   Virtual Impedance Analysis Using Multicurrent Sources

Distributed amplifiers have impedance as a result of the injected signals at each device's output node [6]. Figure 5.2 is a simple schematic of two ideal current sources that combine at a common node connected to a load R. Each current source has a driving impedance (e.g., R_1 and R_2), as shown in Figure 5.2. However, for simplicity, the driving impedances R_1 and R_2 of distributed amplifiers are neglected (as well as in the following analysis). We also make the assumption that the loading effect for medium-power devices is not significant when loaded into the drain line. Thus, the impedance seen by each current source due to injected sources can be called *virtual impedance.*

As shown in Figure 5.2, the current $i(t)$ through the load would be the sum of the two sources $i_1(t)$ and $i_2(t)$. The complex current source and response may be simply represented by applying Euler's identity [42]. The $i_1(t)$ and $i_2(t)$ source thus becomes

$$i_1(t) = I_1 e^{j(wt+\theta_1)} = I_1 \cos(wt + \theta_1) + jI_1 \sin(wt + \theta_1) \tag{5.5}$$

$$i_2(t) = I_2 e^{j(wt+\theta_2)} = I_2 \cos(wt + \theta_2) + jI_2 \sin(wt + \theta_2) \tag{5.6}$$

where I_1 and I_2 represent the magnitude of the complex current source, and θ_1(t) and θ_2(t) are independent phase values, respectively.

The impedance Z_1 looking into the common node with $i_2(t)$ in parallel with R is what the current source vector $i_1(t)$ is loaded with. By simplifying Figure 5.2 and applying the Thevenin theorem, we can show the simplified circuit in Figure 5.3. Thus, impedance Z_1 can be derived as follows:

$$Z_1 = v(t)/i_1(t) = \frac{\left[i_1(t) + i_2(t)\right]}{i_1(t)} R \tag{5.7}$$

$$= R\left[1 + \frac{I_2 e^{j(w(t)+\theta_2)}}{I_1 e^{j(w(t)+\theta_1)}}\right] = R\left[1 + \frac{I_2}{I_1}\left(\cos(\theta_2 - \theta_1) + j\sin(\theta_2 - \theta_1)\right)\right] \tag{5.8}$$

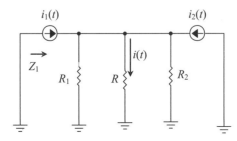

Figure 5.2 Two current sources are combining at a single node.

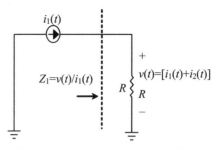

Figure 5.3   Simplification of the circuit from Figure 5.2 to understand Z_1.

If no phase offset or in-phase combining ($\theta_2 = \theta_1$) is applied in the distributed amplifier, the amplitude of the reactive term becomes zero. Thus, virtual impedance Z_1 is positive real. Hence, impedance Z_1 can be simplified as shown below:

$$Z_1 = R\left[1 + \frac{I_2}{I_1}\right] \tag{5.9}$$

From (5.8), the normalized virtual impedance real and imaginary part Z_1/R versus ($\theta_2 - \theta_1$) for a few cases of I_2/I_1 (from 0.1 to 2) is illustrated in Figure 5.4. For the lower ratio of I_2/I_1 (i.e., 0.1), the real Z_1/R is almost 1, and it increases to 3 when I_2/I_1 is 2. The imaginary Z_1/R remains unchanged for any ratio of I_2/I_1. This is evidence that current source properties (i.e., magnitude and phase) can be adjusted to achieve the desired Z_1/R.

Consider multicurrent sources that are combined at a common node connected to a load R. As shown in Figure 5.5, virtual impedance is seen by the current sources in two directions, so the upper and right directions $Z_{u(k)}$ and $Z_{r(k)}$, respectively, can be determined. Note that the Thevenin driving impedances of the current sources are neglected for simplicity, and $k = 1, 2, 3 \ldots n$. The RF excursion swing of voltage and current for each source depends on $Z_{u(k)}$, and $Z_{r(k)}$ will determine the impedance that will be synthesized along the drain artificial transmission line. We begin with $Z_{u(1)}$ by finding the Norton equivalent network across port aa' (see Figure 5.5). The Norton equivalent impedance R_N can be determined by removing all current sources except the first one ($i_2 = i_3 \ldots = i_n = 0$), thus $R_N = R$. The Norton equivalent current i_N across port aa' can be determined by applying the superposition theorem. When finding i_N, port aa' must be shorted, and $i_N = i_2 + i_3 + \ldots + i_n$. Current generator i_1 and the Norton equivalent network of Figure 5.5 can be simplified to Figure 5.6. Thus, the virtual impedance $Z_{u(1)}$ seen by current generator i_1 (from Figure 5.5) can be deduced to

$$Z_{u(1)} = \frac{v}{i_1} = \left[\frac{(i_1 + i_N)R_N}{i_1}\right] = \left[1 + \frac{i_2 + i_3 + \cdots + i_n}{i_1}\right]R \tag{5.10}$$

where $I_1, I_2 \ldots I_n$ represent the magnitude of the complex current source and $\theta_1, \theta_1 \ldots \theta_n$ are independent phase values, respectively.

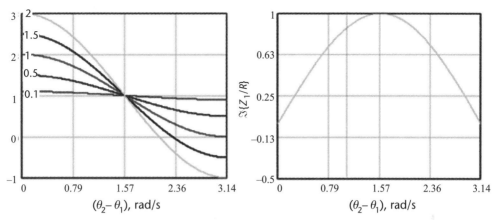

Figure 5.4 The plot of real and imaginary Z_1/R versus $(\theta_2 - \theta_1)$ for a few cases of I_2/I_1 from (5.8).

The ratio I_2/I_1 is varied from 0.1 to 2, and imaginary part does not change with I_2/I_1. By substituting $i_k = I_k e^{j(wt+\theta_k)}$, where $k = 1, 2 \ldots n$, we can derive the virtual impedance:

$$Z_{u(1)} = R\left[1 + \frac{I_2}{I_1}\left(\cos\left(\theta_2 - \theta_1\right) + j\sin\left(\theta_2 - \theta_1\right)\right) + \cdots + \frac{I_n}{I_1}\left(\cos\left(\theta_n - \theta_1\right) + j\sin\left(\theta_n - \theta_1\right)\right)\right]$$

$$(5.11)$$

In a similar manner, $Z_{u(n-1)}$ and $Z_{u(n)}$ seen by i_{n-1} and i_n can be determined as

$$Z_{u(n-1)} = R\left[\begin{array}{l}1 + \dfrac{I_1}{I_{n-1}}\left(\cos\left(\theta_1 - \theta_{n-1}\right) + j\sin\left(\theta_1 - \theta_{n-1}\right)\right) \\[2mm] +\cdots + \dfrac{I_n}{I_{n-1}}\left(\cos\left(\theta_n - \theta_{n-1}\right) + j\sin\left(\theta_n - \theta_{n-1}\right)\right)\end{array}\right] \qquad (5.12)$$

and

$$Z_{u(n)} = R\left[\begin{array}{l}1 + \dfrac{I_1}{I_n}\left(\cos\left(\theta_1 - \theta_n\right) + j\sin\left(\theta_1 - \theta_n\right)\right) \\[2mm] +\cdots + \dfrac{I_{n-1}}{I_n}\left(\cos\left(\theta_{n-1} - \theta_n\right) + j\sin\left(\theta_{n-1} - \theta_n\right)\right)\end{array}\right] \qquad (5.13)$$

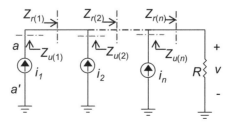

Figure 5.5 The virtual impedance Z_i seen by the current generator in two directions, $Z_{u(k)}$ and $Z_{r(k)}$, respectively, with $k = 1, 2, 3 \ldots n$.

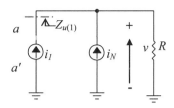

Figure 5.6 Simplified network of current generator i_1 and the Norton equivalent network of Figure 5.5 to determine $Z_{u(1)}$.

It is clear from Figure 5.5, that $Z_{r(1)} = Z_{u(1)}$ and $Z_{r(n)} = R$, but $Z_{r(n-1)}$ is given as

$$Z_{r(n-1)} = \frac{v}{i_1 + \cdots + i_{n-1}} = \left[\frac{(i_1 + i_N)R_N}{i_1 + \cdots + i_{n-1}}\right] = \left[\frac{i_2 + i_3 + \cdots + i_n}{i_1 + \cdots + i_{n-1}}\right]R \quad (5.14)$$

and substituting $i_k = I_k e^{j(wt + \theta_k)}$ into (5.14), it can be deduced as

$$Z_{r(n-1)} = R\left[\frac{1 + \frac{I_2}{I_1}\left(\cos(\theta_2 - \theta_1) + j\sin(\theta_2 - \theta_1)\right) + \cdots + \frac{I_n}{I_1}\left(\cos(\theta_n - \theta_1) + j\sin(\theta_n - \theta_1)\right)}{1 + \frac{I_2}{I_1}\left(\cos(\theta_2 - \theta_1) + j\sin(\theta_2 - \theta_1)\right) + \cdots + \frac{I_{n-1}}{I_1}\left(\cos(\theta_{n-1} - \theta_1) + j\sin(\theta_{n-1} - \theta_1)\right)}\right]$$

$$(5.15)$$

As derived above, $Z_{r(k)}$ and $Z_{u(k)}$ must be satisfied to combine n section current sources to a single load R, assuming a minimum loading effect of the output transistor. Beyer et al. have shown that gain in a conventional distributed amplifier cannot be increased indefinitely by adding more section losses associated with active transistor resistances [1]. Typically, $n = 4$ is applied in a distributed amplifier [1].

From (5.11) through (5.15), for $k = 1, 2, \ldots 4$, the following set of equations is deduced:

$$Z_{u(1)}/R = 1 + \frac{I_2}{I_1}\left(\cos(\theta_2 - \theta_1) + j\sin(\theta_2 - \theta_1)\right) + \frac{I_3}{I_1}\left(\cos(\theta_3 - \theta_1) + j\sin(\theta_3 - \theta_1)\right)$$

$$+ \frac{I_4}{I_1}\left(\cos(\theta_4 - \theta_1) + j\sin(\theta_4 - \theta_1)\right)$$

$$(5.16)$$

$$Z_{u(2)}/R = 1 + \frac{I_1}{I_2}\left(\cos(\theta_1 - \theta_2) + j\sin(\theta_1 - \theta_2)\right) + \frac{I_3}{I_2}\left(\cos(\theta_3 - \theta_2) + j\sin(\theta_3 - \theta_2)\right)$$

$$+ \frac{I_4}{I_2}\left(\cos(\theta_4 - \theta_2) + j\sin(\theta_4 - \theta_2)\right)$$

$$(5.17)$$

$$Z_{u(3)}/R = 1 + \frac{I_1}{I_3}\left(\cos(\theta_1 - \theta_3) + j\sin(\theta_1 - \theta_3)\right) + \frac{I_2}{I_3}\left(\cos(\theta_2 - \theta_3) + j\sin(\theta_2 - \theta_3)\right)$$
$$+ \frac{I_4}{I_3}\left(\cos(\theta_4 - \theta_3) + j\sin(\theta_4 - \theta_3)\right)$$

$$(5.18)$$

$$Z_{u(4)}/R = 1 + \frac{I_1}{I_4}\left(\cos(\theta_1 - \theta_4) + j\sin(\theta_1 - \theta_4)\right) + \frac{I_2}{I_4}\left(\cos(\theta_2 - \theta_4) + j\sin(\theta_2 - \theta_4)\right)$$
$$+ \frac{I_3}{I_4}\left(\cos(\theta_3 - \theta_4) + j\sin(\theta_3 - \theta_4)\right)$$

$$(5.19)$$

$$Z_{r(1)}/R = Z_{u(1)}/R \tag{5.20}$$

$$Z_{r(2)}/R = 1 + \frac{\frac{I_3}{I_1}\left(\cos(\theta_3 - \theta_1) + j\sin(\theta_3 - \theta_1)\right) + \frac{I_4}{I_1}\left(\cos(\theta_4 - \theta_1) + j\sin(\theta_4 - \theta_1)\right)}{1 + \frac{I_2}{I_1}\left(\cos(\theta_2 - \theta_1) + j\sin(\theta_2 - \theta_1)\right)}$$

$$(5.21)$$

$$Z_{r(3)}/R = 1 + \frac{\frac{I_4}{I_1}\left(\cos(\theta_4 - \theta_1) + j\sin(\theta_4 - \theta_1)\right)}{1 + \frac{I_2}{I_1}\left(\cos(\theta_2 - \theta_1) + j\sin(\theta_2 - \theta_1)\right) + \frac{I_3}{I_1}\left(\cos(\theta_3 - \theta_1) + j\sin(\theta_3 - \theta_1)\right)}$$

$$(5.22)$$

$$Z_{r(4)}/R = 1 \tag{5.23}$$

Results of normalized real and imaginary $Z_{u(k)}/R$ ($k = 1, 2, 3$, and 4) versus $(\theta_a - \theta_b)$, are illustrated in Figure 5.7. The term $(\theta_a - \theta_b)$[1] refers to the phase difference, and the a and b refer to any case as given in (5.16) through (5.19). The real and imaginary parts $Z_{u(k)}/R$ and $Z_{r(k)}/R$ are plotted for the same magnitude of current sources, where $I_1 = I_2 = I_3 = I_4$. From Figure 5.7, for in-phase combining $(\theta_a - \theta_b) = 0$, then the real part $Z_{u(k)}/R = 4$ and imaginary part $Z_{u(k)}/R = 0$. As $(\theta_a - \theta_b) = \pi/2$

---

[1] The term $(\theta_a - \theta_b)$ represents the different phase between two current source properties; as an example, $Z_{u(1)}/R$, $\theta_a - \theta_b = \theta_2 - \theta_1 = \theta_3 - \theta_1 = \theta_4 - \theta_1$. The different phases can be determined from (5.16) through (5.19).

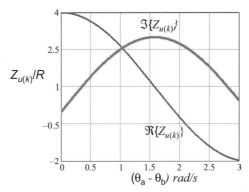

Figure 5.7 Real and imaginary $Z_{u(k)}/R$ (k = 1, 2, 3, and 4) versus ($\theta_a - \theta_b$) for identical current sources, $I_1 = I_2 = I_3 = I_4$. The response $Z_{u(1)}/R = Z_{u(2)}/R = Z_{u(3)}/R = Z_{u(4)}/R$.

rad/s, the real part vanishes to 0 and the imaginary part has the highest reactance. The plot of $Z_{r(k)}/R$ (k = 1, 2, 3, and 4) versus ($\theta_a - \theta_b$), where $I_1 = I_2 = I_3 = I_4$ is shown in Figure 5.8.

The real part $Z_{r(k)}/R$ value is reduced and the imaginary part is null for in-phase combining ($\theta_a - \theta_b$) = 0. Real and imaginary parts $Z_{r(2)}/R$, $Z_{r(3)}/R$, and $Z_{r(4)}/R$ are almost constant for ($\theta_a - \theta_b$) $\leq \pi/2$ rad/s. As ($\theta_a - \theta_b$) increases, the real part vanishes.

An important point to notice is that, to match the standard R of 50Ω, $Z_{u(k)}$ equals 200Ω (k = 1, 2, 3, and 4). However, to achieve $Z_{u(k)}$ < 200Ω (power matching), R has to be lower than 50Ω; therefore, an impedance transformation from R to 50Ω is needed. To avoid impedance transformation, one possibility is to achieve $Z_{u(k)}$ < 200Ω while retaining R = 50Ω, the current source properties (e.g., I_1, I_2 ... I_n and θ_1, θ_2 ... θ_n can be adjusted). Table 5.1 shows tabulated results of $Z_{u(k)}/R$ and $Z_{r(k)}/R$ (k = 1, 2, 3, and 4) for various magnitude and phase selections. From (5.16) through (5.23), the real parts of $Z_{u(k)}/R$ and $Z_{r(k)}/R$ are given in the Table 5.1; the imaginary part is not shown; it is instead absorbed in the input and output transmission lines. For the same magnitude and in-phase combining current source, the $\Re\{Z_{u(k)}/R\}$'s are equal, and $\Re\{Z_{r(k)}/R\}$ is reduced from 4, 2 and to 1, for all k. Figure 5.9 shows $Z_{r(k)}/R$ by each current source and evaluated according to (5.20) through

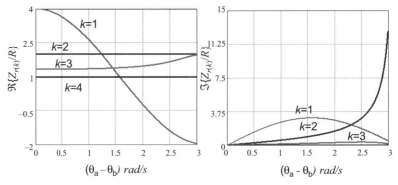

Figure 5.8 Real and imaginary $Z_{r(k)}/R$ (k = 1, 2, 3, and 4) versus ($\theta_a - \theta_b$) for identical current sources, $I_1 = I_2 = I_3 = I_4$. For k = 4, the imaginary part is null.

Table 5.1 $Z_{u(k)}/R$ and $Z_{r(k)}/R$ for Various I_k and θ_k Selections (k = 1, 2, 3 and 4)

I_1	1	1	1	1	1	1	1	1	1	1
I_2	1	1	1	0.7	1	1	1	1	1	0.7
I_3	1	1	0.7	0.4	1	1	1	1	0.6	0.4
I_4	1	0.1	0.1	0.1	1	1	1	0.1	0.1	0.1
θ_1	0	0	0	0	0	0	0	0	0	0
θ_2	0	0	0	0	0	0	$\pi/6$	0	0	$\pi/6$
θ_3	0	0	0	0	0	$\pi/5$	$\pi/4$	0	$\pi/5$	$\pi/4$
θ_4	0	0	0	0	$\pi/4$	$\pi/4$	$\pi/2$	$\pi/4$	$\pi/4$	$\pi/2$
$\Re\{Z_{u(1)}/R\}$	4	3.1	2.8	2.2	3.7	3.5	3.38	3.07	2.55	2.16
$\Re\{Z_{u(2)}/R\}$	4	3.1	2.8	3.14	3.7	3.5	3.82	3.07	2.55	3.22
$\Re\{Z_{u(3)}/R\}$	4	3.1	4	5.5	3.7	3.6	3.79	3.07	3.86	3.67
$\Re\{Z_{u(4)}/R\}$	4	31	28	22	3.1	3.4	3.66	22.2	21.06	20.75
$\Re\{Z_{r(1)}/R\}$	4	3.1	2.8	2.2	3.7	3.5	3.38	3.07	2.25	2.16
$\Re\{Z_{r(2)}/R\}$	2	1.55	1.4	2.9	1.85	1.75	1.93	1.53	1.27	1.38
$\Re\{Z_{r(3)}/R\}$	1.33	1.03	1.03	1.04	1.23	1.29	1.39	1.02	1.03	1.04
$\Re\{Z_{r(4)}/R\}$	1	1	1	1	1	1	1	1	1	1

(5.23). Either adjustment of magnitude or phase of the current source will result in a reduction in $\Re\{Z_{u(k)}/R\}$.

As an example from Table 5.1, for in-phase combining, and when the final current magnitude is reduced by a factor of 10, then $\Re\{Z_{u(1)}/R\} = \Re\{Z_{u(2)}/R\} = \Re\{Z_{u(3)}/R\} = 3.1$, but $\Re\{Z_{u(4)}/R\} = 31$. Therefore, the fourth (or last) section can be placed with a high-f_τ transistor (having a lower device periphery), which may not contribute power to the load and absorb power at a certain frequency, but simply acts as active load matching. In a similar manner, lower $\Re\{Z_{u(k)}/R\}$ can be achieved for the case of fixed magnitude while adjusting the phase. With adjustment of both properties (i.e., the magnitude and phase of the current sources), $\Re\{Z_{u(k)}/R\}$ can be reduced to a lower value. As shown in Table 5.1, when $I_1 = 1$, $I_2 = 0.7$, $I_2 = 0.4$, and $I_2 = 0.1$, and $\theta_1 = 0$, $\theta_2 = \pi/6$, $\theta_2 = \pi/4$, and $\theta_2 = \pi/2$, the value of $\Re\{Z_{u(1)}/R\} = 2.1$, $\Re\{Z_{u(2)}/R\} = 3.2$, $\Re\{Z_{u(3)}/R\} = 3.67$, and $\Re\{Z_{u(4)}/R\} = 20$. To achieve this, the device periphery of the transistors can be tapered [43] while adjusting the gate-line

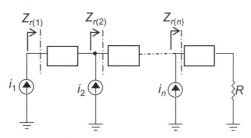

Figure 5.9   The term $Z_{r(k)}$ is evaluated according to (5.20) through (5.23) for designing an artificial transmission line. The output capacitance parallel to the current source, typically known as C_{ds}, will be absorbed in the artificial transmission line design.

characteristics to provide phase conditions. Therefore, it is the designer's choice to adjust the I_k and θ_k to obtain reasonable $Z_{u(k)}/R$ and $Z_{r(k)}/R$.

It is clear that to combine the current from each source at a single node (at load R), the virtual impedance seen by the current sources in two directions ($Z_{u(k)}$ and $Z_{r(k)}$) must be presented to each current source. If current source sees high $Z_{u(k)}$ (e.g., 200Ω), it is not possible to deliver power from each current source due to the RF current swing limitation. Certainly, it is possible to achieve $Z_{u(k)} < 200\Omega$ while retaining $R = 50\Omega$ as long as (5.16) through (5.23) are satisfied. In (5.1), the impedance seen by each current source loaded with a constant-k ladder network transmission line, typically 50Ω impedance, limits the excursion of RF current swing. Nevertheless, to maximize the current swing, it is necessary to match to the optimum power load of each current source (or transistor) [45], and in a distributed amplifier reducing $Z_{u(k)}$ will increase the power from each source. It is important, however, to note that there are boundary limitations to practical realizations to achieve lower $Z_{u(k)}$ while retaining $R = 50\Omega$, such as the device periphery ratio of the transistor, gate-line controlling, and so forth.

An unequal magnitude of the current source can be achieved by using a capacitively coupled technique [13]. The unequal injection of current source is due to different v_{gs} drops along the gate line, as illustrated in Figure 5.10. An external capacitor is coupled in series between the lumped inductance and input parasitic device capacitance C_{gs} and the external capacitor controlling the voltage drop across the C_{gs}. Thus, by varying the voltage drop ratio along the gate line, the input excitation can be tailored to individual transistors. From Figure 5.10, when $\omega C_{gs} R_{gs} \leq 1$, the equivalent capacitance loading the gate line is

$$C_g = \frac{C_a C_{gs}}{C_a + C_{gs}} \tag{5.24}$$

where R_{gs} is the series gate resistance of the active device. For simplicity, R_{gs} is not shown in Figure 5.10. The voltage drop across the C_{gs} junction can be deduced as

$$v_{gs} = \frac{C_a}{C_a + C_{gs}} V_i \tag{5.25}$$

where V_i is voltage that appears as shown in Figure 5.10. For a lossless or ideal gate line, the magnitude of V_i is approximately the same for any nodal stage. Each device transistor injects a current of $g_m v_{gs}$ into each drain of the transistor. Notations g_m and v_{gs} are referred to as the transconductance and gate source voltage of the transistor, respectively.

However, the phase adjustment of the current source is difficult to realize, and keep in mind that the constant-k ladder network behavior changes with respect to frequency. Stengel et al. invented a technique to drive the gate line with a drive generator circuit to produce modulation of the virtual load impedance at each amplifier stage [46] and a new distributed amplifier architecture to provide programmable constructive vector signal combining at a fundamental frequency along with programmable destructive at harmonic frequencies [47]. Stengel et al. [47] have shown that harmonic components of the current sources can be controlled with I/Q baseband

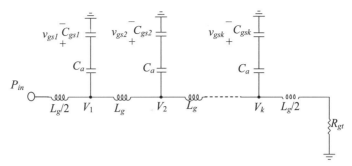

Figure 5.10 Input gate line coupled with external capacitor in series with C_{gs}. Note that the elements may have different values if unequal injection is required.

signal injection to the gate line. A simple approach, like that shown in [30], phase adjustment is achieved with nonuniform gate-line design where gate-line impedance is adaptively reduced to provide better phase synchronization.

5.3   Simulation Analysis of Efficiency Analysis

Validation of the preceding concept for equal magnitude and in-phase combining where $I_1 = I_2 = I_3 = I_4 = 1$ and $\theta_1 = \theta_2 = \theta_3 = \theta_4 = 0$ have been selected (from Table 5.1); $R = 50\Omega$ selection leads to $\Re\{Z_{u(1)}\} = \Re\{Z_{u(2)}\} = \Re\{Z_{u(3)}\} = \Re\{Z_{u(4)}\} = 200\Omega$, and $\Re\{Z_{r(1)}\} = 200\Omega$, $\Re\{Z_{r(2)}\} = 100\Omega$, $\Re\{Z_{r(3)}\} = 67\Omega$, and $\Re\{Z_{r(4)}\} = 50\Omega$, and an imaginary part does not exist. This is illustrated in Figure 5.11. For simplicity, a hypothetical drain terminal for a transistor has been modeled as an ideal current source with parallel resistance (high impedance distributed amplifier, e.g., 1 MΩ) [30]. To additively combine the currents at each junction (as in Figure 5.11), phase synchronization between the current source and the transmission line is crucial. Because transmission lines delays $\theta_{d(k)}$ vary linearly with frequency, making the current source delays also vary with frequency would guarantee delay matching between the sources and the transmission lines. Putting current source delays as $\theta_1 = 0°$, $\theta_2 = -1 * 10° * 1$ GHz/*freq*, $\theta_3 = 2 * 10° * 1$ GHz/*freq*, and $\theta_4 = -3 * 10° * 1$ GHz/*freq*, respectively. The notation *freq* is referring to frequency in an ac simulation (Advanced Design System) and $\theta_{d(k)}$ is simply set to a fixed real value (e.g., 40°). Figure 5.12 shows a vector diagram of magnitude and phase of the current properties for Figure 5.11. At every junction, the resultant magnitude of the sources (i_b, i_d, and i_f) are constructive for in-phase combining between i_a and i_2, i_c and i_3, and i_e and i_4.

Power delivered by each current source is shown in Figure 5.13. It indicates that the calculated power of 100W is achieved across the frequency, where all the sources are presented with exactly the required hypothetical optimum impedance of $200 + j * 0\Omega$ at all frequencies. Phase coherency between the current sources and the phase of the transmission must be well matched to achieve maximum power delivery from the sources to the output line over the entire frequency range. Analogous results are obtained for various selected cases of I_k and θ_k ($k = 1, 2, 3,$ and 4) from Table 5.1.

In reality, however, lumped inductance with parasitic capacitance (of the transistor) will form a constant-k artificial transmission line along the drain line, which is

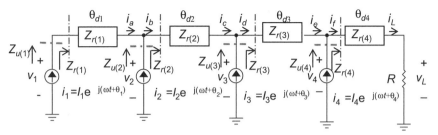

Figure 5.11. Circuit showing multiple-current sources combining at a single load termination, where $R = 50\Omega$.

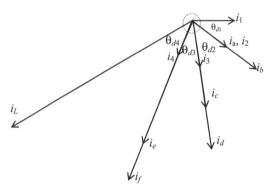

Figure 5.12 Vector diagram of magnitude and phase of the current properties for Figure 5.11, where the magnitude and phase of the current sources are equal and i_1 is set to be a reference.

found to be suitable for the low microwave region [48, 49]. Again, a similar analysis is performed to understand the power delivery behavior in the presence of lumped inductance and capacitance. Figure 5.14 shows a circuit representation for an ideal current source with lumped inductance and capacitance. Cutoff frequency ω_c is set to 2.12 GHz. The transmission line has nonuniform impedance, $Z_{r(1)} = 200\Omega$, $Z_{r(2)} = 100\Omega$, $Z_{r(3)} = 67\Omega$, and $Z_{r(4)} = 50\Omega$, and an imaginary part does not exist. Careful phase consideration of each current source must be taken into account due to the delay of each L-C combination with respect to the frequency. Therefore, the delay

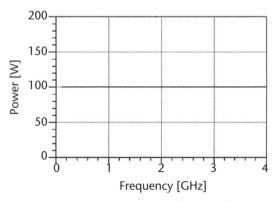

Figure 5.13 Power delivered by each current source with matched delay values connected to the output transmission line. Note that all sources deliver maximum power and have the same energy level over the frequency range of interest.

or phase velocity equation of the constant-k artificial transmission line is defined for the current source delay $\theta_{d(k)}$ as given by

$$\theta_{d(k)} = \cos^{-1}\left[1 - \frac{2\omega^2}{\omega_c^2}\right] \tag{5.26}$$

The $\theta_1 = 0°$, $\theta_2 = -1 * \theta_{d(1)}$, $\theta_3 = -2 * \theta_{d(2)}$, and $\theta_4 = -3 * \theta_{d(4)}$, respectively. Figure 5.15 shows the power response of the current sources across a wide frequency range. At low frequencies (close to dc), maximum power is delivered by each current source. The power response is being degraded as the frequency increases toward ω_c. Due to the fact that the phase delay changes with frequency, it is difficult to achieve phase coherency for optimum current combining at each junction (as in Figure 5.11). The further the location of the current source (closer to the load termination R), the earlier the power degraded with respect to frequency, and at a certain frequency it absorbs power from the line (i.e., negative impedance). Keep in mind that power delivered by each transistor follows the real part impedance. Note that a strong peaking occurred the closer the signal got to ω_c, especially at the first and last section. In the following section, a technique to compensate phase delay by designing a nonuniform gate line to improve power delivery from each transistor is discussed [30].

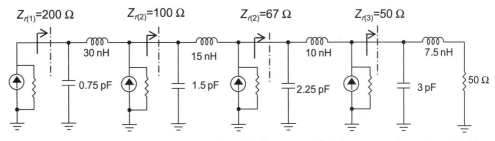

Figure 5.14 Circuit showing an analysis of a drain line tapered with constant-k terminated with 50Ω load termination. All current sources are loaded with a parallel resistor of 1 MΩ.

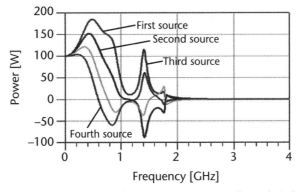

Figure 5.15 Power delivered by each ideal current source nonuniform drain line with lumped elements, which consist of the first source, second source, third source, and fourth source.

5.4   Design Example of High-Efficiency Distributed Amplifier

The design methodology (device selection, synthesizing gate/drain-line elements from device packaged values, until full-wave simulation/layout optimizations) discussed in Section 4.5 is applied in this section. The basic design goal of the work is to achieve high efficiency for SDR driver power amplifier applications, with a power operation of ~27 dBm. Therefore, a medium-power device, for example, a pHEMT device (ATF511P8) [50], is suitable. Low dc supply operation is required for the device, which is typically about 4.5V. Breakdown voltage V_{bk} of the device is ~16V, which is slightly lower than the computation value obtained from (4.47). The drain loading effect of the device is not significant, but $X_{opt}(\omega)$ is important because it determines the drain-line cutoff frequency ω_c. A value of 4 pF for C_{opt} is extracted by means of device modeling (with inclusion of package properties). Therefore, effective drain-line elements L_i are synthesized according to (4.58) to obtain the desired ω_c (~1.9 GHz). Dummy drain termination is eliminated to improve the efficiency performance [30].

A simplified design schematic for a four-section pHEMT distributed amplifier applying a nonuniform drain line is shown in Figure 5.16. The m-derived section is implemented at both terminations of the gate line. Each device is fed with a 5V drain supply voltage. A bias voltage of 0.44V is applied to each gate, resulting in Class AB operation, with quiescent current I_{dq} of ~90 mA (~10% of I_{dss}). A series gate resistor $R_{g(k)}$ of 5Ω is used at each section for stability purposes. Power performance due to load termination R for $n = 4$ is investigated. The virtual impedance seen by the transistor in both directions depends on selection as illustrated in Section 5.2 and can be computed using (5.16) through (5.23). Thus, computation of the virtual impedance must be repeated for the different cases of load termination R. All the passive elements are slightly tuned at the simulation level for optimum results, and the summary elements are tabulated in Table 5.2. As given in Table 5.2, $C_{d(k)}$ refers to the external capacitance required to form the desired ω_c with combination of C_{opt}. The gate-line impedance network remains unchanged for the analysis. The supply voltage and gate bias voltage are fixed as previous values.

Simulated results input return loss S_{11} and output return loss S_{22}, and power performances (PAE and gain) for various load terminations R are shown in Figures 5.17 and Figure 5.18, respectively. Figure 5.17 shows that the output return loss (S_{22})

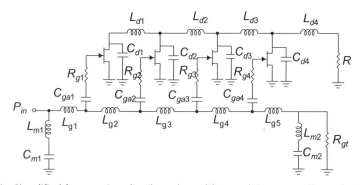

Figure 5.16   Simplified four-section distributed amplifier applying nonuniform drain-line impedance. The m-derived section is implemented close to the termination point of the gate line.

Table 5.2 Summary Drain-Line Network Elements for Various Load Terminations R^*

Elements	$R = 5\Omega$	$R = 8\Omega$	$R = 12.5\Omega$	$R = 25\Omega$	$R = 50\Omega$
L_{d1}	3 nH	4.8 nH	7.5 nH	15 nH	30 nH
L_{d2}	1.5 nH	2.4 nH	3.75 nH	7.5 nH	15 nH
L_{d3}	1 nH	1.6 nH	2.5 nH	5 nH	10 nH
L_{d4}	0.75 nH	1.2 nH	1.88 nH	3.75 nH	7.5 nH
C_{d1}	6.6 pF	3.8 pF	2.1 pF	0.6 pF	n/a
C_{d2}	14.1 pF	8.48 pF	5.1 pF	2.1 pF	0.6 pF
C_{d3}	21.6 pF	13.2 pF	8.1 pF	3.6 pF	1.35 pF
C_{d4}	29.1 pF	17.9 pF	11.1 pF	5.1 pF	2.1 pF

* The n section is 4 and $C_{d(k)}$ is the external shunt capacitance placed parallel to the transistor.

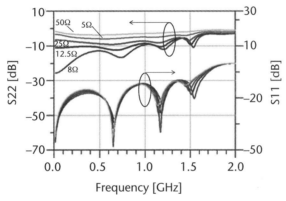

Figure 5.17 Simulated analysis of S_{11} and S_{22} for various load terminations R ($R = 5\Omega$; $R = 8\Omega$; $R = 12.5\Omega$; $R = 25\Omega$; $R = 50\Omega$) for the circuit shown in Figure 5.16.

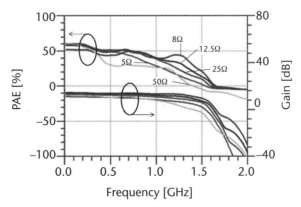

Figure 5.18 Simulated analysis of PAE and gain for various load terminations R ($R = 5\Omega$; $R = 8\Omega$; $R = 12.5\Omega$; $R = 25\Omega$; $R = 50\Omega$) for the circuit shown in Figure 5.16.

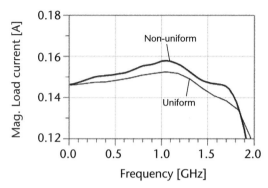

Figure 5.19 Magnitude of load current for nonuniform and uniform gate-line design across the frequency range of 10 to 1800 MHz.

is dependent on R. Right selection of load termination R is important for optimum matching and power performance across a wide frequency range. For example, the R's within the range of 8Ω and 12.5Ω are the optimum case. For the best efficiency result (PAE >40%) across the frequency range, R of 12.5Ω is the optimum case. A constant-k nonuniform drain-line network with lower R (e.g., 5Ω) is not effective. For the gate-line termination R_{gt} is 50Ω; optimum R is within range of 8Ω to 12.5Ω for the case $n = 4$. For an R of 50Ω, the power performance is degraded beyond 1400 MHz.This analysis indicated that R is important to provide better phase equalization between gate and drain line.

Phase velocity synchronization between gate and drain line is achieved with a nonuniform gate-line design as well [30]. A simple adjustment of $C_{ga(k)}$ and $L_{g(k)}$ (for $k = 4$, as shown in Figure 5.10) offers a power and efficiency improvement. The gate-line impedance and dummy termination R_{gt} are adaptively reduced at the simulation level. When R of 50Ω and R_{gt} of 41Ω are selected, the magnitude current of load termination is improved by ~15 mA at the simulation level compared with uniform gate-line impedance across a frequency range. Figure 5.19 shows current for a load termination comparison for the cases of with and without nonuniform gate-line impedance. In a similar manner, for R of 12.5Ω, the gate-line impedance and dummy termination are optimized in at the simulation level. It is important to note that the gate-line properties are dependent on the termination R value. Detailed analysis of a nonuniform gate-line network for phase synchronization with a nonuniform drain line is not presented but it is recommended for future work.

An R of 12.5Ω would be the optimum selection due to the resulting good power performance and lower transformation impedance ratio to 50Ω compared with an R of 8Ω. Wideband impedance transformer design to match an impedance of 12.5Ω to 50Ω will be discussed in the following section. A high-efficiency distributed amplifier having nonuniform drain and gate-line networks terminated with an R of 12.5Ω and 50Ω, respectively, is chosen to proof the concept at the measurement level (see discussion at the end of this chapter). A wideband impedance transformer ratio of 1:4 based on asymmetric parallel coupled line [52, 53] is used in this design to achieve impedance transformation from 12.5Ω to 50Ω, covering a frequency range of 10 to 1800 MHz. The detailed design of the transformer is covered in the following section.

5.5   Broadband Impedance Transformer Design

Recent works [51, 54–58] have demonstrated an impedance transformer for microwave applications. Coupled transmission lines have been suggested as a matching element due to greater flexibility and compactness in comparison to quarter-wave transmission lines [55, 56]. The quarter-wave transformer is simple and easy to use, but it has no flexibility beyond the ability to provide a perfect match at the center frequency for a real valued load. The coupled line section provides a number of variables that can be utilized for matching purposes and these variables are the even- and odd-mode impedance distributed amplifier and loading of the through and coupled ports [52]. Recent work [52] demonstrated a fractional bandwidth of more than 100% for a −20-dB reflection with asymmetric coupled lines implemented in a nonhomogenous medium. A microstrip is one of the most commonly used classes of transmission lines in nonhomogenous media. Figure 5.20 shows general coupled line configuration, and coupled and through ports can be loaded with an external impedance termination to extend the bandwidth [52]. Figure 5.20 can be represented as a two-port network, as shown in Figure 5.21. The magnitude of the reflection coefficient at port 2 [53] is equal to

$$|S_{22}|_{dB} = 20 \; \log\left(\left|\frac{Z_{IN}\left(Z_{ij}, Z_{ij}'', Z_L - Z_g\right)}{Z_{IN}\left(Z_{ij}, Z_{ij}'', Z_L + Z_g\right)}\right|\right) \tag{5.27}$$

where Z_{IN} is the input impedance of the transformer, which is a function of the load Z_L, impedance matrix elements of coupled lines Z_{ij}, and arbitrary load Z_{ij}'' (i and j are the indexes of the matrix elements). Design equations for the asymmetric coupled ports in nonhomogenous medium can be obtained from [52].

When loaded the coupled and through ports with stepped impedance transmission line the operating bandwidth of the transformer has increased by three times in comparison to the traditional quarter-wave transformer [52]. However, in this

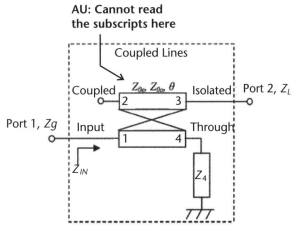

Figure 5.20 General coupled line configuration. The four-port section is reduced to a two-port with coupled and through ports that can be loaded with an external impedance termination [52]. In this figure, the through port is terminated with termination impedance Z_4.

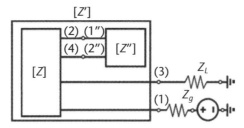

Figure 5.21 Two-port network representation for a coupled line transformer [52].

work, bandwidth is further increased by creating an impedance transformer using more coupled line sections connected in series [28, 51]. It consists of two coupled transmission line sections, as illustrated in Figure 5.21. Each section is a quarter-wavelength long at the center frequency of operation. The impedance transformation of 12.5Ω to 50Ω is achieved between two diagonal ports of the section. The remaining ports are interconnected by a stepped impedance transmission line. This interconnection widens the operating frequency band and compensates for the differences in electrical lengths of the coupled lines for the c and π modes [59] in a nonhomogeneous medium (assuming microstrip realization).

From Figure 5.22, an analysis of the series stepped impedance of port $1''2''$ is carried out based on [52]. Thus, the impedance matrix elements are found to be

$$Z_{11}'' = Z_{11}^1 - \frac{\left(Z_{12}^{(1)}\right)^2}{Z_{11}^{(2)} + Z_{11}^{(1)}} = Z_{01}\coth\left(\gamma_1 l_1\right) - \frac{Z_{01}^2}{\left(Z_{01}\coth\left(\gamma_1 l_1\right) + Z_{02}\coth\left(\gamma_2 l_2\right)\right) \cdot \sinh^2\left(\gamma_1 l_1\right)}$$

(5.28)

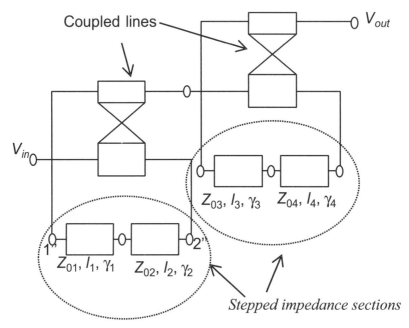

Figure 5.22 Proposal circuit of the compact impedance transformer, and the impedance transformation from 12.5Ω to 50Ω.

$$Z''_{12} = Z''_{21} - \frac{Z^{(1)}_{12} Z^{(2)}_{12}}{Z^{(2)}_{11} + Z^{(1)}_{11}} = \frac{Z_{01} Z_{02}}{\left(Z_{01} \coth(\gamma_1 l_1) + Z_{02} \coth(\gamma_2 l_2) \right) \cdot \sinh(\gamma_1 l_1) \sinh(\gamma_2 l_2)}$$

(5.29)

$$Z''_{22} = Z^{(2)}_{11} - \frac{\left(Z^{(2)}_{12} \right)^2}{Z^{(2)}_{11} + Z^{(1)}_{11}} = Z_{02} \coth(\gamma_2 l_2) - \frac{Z^2_{02}}{\left(Z_{01} \coth(\gamma_1 l_1) + Z_{02} \coth(\gamma_2 l_2) \right) \cdot \sinh^2(\gamma_2 l_2)}$$

(5.30)

where transmission lines with characteristic impedances Z_{01}, Z_{02}, length l_1, l_2, and propagation constants γ_1, γ_2 [58] are given by

$$\left[Z^1 \right] = \begin{bmatrix} Z^{(1)}_{11} & Z^{(1)}_{12} \\ Z^{(1)}_{21} & Z^{(1)}_{22} \end{bmatrix} = \begin{bmatrix} Z_{01} \coth(\gamma_1 l_1) & \dfrac{Z_{01}}{\sinh(\gamma_1 l_1)} \\ \dfrac{Z_{01}}{\sinh(\gamma_1 l_1)} & Z_{01} \coth(\gamma_1 l_1) \end{bmatrix}$$

(5.31)

and

$$\left[Z^1 \right] = \begin{bmatrix} Z^{(2)}_{11} & Z^{(2)}_{12} \\ Z^{(2)}_{21} & Z^{(2)}_{22} \end{bmatrix} = \begin{bmatrix} Z_{01} \coth(\gamma_2 l_2) & \dfrac{Z^2}{\sinh(\gamma_2 l_2)} \\ \dfrac{Z_{02}}{\sinh(\gamma_2 l_2)} & Z_{02} \coth(\gamma_2 l_2) \end{bmatrix}$$

(5.32)

The compensation allows for uniform distribution of the reflection zeros in the frequency domain. Each stepped impedance transmission line consists of two $\lambda_g/8$ length transmission lines. The total electrical length of the transformer is equal to half a wavelength at the center frequency. The return loss response S_{22}, from (5.27) of the transformer is shown in Figure 5.23. For comparison, the transformer discussed in [52] is included. The transformer exhibits six minima (zero reflection) in the spectrum of the reflection coefficient across the frequency range. The achieved fractional matching bandwidth is beyond a decade at the −20-dB reflection coefficient level with the new approach. In addition, the distance between the minima location $\Delta\sigma$ (as shown in Figure 5.23) can be widened to improve low-frequency matching by adjusting the parameters of the structure, that is, the step impedance sections (width and length).

Performance of the broadband impedance transformer is verified with a 3D-EM full-wave simulator (CST). The layout is imported from Cadence (in ODB++ format) as shown in Figure 5.24(a). Port 1 is terminated with waveguide port (12.5Ω), and port 2 is connected with a 50Ω SMA connector. A prototype board is fabricated using Rogers three-layer PCB material, which has a dielectric constant ε_r of 3.6 and a thickness h of 0.762 mm [see Figure 5.24(b)]. Since port 1 will be connected to a

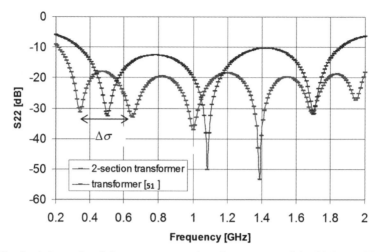

Figure 5.23 Analysis results of the output return loss S_{22} response of the 12.5Ω to 50Ω impedance transformer shown in Figure 5.22. A six zero (minimum) reflections exist across the frequency range. For comparison, the transformer shown in [51] is included.

50Ω SMA connector for real measurements, the two-port S-parameter data of the connector must be de-embedded for accurate results [60].

Simulated (line with triangle) and measured results (line with circle) are given for the connector across the bandwidth). Figures 5.26 and Figure 5.27 present simulated and measured results for the complete broadband transformer. Insertion loss S_{21} is acceptable from 10 to 1800 MHz, which is less than 1.5 dB at the simulation level. From the analytical (as shown in Figure 5.23), there is a six-zero reflection; however, only three- and four-zero reflections (S_{22}) exist in the simulation response of CST and measured data, respectively, within the passband (10 to 1800 MHz). As predicted in CST, there is no zero reflection occurrence at the center frequency (1000 MHz). The measured result is acceptable below 1700 MHz (i.e., insertion loss lower about 2 dB), and four-zero reflection occurred. The results demonstrated are adequate to prove the high-efficiency distributed amplifier concept although the insertion loss above 1500 MHz is more than 1.5 dB.

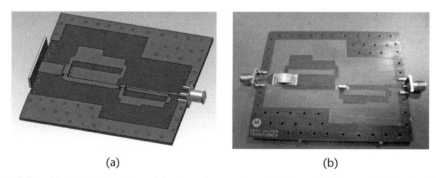

(a)   (b)

Figure 5.24 (a) Layout structure of the transformer imported from Cadence with the simulation performed in CST and (b) an actual prototype board.

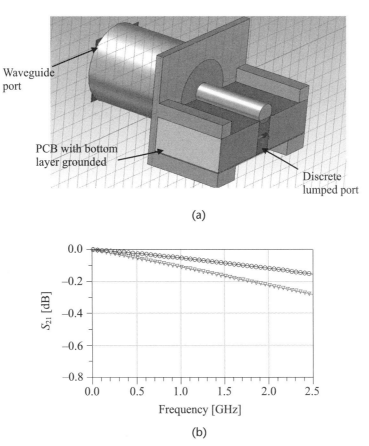

(a)

(b)

Figure 5.25   (a) Connector modeling in CST, where waveguide and lumped ports are used, and (b) simulated (line with triangle) and measured (line with circle) results of the connector across the bandwidth.

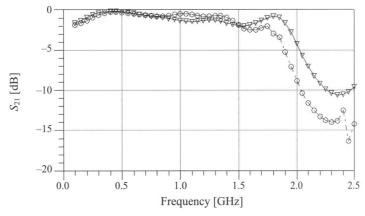

Figure 5.26   Simulated versus measured results of insertion loss S_{21} of the transformer. The simulation response of the broadband impedance transformer is performed with CST (line with triangle) and the measured result (line with circle).

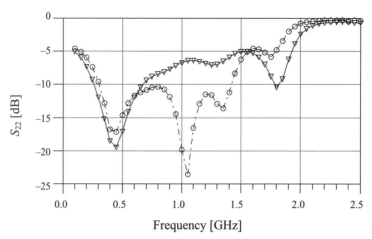

Figure 5.27 Simulated versus measured results of output return loss S_{22} of the transformer. The simulation response of the broadband impedance transformer is performed with CST (line with circle) and the measured result (line with triangle).

To experimentally validate the concept of a high-efficiency distributed amplifier, a prototype board is fabricated using a Rogers three-layer PCB material that has a permittivity ε_r of 3.66 and a thickness h of 0.762 mm. The first layer is used for RF and dc line routing, where all the component placement takes place on top of layer 1. General layout guidelines (e.g., component placements, RF and dc routing, and general layout rules) are still applied in this work.

In addition to a series gate resistor, a ferrite bead is implemented at the drain line together with a 33-μF tantalum capacitor[2] and 1.8-nF ceramic capacitor[3] as a precaution of low frequency oscillation. The dc gate biasing terminals are bypassed to ground with multiple chip and tantalum capacitors (e.g., 100 pF, 33 nF, 10 μF). The dc feeding lines for the final stage are connected to a high-Q air-wound coil (from Taito Yuden, Inc.) and high-Q chip inductors (value of 220 nH) from Coilcraft, Inc. The photograph of the high-efficiency distributed amplifier board is shown in Figure 5.28. The effective distributed amplifier size area is 27 mm × 24 mm. Two measurements, R of 12.5Ω and 50Ω, are performed in the same board outline without repeating another layout work in Cadence, therefore, point a (in Figure 5.28) will be used to tap the RF output directly and the transformer section is bypassed when R of 50Ω is selected. The full transformer section is utilized when R of 12.5Ω is selected. In other words, two different board designs are used but both boards have the same outline. The drain and gate-line elements value for both cases of R are different and optimized for best performance.

Measured results of small-signal S-parameters and PAE for a four-section high-efficiency distributed amplifier when terminated at $R = 50\Omega$ are given in Figures 5.29 and Figure 5.30. Each device is fed with a 5V drain supply voltage. Bias voltage

---

[2] The tantalum capacitor part number is T491D22K016AT, from KEMET, Inc. The rated voltage is 16V at 85°C.

[3] The capacitors used were 600S Series ultralow ESR, high-Q microwave capacitors, from ATC Inc. Other capacitors in the series are 545-L ultrabroadband high-Q capacitors, from Murata, Inc.

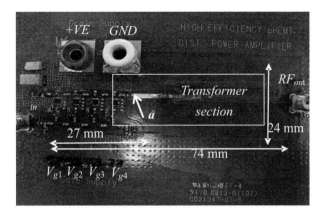

Figure 5.28 Photograph of the high-efficiency distributed amplifier prototype board. The terminal *a* will be tap when $R = 50\Omega$ output or the transformer will be used if $R = 12.5\Omega$. The effective distributed amplifier size area is 27 mm × 24 mm.

of 0.44V is applied to each gate, resulting in Class AB operation, with quiescent current I_{dq} of ~90 mA (10% of I_{max}). Input return loss S_{11} is below −10 dB, but output return loss S_{22} is approximately −5 dB across the 10- to 1800-MHz frequency range (see Figure 5.29). For power performance (see Figure 5.30), the output power of ~500 mW, gain of 8 dB, and PAE of 30% are achieved throughout the 10- to 1800-MHz frequency range. Good agreement between simulations and measurement results is obtained. The measured PAE result and gain for a few cases of bias and supply voltage V_g and V_d, respectively, are recorded in Figure 5.31. With a high supply voltage (5.4V) and applied bias close to pinch-off ($Vg = 0.4V$ where I_{dq} ~5% of I_{max}), PAE at a low frequency is increased by 10% with minimum changes in gain with comparison to the result shown in Figure 5.31. There is no improvement

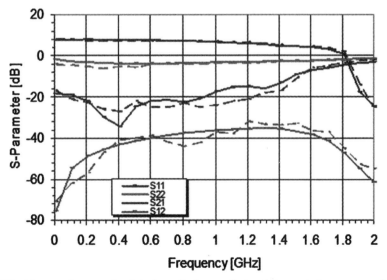

Figure 5.29 Measured versus simulated results for small-signal *S*-parameters across a frequency range of 10 to 1800 MHz when terminated $R = 50\Omega$. *Legend:* solid line: simulated results; dashed line: measured results.

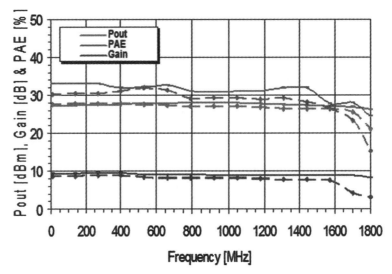

Figure 5.30 Measured versus simulated results of PAE, output power (or P_{out}), and gain across a frequency range of 10 to 1800 MHz when terminated $R = 50\Omega$. *Legend:* solid line: simulated results; dashed-dotted line: measured results.

in PAE with a supply adjustment. The performance of ω_c has limitations on gate-line characteristics.

Due to the high insertion loss of the transformer beyond 1.5 GHz, the measured results of a four-section high-efficiency distributed amplifier when terminated at $R = 12.5\Omega$ are degraded as the frequency increases. The same supply voltage and bias voltage are applied to each gate of the transistor (as applied for the case of $R = 50\Omega$). Measured results of the power performance of a four-section high-efficiency distributed amplifier when $R = 12.5\Omega$ are shown in Figure 5.32. At a low-frequency output power of ~700 mW, a gain of 10 dB and PAE of ~38% is achieved. At higher frequencies (e.g., beyond 1.4 GHz), the performance degraded.

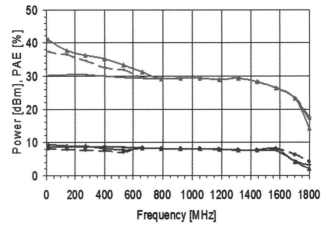

Figure 5.31 Measured results of PAE and gain for three different cases when terminated $R = 50\Omega$. *Legend:* solid line: $V_g = 0.44V$ and $V_d = 5V$; dashed line: $V_g = 0.4V$ and $V_d = 4.6V$; solid line with triangles: $V_g = 0.44V$ and $V_d = 5.4V$.

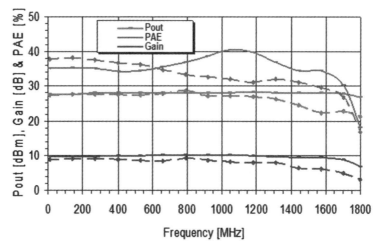

Figure 5.32   Measured versus simulated results of PAE, output power (or P_{out}), and gain across the frequency range of 10 to 1800 MHz when terminated $R = 12.5\Omega$. *Legend:* solid line: simulated results; dashed-dotted line: measured results.

It is necessary to achieve good performance from the broadband impedance transformer at the measurement level in order to demonstrate the best efficiency results (as shown in Figure 5.32, where PAE > 40% at the simulation level). This proves that the PAE > 38% at low frequencies with $R = 12.5\Omega$ at the measurement level. Efficiency of more than 30% is recorded across the bandwidth operation. It is strongly evident from the measurement results that the concept to achieve high efficiency is well demonstrated. Nevertheless, the distributed amplifier performance at higher frequency operation can be improved further with transformer optimization.

References

[1]   J. B. Beyer et al., "MESFET Distributed Amplifier Design Guidelines," *IEEE Trans. Microwave Theory and Techniques*, Vol. MTT-32, No. 3, pp. 268–275, Mar. 1984.

[2]   B. Kim, H. Q. Tserng, and H. D. Shih, "High Power Distributed Amplifier Using MBE Synthesized Material," *IEEE Microwave and Millimeter Monolithic Integrated Circuits Symposium*, pp. 35–37, 1985.

[3]   P. H. Ladbrooke, "Large Signal Criteria for the Design of GaAs FET Distributed Power Amplifier," *IEEE Trans. Electron Devices*, Vol. ED-32, No. 9, pp. 1745–1748, Sep. 1985.

[4]   B. Kim and H. Q. Tserng, "0.5W 2–21 GHz Monolithic GaAs Distributed Amplifier," *Electronics Letters*, Vol. 20, No. 7, pp. 288–289, Mar. 1984.

[5]   M. J. Schindler et al., "A 15 to 45 GHz Distributed Amplifier Using 3 FETs of Varying Periphery," *IEEE GaAs IC Symposium Technical Digest*, pp. 67–70, 1986.

[6]   L. Zhao et al., "A 6 Watt LDMOS Broadband High Efficiency Distributed Power Amplifier Fabricated Using LTCC Technology," *IEEE MTT-S Int. Microwave Symp. Dig.*, pp. 897–900, June 2002.

[7]   J. Gassmann et al., "Wideband High Efficiency GaN Power Amplifiers Utilizing a Non-Uniform Distributed Topology," *IEEE MTT-S Int. Microwave Symp. Dig.*, pp. 615–618, June 2007.

[8]   K. Narendra et al., "Vectorially Combined pHEMT/GaN Distributed Power Amplifier for SDR Applications," *IEEE Trans. Microwave Theory and Techniques*, Vol. 60, No. 10, pp. 3189–3200, Oct. 2012.

[9]   K. Narendra et al., "Vectorially Combined Distributed Power Amplifier with Load Pull Determination," *Elect. Letters*, Vol. 46, No. 16, pp. 1137–1138, Aug. 2010.

[10] E.W. Strid and K. R. Gleason, "A DC-12 GHz Monolithic GaAsFET Distributed Amplifier," *IEEE Trans. Electron Devices*, Vol. ED-29, No. 7, pp. 1065–1071, July 1982.

[11] K. Krishnamurthy et al., "Broadband GaAs MESFET and GaN HEMT Power Amplifiers," *IEEE Journal Solid State Circuits*, Vol. 35, No. 9, pp. 1285–1292, 2000.

[12] H. Amasuga et al., "A High Power and High Breakdown Voltage Millimeter-Wave GaAs pHEMT with Low Nonlinear Drain Resistance," *IEEE MTT-S Int. Microwave Symp. Dig.*, pp. 821–824, June 2007.

[13] Y. Ayasli et al., "Capacitively Coupled Traveling Wave Power Amplifier," *IEEE Trans. Microwave Theory Techniques*, Vol. MTT-32, No. 12, pp. 1704–1709, Dec. 1984.

[14] A. Ayasli et al., "A Monolithic GaAs 1–13 GHz Traveling Wave Amplifier," *IEEE Trans. Microwave Theory Techniques*, Vol. 30, No. 12, pp. 976–981, July 1982.

[15] K. B. Niclas, R. R. Pereira, and A. P. Chang, "On Power Distribution in Additive Amplifiers," *IEEE Trans. Microwave Theory and Techniques*, Vol. 38, No. 3, pp. 1692–1699, Nov. 1990.

[16] J. L. B. Walker, "Some Observations on the Design and Performance of Distributed Amplifiers," *IEEE Trans. Microwave Theory and Techniques*, Vol. 38, pp. 164–1698, Jan. 1992.

[17] R. Halladay, M. Jones, and S. Nelson, "2–20 GHz Monolithic Distributed Power Amplifiers," *IEEE Microwave and Millimeter-Wave Monolithic Integrated Circuits Symp.*, pp. 35–37, 1985.

[18] S. N. Prasad, J. B. Beyer, and I. S. Chang, "Power-Bandwidth Considerations in the Design of MESFET Distributed Amplifiers," *IEEE Trans. Microwave Theory Techniques*, Vol. 36, pp. 1117–1123, July 1988.

[19] S. N. Prasad, S. Reddy, and S. Moghe, "Cascaded Transistor Cell Distributed Amplifiers," *Microwave and Optical Technology Lett.*, Vol. 12, No. 3, pp. 163–167, June 1996.

[20] K. Narendra et al., "Dual Fed Distributed Power Amplifier with Controlled Termination Adjustment," *Progress in Electromagnetic Research*, Vol. 139, pp. 761–777, May 2013.

[21] K. Narendra et al., "Discrete Component Design of a Broadband Impedance Transforming Filter for Distributed Power Amplifier," *10th IEEE Microwave Mediterranean Symp.*, pp. 292–295, Aug. 2010.

[22] W. S. Percival, "Improvements In and Relating to Thermionic Valve Circuits," British Patent 460,562, 1937.

[23] E. L. Ginzton et al., "Distributed Amplification," *Proc. IRE*, pp. 956–969, 1948.

[24] T. T. Y. Wong, *Fundamentals of Distributed Amplification*, Norwood, MA: Artech House, 1993.

[25] Y. Ayasli et al., "2-to-20 GHz GaAs Traveling Wave Power Amplifier," *IEEE Trans. Microwave Theory Techniques*, Vol. 32, No. 3, pp. 290–295, Mar. 1984.

[26] A. W. Podgorski and L. Y. Wei, "Theory of Traveling-Wave Transistors," *IEEE Trans. Electron Devices*, Vol. ED-29, No. 12, pp. 1845–1853, Dec. 1982.

[27] K. Narendra et al., "Cascaded Distributed Power Amplifier with Non-Identical Transistors and Inter-Stage Tapered Impedance," *40th European Microwave Conf.*, pp. 549–522, Sep. 2010.

[28] K. Narendra et al., "Design Methodology of High Power Distributed Amplifier Employing Broadband Impedance Transformer," *IEEE Int. Conf. of Antenna, Systems and Propagation 2009*, Sep. 2009.

[29] K. Narendra et al., "PHEMT Distributed Power Amplifier Adopting Broadband Impedance Transformer," *Microwave Journal*, Vol. 56, pp. 76–82, 2013.

[30] K. Narendra et al., "High Efficiency Applying Drain Impedance Tapering for 600mW pHEMT Distributed Power Amplifier," *IEEE Int. Conf. on Microwave and Millimeter Wave Technology*, pp. 1769–1772, Apr. 2008.

[31] P. H. Ladbrooke, "Large Signal Criteria for the Design of GaAs FET Distributed Power Amplifier," *IEEE Trans. Electron Devices*, Vol. ED-32, No. 9, pp. 1745–1748, Sep. 1985.

[32] R. Halladay, M. Jones, and S. Nelson, "2–20 GHz Monolithic Distributed Power Amplifiers," *IEEE Microwave and Millimeter-Wave Monolithic Integrated Circuits Symp.*, pp. 35–37, 1985.

[33] S. N. Prasad, S. Reddy, and S. Moghe, "Cascaded Transistor Cell Distributed Amplifiers," *Microwave and Optical Technology Lett.*, Vol. 12, No. 3, pp. 163–167, June 1996.

[34] R. W. Chick, "Non-Uniformly Distributed Power Amplifier," U.S. Patent 5,485,118, Jan. 1996.

[35] M. Campovecchio et al., "Optimum Design of Distributed Power FET Amplifiers: Application to 2–18 GHz MMIC Module Exhibiting Improved Power Performances," *IEEE MTT-S Int. Microw. Symp. Dig.*, San Diego, pp. 125–128, June 1994.

[36] C. Duperrier et al., "New Design Method of Uniform and Nonuniform Distributed Power Amplifiers," *IEEE Microwave Theory Techniques*, Vol. 49, No. 12, pp. 2494–2499, Dec. 2001.

[37] C. Xie and A. Pavio, "Development of GaN HEMT Based High Power High Efficiency Distributed Power Amplifier for Military Applications," *IEE MILCOM 2007*, pp. 1–4, Oct. 2007.

[38] A. Martin et al., "Balanced AlGaN/GaN Cascode Cells: Design Method for Wideband Distributed Amplifiers," *Electronics Letters*, Vol. 44, No. 2, pp. 116–117, Jan. 2008.

[39] P. Heydari, "Design and Analysis of a Performance-Optimized CMOS UWB Distributed LNA," *IEEE Journal Solid-State Circuits*, Vol. 42, No. 9, pp. 1892–1905, Sep. 2007.

[40] W. H. Horton, J. H. Jasberg, and J. D. Noe, "Distributed Amplifiers: Practical Considerations and Experimental Results," *Proc. IRE*, Vol. 38, pp. 748–753, July 1950.

[41] S. N. Prasad and A. S. Ibrahim, "Design Guidelines for a Novel Tapered Drain Line Distributed Power Amplifier," *Proc. 36th European Microwave Conf.*, pp. 1274–1277, Sep. 2006.

[42] K. Krishnamurthy, S. I. Long, and M. J. W. Rodwell, "Cascode Delay Matched Distributed Amplifiers for Efficient Microwave Power Amplification," *IEEE MTT-S Int. Microw. Symp. Dig.*, pp. 819–821, June 1999.

[43] M. E. V. Valkenburg, *Network Analysis*, 3rd ed., Upper Saddle River, NJ: Prentice Hall, 1974.

[44] S. D. Agostino and C. Paoloni, "Design of a Matrix Amplifier Using FET Gate-Width Tapering," *Microwave Opt. Technol. Letters*, Vol. 8, pp. 118–121, Feb. 1995.

[45] S. C. Cripps, *Power Amplifier for Wireless Communications*, 2nd ed., Norwood, MA: Artech House, 2006.

[46] B. Thompson et al., "Distributed Power Amplifier with Electronic harmonic Filtering," *IEEE Radio Frequency Integrated Circuits Symp.*, pp. 241–244, June 2009.

[47] B. Thompson and R. E. Stengel, "System and Method for Providing an Input to a Distributed Power Amplifier System," U.S. Patent 7,233,207, June 2007.

[48] T. S. Tan, M. F. Ain, and S. I. S. Hassan, "Large Signal Design of Distributed Power Amplifier with Discrete RF MOSFET Devices," *2006 Int. RF & Microwave Conf. Proc.*, pp. 58–61, Sep. 2006.

[49] T. S. Tan, M. F. Ain, and S. I. S. Hassan, "100–500MHz, 1 Watt Distributed Power Amplifier with Discrete MOSFET Devices," *Asia Pacific Conference on Applied Electromagnetics Proc.*, pp. 198–203, Dec. 2005.

[50] Application Notes ATF511P8, www.avagotech.com.

[51] K. Narendra and S. Pragash, "Distributed Amplifier with Multi-Section Transformer," www.ip.com, Sep. 2009.

[52] T. Jensen et al., "Coupled Transmission Line as Impedance Transformer," *IEEE Microw. Theory Tech.*, Vol. 55, No. 12, pp. 2957–2965, Dec. 2007.

[53] V. Zhurbenko, V. Krozer, and P. Meincke, "Broadband Impedance Transformer Based on Asymmetric Coupled Transmission Lines in Nonhomogeneous Medium," *IEEE MTT-S Int. Microwave Symp.*, 2007.

[54] J. Horn, M. Huber, and G. Boeck, "Wideband Balun and Impedance Transformers Integrated in a Four-Layer Laminate PCB," *35th European Microwave Conf.*, Sep. 2005.

[55] J. Chramiec and M. Kitlinski, "Design of Quarter-Wave Compact Impedance Transformers Using Coupled Transmission Line," *Electronics Letters*, Vol. 38, No. 25, pp. 1683–1685, Dec. 2002.

[56] K. S. Ang, C. H. Lee, and Y. C. Leong, "Analysis and Design of Coupled Line Impedance Transformers," *IEEE MTT-S Int. Microwave Symp. Dig.*, pp. 1951–1954, June 2004.

[57] K. S. Ang, C. H. Lee, and Y. C. Leong, "A Broad-Band Quarter-Wavelength Impedance Transformer with Three Reflection Zeroes Within Passband," *IEEE Microw. Theory Tech.*, Vol. 52, No. 12, pp. 2640–2644, Dec. 2004.

[58] G. Jaworski and V. Krozer, "Broadband Matching of Dual-Linear Polarization Stacked Probe-Fed Microstrip Patch Antenna," *Electron Lett.*, Vol. 40, No. 4, pp. 221–222, 2004.

[59] V. Tripathi, "Asymmetric Coupled Transmission Line in Inhomogeneous Medium," *IEEE Trans. Microwave Theory Techniques*, Vol. MTT 32, No. 9, pp. 734–739, Sep. 1975.

[60] Agilent report de-embed technique.

CHAPTER 6

Stability Analysis of Distributed Amplifiers

Hardware fabrication of microwave circuits is expensive, especially power amplifier circuits; therefore, it is necessary to use a good technique that allows designers to correctly predict the behavior of microwave circuits with the purpose of detecting and correcting possible problems during the design stage. Manufacturing tolerances cause additional variations of input and output termination impedance. Stability, as well as circuit performance, has to be guaranteed in spite of impedance variations. Unconditional stability criteria [1–14] can be applied to the overall amplifier. Note, however, that unconditional stability can be too stringent of a requirement for the overall amplifier and can prevent the achievement of the desired performance. If maximum variations of input and output terminations can be estimated early on, the amplifier can be designed to be stable for all the input and output termination impedances. The stability of a two-port network for different values of the input and output loads is investigated in [15, 16].

Stability criteria can be used to study the stability behavior of a two-port network made up of a single active device as highlighted in [17–19], but this is not adequate, and a Nyquist analysis of all the internal feedback loops is required. These techniques, as well as stability circles and other graphical methods, require us to visually inspect the polar plots of properly chosen functions. Therefore, they allow the stability check to be performed only at the end of the design phase, often resulting in a tedious trial-and-error design process. A synthesis-oriented criterion based on a stability factor is required to make use of optimization routines provided by computer-aided design (CAD) tools in the design phase. The criterion allows for a guarantee of stability in circular regions of the input and output reflection coefficient planes surrounding the nominal purely resistive input and output loads [20, 21]. In [21], such a criterion has been extended to ensure the stability of circuits with nominal complex termination impedances.

6.1 Motivation for Conducting Stability Analyses

Originally, Rollet [22] deduced the basic results for the stability of linear two-port RF amplifiers as a function of the Z-, Y-, G-, or H-parameters of a network. However, Woods shows simple examples of circuits with negative resistances that do not fulfill the Rollet condition and whose stability cannot be correctly deduced using the conditions [23]. Platzker also paid attention to this fact illustrating real examples of circuits with various active elements in which using the K-factor led to erroneous conclusions [24]. Instabilities at low frequencies due to bias circuits cannot be detected using the K-factor [25]. A study by Platzker et al. [24] showed that the K-factor is not sufficient to analyze the stability criteria of multistage amplifiers due to the fact that possible feedback between stages is not taken into account. In power-combining structures with n transistors in parallel, n possible modes of oscillation exist [25, 26], and the K-factor only allows for the detection of even modes of oscillation. As an alternative, Freitag [27] proposed an analysis of the K-factor capable of detecting odd-mode oscillations using an ideal transformer that forces the odd mode of oscillation in the circuit.

Ohtomo proposes a rigorous method for evaluating the stability of a dc solution in microwave circuits with multiple active devices that can contain multiple feedback loops [28]. The method is valid for the detection of odd- and even-mode oscillations and can be applied to any linear or nonlinear circuits with a linear equivalent around the bias point. Centurelli et al. [20, 21] present the necessary and sufficient conditions for the stability of circuits with multiple active devices in agreement with the procedure proposed by Ohtomo [28], but also guarantee a stability margin in circular regions around complex terminations. Ohtomo's method is based on the Nyquist criterion and requires observing the $2N$ transfer functions associated with each port. However, Centurelli et al. define the margins of gain and phase that allow determination of the stability of the circuit.

Kassakian and Lau emphasize the possibility of the appearance of odd-mode oscillations with voltage-combining structures [29]. Postulating a priori the mode of oscillation of the circuit and applying the Routh-Hurwitz [31] criterion to the zeros of the characteristic equation, an odd-mode oscillation in a power amplifier can be detected. Freitag proposes a method for the detection of odd-mode oscillations based on the representation of Z-parameters of the circuit [26]. It identifies the different modes of operation of the multistage amplifiers. The drawback of this method is that a characteristic matrix of the system is needed; therefore, it is not suitable for implementation for commercial simulators.

Ramberger and Merkle [31] apply Ohtomo's method [28] to circuits with voltage-combining structures. Instead of applying the method to the complete circuit, they use an equivalent circuit with a single branch, which remarkably reduced the simulation time. Costantini et al. [32] also proposed a method that allows the detection of odd- and even-mode oscillations, and the circuit must be analyzed using as many equivalent circuits as modes of oscillation that can possibly exist in the circuit.

6.2   Method of Stability Analysis

In this chapter, some stability analyses—K-factor for a two-port network, feedback and NDF factor, and finally pole-zero identification method—are discussed. This will give the reader a good understanding of stability analysis.

6.2.1   K-Factor Stability of a Two-Port Network

A two-port network is unconditionally stable if no combination of passive source and load impedance exists that can cause the circuit to oscillate. For it, it is a necessary condition that the real part of the admittance (impedance or admittance) observed at the input of each of the ports remains positive with any passive termination that is connected to the other port. This is given in the following conditions from [33], where

$$K = \frac{2\operatorname{Re}(\gamma_{11})\operatorname{Re}(\gamma_{22}) - \operatorname{Re}(\gamma_{12}\gamma_{21})}{|\gamma_{12}\gamma_{21}|} \geq 1 \tag{6.1}$$

$$\operatorname{Re}(\gamma_{11}) \geq 0 \tag{6.2}$$

$$\operatorname{Re}(\gamma_{22}) \geq 0 \tag{6.3}$$

where γ_{ij} represents the element (i, j) of any of the Z-, Y-, G-, or H-parameter matrices. Equation (6.1) can be rewritten as a function of the S-parameters, which are more adequate for describing RF and microwave circuits, obtaining, in this manner, a set of conditions equivalent to (6.1) for unconditional stability [22]. If in the whole range of frequencies for which the device shows the following equations are fulfilled

$$K = \frac{1 - |S_{11}|^2 - |S_{22}|^2 + |\Delta_s|^2}{2|S_{12}S_{21}|} > 1 \tag{6.4}$$

together with one of the following oscillation conditions,

$$|\Delta_s| = |S_{11}S_{22} - S_{12}S_{21}| < 1 \tag{6.5}$$

$$B_1 = 1 + |S_{11}|^2 - |S_{22}|^2 - |\Delta_s|^2 > 0 \tag{6.6}$$

$$B_2 = 1 - |S_{11}|^2 + |S_{22}|^2 - |\Delta_s|^2 > 0 \tag{6.7}$$

$$1 - |S_{11}|^2 > |S_{12}S_{21}| \tag{6.8}$$

$$1 - |S_{22}|^2 > |S_{12}S_{21}| \tag{6.9}$$

The unconditional stability of the two-port network is guaranteed only if the Rollet condition is fulfilled, that is, whenever the unloaded circuit does not have poles in the right half plane of the s-plane. The condition for K and the auxiliary conditions can be replaced by a single figure of merit μ [34]. The new condition for unconditional stability is

$$\mu = \frac{1 - |S_{11}|^2}{|S_{22} - S_{11}^*\Delta| + |S_{21}S_{12}|} > 1 \tag{6.10}$$

whenever the Rollet condition is fulfilled. The parameter μ, in addition to evaluating the unconditional stability of a two-port network, allows estimation of its degree of potential instability, since it can be geometrically interpreted as the minimum distance between the origin of the unit Smith chart and the unstable region. Nevertheless, the parameter μ has not been able to replace the K-factor, which continues to be mainly used by microwave circuit designers.

If the conditions of (6.1) through (6.9) or (6.10) are not fulfilled at all frequencies, it is said that the two-port network is conditionally stable. It is important to study the impedance that, when connected to the input or output, can make the circuit oscillate; that is, the source Z_S and load Z_L *impedance* for which

$$\left|\Gamma_{\text{in}}\right| > 1 \tag{6.11}$$

$$\left|\Gamma_{\text{out}}\right| > 1 \tag{6.12}$$

where Γ_{in} and Γ_{out} are the reflection coefficients at the input and output of the two-port network (Figure 6.1) that are given by the following expressions:

$$\left|\Gamma_{\text{in}}\right| = S_{11} + \frac{S_{12}S_{21}\Gamma_L}{1 - S_{22}\Gamma_L} \tag{6.13}$$

$$\left|\Gamma_{\text{out}}\right| = S_{22} + \frac{S_{12}S_{21}\Gamma_S}{1 - S_{11}\Gamma_S} \tag{6.14}$$

In a circuit that is not unconditionally stable, it is necessary to carefully select the load impedance to avoid the presence of undesired oscillations. For it, it is convenient to find the limit between the unstable and stable region. At the input, the location of the points of Γ_L that make $\left|\Gamma_{\text{in}}\left(\Gamma_L\right)\right|$ can be drawn in the plane of the reflection coefficient of the load. The result is the output stability circle [35] that is characterized by its center C_L,

$$C_L = \frac{S_{22}^* - \Delta_S^* S_{11}}{\left|S_{22}\right|^2 - \left|\Delta_S\right|^2} \tag{6.15}$$

and its radius r_L,

$$r_L = \left|\frac{S_{12}S_{21}}{\left|S_{22}\right|^2 - \left|\Delta_s\right|^2}\right| \tag{6.16}$$

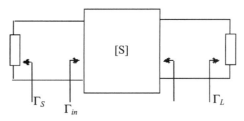

Figure 6.1 Two-port network terminated with input and output impedance.

It contains the values of Γ_L that are found at the limit of the stability region. To find the limit between the unstable and the stable regions, the location of the points of Γ_S that make $\left|\Gamma_{\text{out}}(\Gamma_S)\right|$ can be drawn in the plane of the reflection coefficient of the source [35]. The result is the input stability circle that is also characterized by its center C_S,

$$C_S = \frac{S_{11}^* - \Delta_S^* S_{22}}{\left|S_{11}\right|^2 - \left|\Delta_S\right|^2} \tag{6.17}$$

and its radius r_S,

$$r_S = \left|\frac{S_{12}S_{21}}{\left|S_{11}\right|^2 - \left|\Delta_s\right|^2}\right| \tag{6.18}$$

It contains the values of Γ_S that are found at the limit of the stability region. Equations (6.15) through (6.18) determine the limits of the stability region, but we also need to determine if the stable region is in the interior or outside the stability circles. In analyzing the output stability circles, we will use $\left|\Gamma_{\text{in}}\right| < 1$ and on the other side $\left|\Gamma_{\text{in}}\right| > 1$. If the load impedance is in the center of the Smith chart ($\Gamma_L = 0$), from (6.13) we obtain $\left|\Gamma_{\text{in}}\right| = S_{11}$. In this way, if $\left|S_{11}\right| < 1$, the center of the Smith chart is in the stable region and load $\left|S_{11}\right| > 1$, which places it in the unstable region. Also, the input stability circles delimit the regions for $\left|\Gamma_{\text{out}}\right| < 1$ and $\left|\Gamma_{\text{out}}\right| > 1$ is fulfilled. From (6.14) it is deduced that if $\left|S_{22}\right| < 1$, the center of the Smith chart, which corresponds to $Z_S = Z_0$ and $\Gamma_S = 0$, is in the stable region and if $\left|S_{22}\right| > 1$, it is in the unstable region.

An extensive K-factor still leads to erroneous conclusions since internal feedback loops can exist. Woods [23] shows simple examples of circuits with negative resistances that do not fulfill the Rollet condition and whose stability cannot be correctly deduced using conditions (6.1) through (6.10). Platzker [24] also paid attention to this fact illustrating real examples of circuits with various active elements in which using the K-factor led to erroneous conclusions. The fulfillment of conditions (6.1) through (6.10) at all frequencies only indicates that a stable circuit will continue to be stable when loading it with passive external loads at the output or input. To guarantee the stability of the design, it is necessary to also evaluate the Rollet condition [22]. Hence, it is necessary to verify the stability of the circuit without loading for which the use of alternative methods is indispensable.

6.2.2   Feedback and NDF Factor

Platzker et al. [24] established the inconvenience of applying the Rollet stability criteria [22] in the linear networks that present a pole in the right half plane of the complex plane. In addition, these authors proposed a technique to determine whether a circuit has or does not have a pole in the right half complex plane before applying the Rollet stability criterion [36]. This technique is based on the diagram

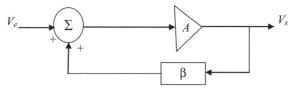

Figure 6.2 Block diagram of a system with a single feedback loop.

or location of a normalized determinant function (NDF) of the circuit in the complex plane and is a generalization of the works of Bode [30] to circuits with multiple active elements.

Taking into account the multiple internal feedbacks that are produced in a circuit with various active elements, the study of stability can be carried out using the theory of feedback systems [30] and thus overcoming the inherent limitations of the K-factor. The Bode theory [30] deals with the systems with a single feedback loop where the active element is included in a passive feedback network. Figure 6.2 shows the block diagram of a system with a single feedback loop.

The transfer function of the complete system can be expressed in the following form:

$$G = \frac{V_s}{V_e} = \frac{A}{1 - A\beta} \tag{6.19}$$

Function $A\beta$ is called the open-loop transfer function and is obtained in the absence of the input voltage ($V_e = 0$) simply cutting the feedback loop, introducing a unit amplitude signal at the input of the active element, and measuring the return signal at the breakpoint (Figure 6.3).

In Bode's works, functions $-A\beta$ and $1 - A\beta$ are designated as return level and return difference or feedback factor, respectively:

$$RR = -A\beta \tag{6.20}$$

$$F = 1 + RR = 1 - A\beta \tag{6.21}$$

To measure the open-loop transfer function of a circuit with a single active element like the one in Figure 6.3, the dependent source $i = g_m v_{in}$ must be replaced by an equivalent dependent source of an auxiliary generator V_{ext} of variable frequency, as in Figure 6.4.

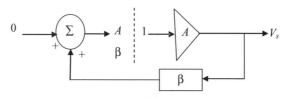

Figure 6.3 Obtaining the open-loop transfer function.

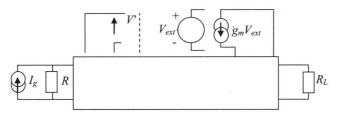

Figure 6.4   Obtaining the return level in a circuit with a single active element.

The return level associated with the voltage-dependent current source is defined as

$$RR = -\frac{V'}{V_{\text{ext}}} \qquad (6.22)$$

and can easily be obtained using the scheme given in Figure 6.4.

The feedback factor F with respect to a transfer parameter g_m can be expressed as [37]

$$F = \frac{\Delta}{\Delta_0} \qquad (6.23)$$

where Δ represents the determinant of the complete circuit (including the terminations of each port) and Δ_0 represents the determinant of the passive network that results from setting the only dependent source of the circuit to zero ($g_m = 0$). Any matrix of Y-, Z- or H-parameters that describes the linear circuit can be used to calculate the determinants Δ and Δ_0. Applying the Nyquist criterion to the function F, we obtain

$$F = 1 + RR = 1 - \frac{V'}{V_{\text{ext}}} \qquad (6.24)$$

for which the stability of the closed-loop circuit can be determined.

For the correct application of the Nyquist criterion, it is important to emphasize that the denominators of functions Δ and Δ_0 are the same; therefore they cancel each other out. In this way, the zeros of function F are the zeros of the characteristic determinant of the system or, in other words, the poles of the closed-loop transfer function. Also, since Δ_0 is calculated from a passive circuit [37], it cannot have zeros with a real positive part; hence, function F cannot have poles with a real positive part by analysis. On the other hand, since $\Delta(\sigma + j\omega)$ and $\Delta_0(\sigma + j\omega)$ are of the same order, when $\omega \to \infty$ or $\sigma \to \infty$, function F tends to one. Lastly, since the response of the circuit is a real function, $F(-j\omega) = F^*(j\omega)$.

In this manner, because function F does not have unstable poles and its zeros represent natural frequencies of the circuit, if the Nyquist trace of function

$$F = \frac{\Delta(j\omega)}{\Delta_0(j\omega)} \qquad (6.25)$$

varying from 0 to ∞ does not encircle the origin, the system is stable. In contrast, the appearance of instability is characterized by the clockwise encirclement of the Nyquist trace around the origin. In this case, the frequency crossing the negative real axis provides an approximation of the starting frequency of the oscillation [38].

6.2.3 Pole-Zero Identification Method

The pole-zero identification method is based on a transfer function approach, in which the function of the system linearized about the steady-state solution is obtained to extract the stability information [39–43]. By introducing a small-signal RF current generator i_{in} in node n of an electric circuit (Figure 6.5), there exists a direct equivalent between the circuit and the system given in (6.26). To obtain the frequency response of the circuit required for stability analysis of a dc solution, it is enough to introduce a small-signal RF current generator i_{in} in any node of the circuit fed solely by the bias sources. The frequency response $H(j\omega)$ is obtained by means of a linear analysis of the impedance observed by the current generator to its operating frequency f_s while the frequency f_s is being swept [40]:

$$H\left(j\omega_s\right) = \frac{v_{out}}{i_{in}} = \frac{Z_2^n}{1 + Z_2^n \dfrac{1}{Z_1^n}} = \frac{Z_1^n Z_2^n}{Z_2^n + Z_1^n} = Z_1^n \| Z_2^n = Z_T^n \left(j\omega_s\right) \qquad (6.26)$$

This is a general result, thus the frequency response in any node n of the linearized circuit can be easily calculated by introducing a small-signal current generator in that node n and measuring the impedance $Z_T^n\left(j\omega_s\right)$ observed by the current source as the frequency ω_s of the current source is swept. It is important to emphasize that, as a result of its parallel connection, the introduced current generator will not have influence on frequencies different from its own operating frequency. The frequency responses associated with the current source generator that displays a low impedance path to ground are not suitable for the analysis, but introducing a voltage generator in that branch to determine admittance $Y_T^m\left(j\omega_s\right)$ is rather effective [42] (see Figure 6.6).

The next step of the technique of stability analysis consists of extracting the information relative to the stability from the frequency response of the circuit. In the case of dc stability analysis (as for a large signal), the information is contained in the denominator of the transfer function associated with the frequency response. Furthermore, in either of the cases there is no guarantee that the associated transfer

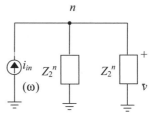

Figure 6.5 General diagram of an electric circuit with a current generator in parallel. The $H(j\omega_s)$ is determined as the ratio of v_{out}/i_{in}.

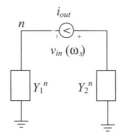

Figure 6.6 General diagram of an electric circuit with a voltage generator in series. The $H(j\omega_s)$ is determined as the ratio of v_{out}/i_{in}.

functions lack unstable zeros, which is the reason why the simplified Nyquist analysis is not, in general, valid [44]. The utilization of system identification techniques like the general methodology for the extraction of information relative to the stability from the frequency response of the circuit is proposed. The techniques of system identification allow one to obtain the transfer function $H(s)$ associated with the frequency response $H(j\omega)$ of the circuit.

$$H(j\omega) = \frac{\text{identification}}{\text{process}} \Rightarrow H(s) = \frac{\prod\limits_{i=1}^{N_z}(s - z_i)}{\prod\limits_{i=1}^{N_p}(s - p_i)} \tag{6.27}$$

where z_i and p_i are the zeros and poles, respectively, of the transfer function $H(s)$ of the system.

Once the poles and zeros of $H(s)$ have been obtained, we proceed to the analysis of the poles to determine the stability of the analyzed steady state. In the case of a frequency response associated with a dc state of the circuit, the poles of the identified transfer function are adjusted to the eigenvalues of the Jacobian matrix [39]. In the case of a frequency response associated with a periodic state, the poles correspond to the Floquet exponents of the system [39]. Therefore, in both cases, the existence of a pair of conjugate complex poles with a positive real part in the transfer function $H(s)$ predicts the instability of the system. In other words, it indicates that the analyzed steady state is unstable and that an oscillation of increasing amplitude to an independent spurious frequency is generated. The autonomous oscillation initiating at ω_a is determined by the magnitude of the imaginary part of these poles [43].

The identification tools available in Scilab[1] are transfer functions that have an excess pole-zero null; therefore, the order of the transfer function is equal to the number of poles and the number of zeros. It is important to emphasize that if we attempt to identify a frequency response with a transfer function order that is higher than necessary, good identification results will be obtained. Nevertheless, there will probably be pole-zero quasi-cancellations that can be eliminated using a lower transfer function order [39]. If the quasi-cancellations are stable, they do not repre-

---

[1] The Scilab program can be downloaded at www.scilab.org.

sent a problem; but if they are in the right half plane, it will be necessary to verify whether a precise identification without the appearance of these quasi-cancellations can be obtained [39].

6.3   Analysis and Conditions of Stability in Distributed Amplifiers

The S-parameters of distributed amplifiers without feedback capacitance C_{gd} have been calculated by Niclas et al. [45], and the analysis of such amplifiers becomes complicated with the inclusion of C_{gd} in an active device. Gamand [46] showed that the occurrence of oscillations in a distributed amplifier (with simplified transistor model) can be caused by the high transconductance g_m and gate-drain capacitance C_{gd} of the transistor. Furthermore, the approach highlighted in [46] indicates that the impedance of the left- and right-hand parts of the circuits from any arbitrary reference plane (node x, y, and so on) are not sufficient to illustrate the oscillation phenomenon due to the fact the line impedances are identical (symmetrical structure), as shown in Figure 6.7. However, in [46], Gamand has explained that the origin of the oscillation can be found in the loop constituted by nodes a, b, c, and d, as in Figure 6.7, and the amplifier tends to oscillate when the gain within the loop becomes too high.

From circuit theory we know that oscillation occurs when a network has a pair of complex conjugate poles on the imaginary axis. If the closed-loop gain (Barkhusien) has a pair of complex conjugate poles in the right half plane (RHP), close to the imaginary axis, due to the ever present noise voltage generated by thermal noise in the network, a growing sinusoidal output voltage appears [35]. As the amplitude of the noise-induced oscillation increases, the amplitude-limiting capabilities of the amplifier produce a change in the location of the poles. The origin of the oscillation in a basic distributed amplifier structure associated with the critical poles can be found in the loop constituted by node a, b,..., and so forth.

To further the analysis, it is convenient to consider a basic single-section distributed amplifier as a basic feedback oscillator circuit and to use a simplified transistor model. The transistor is assumed to be a voltage-controlled current source (VCVS). The real part R_{ds} is included in the model. The basic single-section distributed amplifier and the simplified transistor model are given in Figure 6.8. The transformation of the single-section distributed amplifier to the feedback oscillator circuit is shown in Figure 6.9, where a basic Hartley oscillator is formed. As shown in Figure 6.8(b),

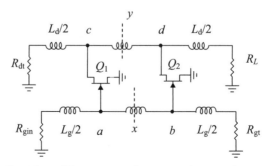

Figure 6.7   Basic distributed amplifier structure in two-section transistors.

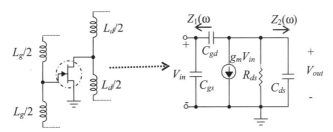

Figure 6.8 (a) Basic single-section distributed amplifier design and (b) simplified transistor model.

$Z_1(\omega)$ and $Z_2(\omega)$ are impedances seen by gate and drain points, respectively. In a basic Hartley oscillator, $Z_1(\omega)$ and $Z_2(\omega)$ can be replaced with single inductance element, for example, L_g and L_d, respectively. For the oscillation to occur, according to [35], the loop resistance must be zero, thus one can define a gain condition as

$$g_m R_{ds} = \frac{L_d}{L_g} \qquad (6.28)$$

where g_m is device transconductance and the frequency of oscillation ω_o is given by

$$\omega_o = \frac{1}{\sqrt{L_T C_{gd}}} \qquad (6.29)$$

where $L_T = L_g + L_d$.

Pole-zero identification of a linearized frequency response is used here for stability analysis [39]. By introducing a small-signal RF current generator $i(f_s)$ to node V_{in} of the circuit shown in Figure 6.9, the frequency response $H(j\omega)$ is obtained by means of a linear analysis of the impedance distributed amplifier observed by the current generator to its operating frequency f_s while the frequency f_s is being swept.

By selecting $C_{gd} = 2.2$ pF, $L_g = 8$ nH, $L_d = 16$ nH, $R_{ds} = 200\Omega$, and $g_m = 10$ mS as an example, f_o is computed from (6.29) and the network oscillates around 694 MHz. The selective elements are determined based on a constant-k network to show the oscillation frequency below 800 MHz (within the passband). Two peaks occur

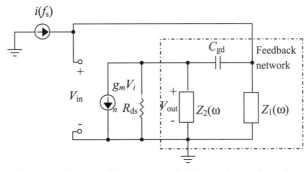

Figure 6.9 A basic single-section distributed amplifier is transformed to a basic feedback oscillator (e.g., a Hartley oscillator). A small-signal RF current generator $i(f_s)$ is introduced to node V_{in}.

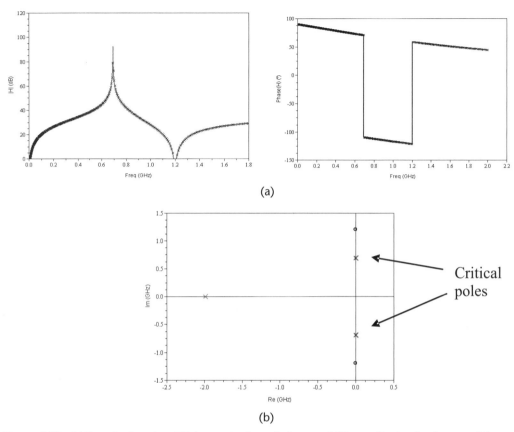

Figure 6.10 (a) Transfer function $H(j\omega)$ magnitude and phase and (b) coordinate of pole-zero of the $H(j\omega)$.

in $H(j\omega)$ as shown in Figure 6.10(a), at the unstable poles (with positive peaking and negative slope phase) and zeros (with negative peaking and positive slope phase) at 694 MHz and 1.2 GHz, respectively. It is clear that the critical pole close to the imaginary axis is located at 694 MHz, as shown in Figure 6.10(b). However, by arranging the gate and drain line to have a cutoff frequency f_c around 800 MHz and a line impedance of 50Ω, no indication of oscillation is observed.

To understand the origin of the oscillation in a distributed amplifier having multiple loops, let's consider a two-section distributed amplifier (from Figure 6.7), where it can be simplified to the Hartley oscillator configuration (as given in Figure 6.9). The feedback network $Z_1(\omega)$ and $Z_2(\omega)$, respectively, is formed by a multiple-loop arrangement (Figure 6.11). The impedances $Z_1(\omega)$ and $Z_2(\omega)$ rely strictly on the loop associated due to g_{m2}, C_{gd2}, L_g, C_{gs2}, L_d, C_{ds2}, terminations, and so forth. An important point to be noticed is that the poles become stable when the primary loop (in loop 1, with $C_{gd2} = 0$) and device 2 have a very low g_{m2} effect (or negligible). Although the primary loop is not connected, the oscillation is still exists when g_{m2} is increases to adequate value due to the fact other loops (in loop 2) still present. Few cases are examined.

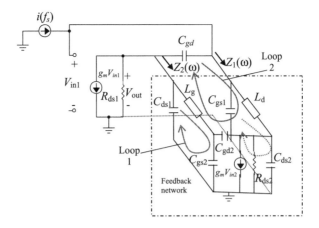

Figure 6.11   In this basic two-section distributed amplifier, the main section is formed with a Hartley oscillator configuration and the feedback network consists of a multiple-loop arrangement.

The following cases of the feedback network have been examined:

Case (a)] $g_{m2} \neq 0$, $C_{gd2} \neq 0$, sweep either g_{m1} or C_{gd2}. The poles are unstable, and the pole evolution is shown in Figure 6.12(a) by sweeping g_{m2}.

Case (b)] $g_{m2} \neq 0$, $C_{gd2} = 0$, sweep g_{m2}. The poles are unstable, and the pole evolution is shown in Figure 6.12(b) by sweeping g_{m2}.

Case (c)] $g_{m2} = 0$, $C_{gd2} = 0$. The poles are stable.

Case (d)] $g_{m2} = 0$, $C_{gd2} \neq 0$, sweep C_{gd2}. The poles are stable.

One should bear in my mind that to have an unstable condition, it necessary for $Z_1(\omega)$ and $Z_2(\omega)$ to be inductive while C_{gd} is present, and this is simply a Hartley oscillator. Therefore, the gate and drain transmission line can be investigated. For instance, look at Figure 6.13(a), where the gate line from Figure 6.8 is redrawn. The imaginary part of Z_{in} behaves in a capacitive manner over a wide frequency range [see Figure 6.13(b)], and it is not possible for oscillation to be present.

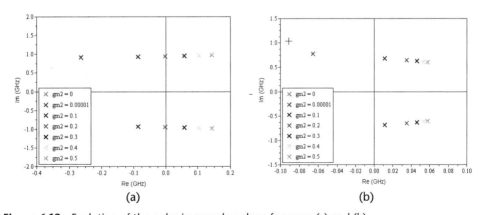

Figure 6.12   Evolution of the poles in complex plane for cases (a) and (b).

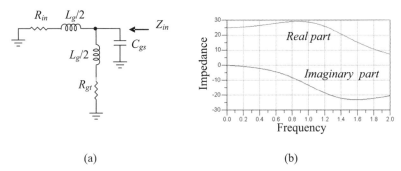

(a)   (b)

Figure 6.13 (a) Gate-line transmission line and (b) plot response of both real and imaginary parts over a wide frequency range.

Let's extend the analysis for the two-section distributed amplifier, as shown in Figure 6.14. The distributed amplifier is analyzed for a transient condition, and an initial voltage condition is introduced in the circuit. By selecting $C_{gd} = 2.2$ pF, $C_{gs} = 5$ pF, $C_{ds} = 5$ pF, $L_g = 10$ nH, $L_d = 10$ nH, $R_{ds} = 200\Omega$, and $g_m = 100$ mS, an odd-mode oscillation takes place (mode +,−) at a frequency of 600 MHz, and Q_1 oscillates 180° out of phase with respect to Q_2; for the virtual ground in the middle, refer to the voltage plot shown in Figure 6.15.

The symmetry of the distributed amplifier topology facilitates the accomplishment of the oscillation conditions for odd-mode oscillations [47]. If the symmetry is not perfect, the mode is not pure. The equivalent circuit of the two-section distributed amplifier for odd-mode oscillation can be simplified to Figure 6.16. The same oscillation condition can still be obtained.

The gate and drain transmission line of Figure 6.16 is investigated. For simplicity, review Figure 6.17(a), where the gate line is redrawn. The imaginary part of Z_{in}

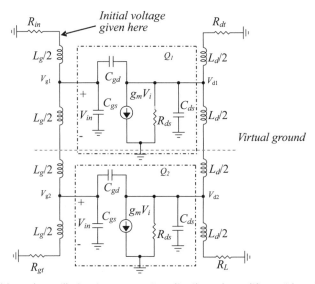

Figure 6.14 Odd-mode oscillation in two-section distributed amplifier, with a virtual ground in the middle.

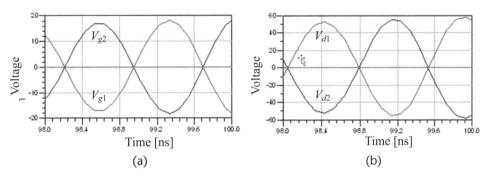

Figure 6.15   Analysis plot of active nodal (i.e., V_{g1}, V_{g2}, V_{d1}, and V_{d2}). (a) The plot of the gate lines: V_{g1} and V_{g2}. (b) The plot of the drain lines: V_{g1} and V_{g2}.

behaves as inductance over a wide frequency range, where the Hartley oscillation condition is fulfilled. The important point to be noticed is that the gate and drain transmission line can be modified by inserting a positive real part element (compare with Figure 6.13) to improve stability in a distributed amplifier having more than a single section.

Our analysis can be extended to a three-section distributed amplifier, as shown in Figure 6.18, which allows different oscillation modes to coexist [47]. They are

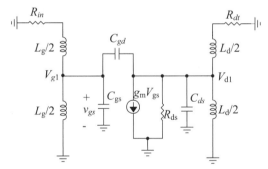

Figure 6.16   Equivalent circuit model of two-section distributed amplifier for odd-mode oscillation. The middle reference plane is the ground plane.

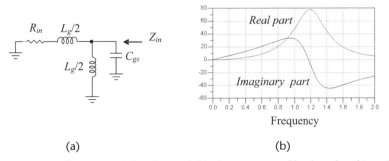

(a)   (b)

Figure 6.17   (a) Gate-line transmission line and (b) plot response of both real and imaginary parts of Z_{in} over a wide frequency range.

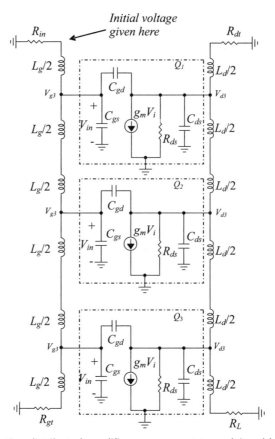

Figure 6.18 Three-section distributed amplifier arrangement to explain odd-mode oscillation.

dependent on the initial conditions. For example, for mode $(+,0,-)$ mode, Q_1 oscillates out of phase with Q_3. The active node, that is, the gate and drain point of the middle section, behaves as a virtual ground; see the analysis plot shown in Figure 5.19(a). On the other hand, there is the $(+,-,+)$ mode, in which Q_1 oscillates in phase with Q_3, and out of phase with Q_2. In a practical sense, odd-mode oscillation

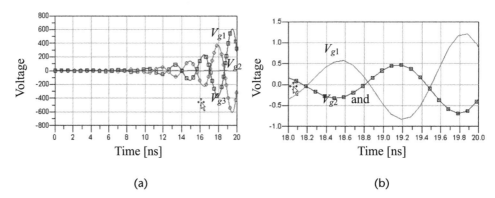

(a)   (b)

Figure 6.19 Analysis plot of active nodal (i.e., V_{g1}, V_{g2}, and V_{g3}: (a) odd-mode oscillation condition for $(+,0,-)$ mode and (b) $(+,-,+)$ mode.

exists when active devices are combined in a parallel configuration (e.g., push-pull or balanced amplifier) [48].

6.4   Parametric Oscillation Detection in Distributed Amplifiers

The multisection nature of distributed amplifiers makes them prone to spurious oscillations due to the presence of multiple nonlinear elements and feedback loops. In fact, under certain conditions, distributed amplifiers can exhibit parametric oscillations, that is, spurious responses of an autonomous nature that are a function of the input power or input frequency. Several stability analyses based on small-signal S-parameters can be found in the literature for distributed amplifiers [48–53]. However, small-signal stability techniques are unable to detect parametric oscillations and thus cannot guarantee circuit stability under large-signal regimes [54–60].

A procedure that includes both small-signal and large-signal stability analyses is used to, first, detect and then, second, eliminate undesired oscillations, including those of a parametric nature. The stability analysis is based on determining the poles of a closed-loop transfer function resulting from linearizing the system around a steady-state regime [54]. The large-signal stability results are used to understand the origin of the instability in the distributed amplifier and to determine an optimum place and value of stabilization resistors that guarantee sufficient stability margins with minimum degradation of performance. The approach is illustrated through the stabilization of a high-efficiency LDMOS distributed amplifier (100 to 700 MHz). The original circuit exhibited a parametric oscillation below 700-MHz deteriorating in-band performance. A second version of the distributed amplifier with the stabilization circuit has been successfully fabricated and no stability problems have been reported.

6.4.1   Stability Analysis of Distributed Amplifiers

Pole-zero identification of a linearized frequency response is again considered here for stability analysis [54]. This technique has the benefit of being applicable to dc small-signal and large-signal stability analyses within a similar methodology and from simulations obtained using commercial CAD tools.

The stability verification of the circuit under study begins with a small-signal stability analysis. The small-signal current probe required to obtain the frequency response is introduced at the gate terminal of the third section. Note that, except for exact pole-zero cancellations, the same stability information can be extracted if the current source is injected at any other circuit node, as described in [39]. The resulting pole-zero map is plotted in Figure 6.20(a). Figure 6.20(b) shows pole-zero map corresponding to the small-signal stability analysis of the stabilized circuit. Because no poles with a positive real part are present, we can conclude that the circuit is stable under small-signal conditions. However, even though all poles are stable, Figure 6.20(a) shows the presence of a couple of conjugate poles with a very small absolute value for their real part. This couple of critical poles is dangerously close to the RHP and reflects the existence of a risky resonance about the frequency given by its imaginary part, about 760 MHz. This means that, although stable in the small-signal regime, the circuit exhibits a low stability margin at that frequency. Such high-frequency critical resonance is commonly found in distributed amplifiers and has its origin in the parasitic loop formed in the distributed structure through the gate drain capacitances of the FET devices [55].

This resonance can become unstable if the gain of the individual section is increased as described in [39]. To ensure circuit stability under large-signal operation, it seems necessary to extend the analysis, studying the evolution of these critical poles versus input power P_{in}. To achieve this goal, a large-signal stability analysis is required. The small-signal current probe is maintained at the same node, and mixer-like harmonic-balance simulations, based on a conversion matrix algorithm, are performed in order to obtain the linearized frequency responses, as in [54], for each power level of the input drive P_{in}.

An input drive at 200 MHz has been arbitrarily chosen for the analysis. The evolution of the real part of the critical poles versus P_{in} is plotted in Figure 6.21, where it can be observed that for $P_{in} > 11$ dBm the poles have a positive real part, indicating an unstable behavior that gives rise to a parametric oscillation. Eventually, the circuit becomes stable again for P_{in} higher than 17.5 dBm. Analogous results are obtained for other frequencies of the input drive.

The results obtained from this large-signal stability analysis confirm that the circuit, originally stable under dc or small-signal regimes, exhibits an undesired parametric oscillation as the input drive increases. To understand the origin of this instability, the evolution of the small-signal gain at 760 MHz versus the input drive (at 200 MHz) has been calculated with conversion matrix simulations. This is equivalent to calculating the S_{21} at 760 MHz in the presence of a large signal at 200 MHz. The result is superimposed in Figure 6.21 where a gain expansion phenomenon is clearly noticeable. This gain expansion is the consequence of the deep Class AB bias required by the circuit in order to achieve high efficiency. The gain expansion at 760 MHz correlates with the evolution of the real part of the critical poles and it is responsible for the undesired oscillation.

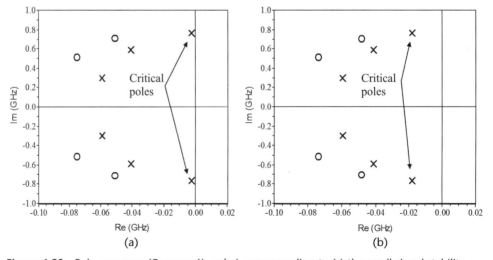

Figure 6.20 Pole-zero map (O: zeros; X: poles) corresponding to (a) the small-signal stability analysis of the original circuit and (b) the small-signal stability analysis of the stabilized circuit.

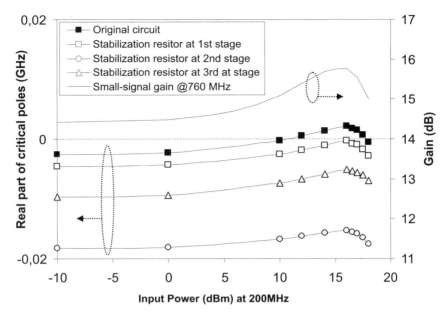

Figure 6.21 Evolution of the real part of critical poles versus P_{in} for the original circuit and the three stabilization configurations; evolution of the small-signal gain at 760 MHz versus P_{in}, and the frequency of the input drive is 200 MHz.

On the one hand, a frequency division by 2 is encountered for f_{in} at about 500 MHz and particular load conditions and input drive levels. As an example, Figure 6.22(a) shows a pole-zero map for $f_{in} = 500$ MHz, $P_{in} = 17.1$ dBm, and $\Gamma_L = -0.75$. The presence of a couple of RHP complex conjugate poles at $f_{in}/2$ reveals the frequency division instability. Figure 6.22(b) plots on the Smith chart the values of load termination Z_L that imply frequency division for $f_{in} = 500$ MHz and $P_{in} = 17.1$ dBm. The detected instabilities have been experimentally verified through

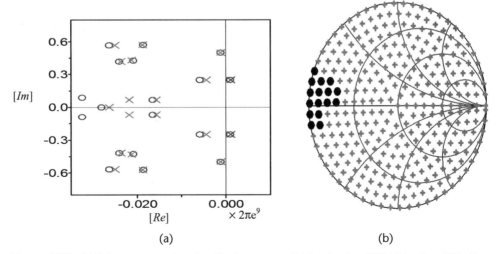

Figure 6.22 (a) Pole-zero map showing the frequency division for $f_{in} = 500$ MHz, $P_{in} = 17.1$ dBm, and $\Gamma_L = 0.75 \angle 180°$. (b) Unstable (circles) and stable (crosses) values of Γ_L for $f_{in} = 500$ MHz and $P_{in} = 17.1$ dBm.

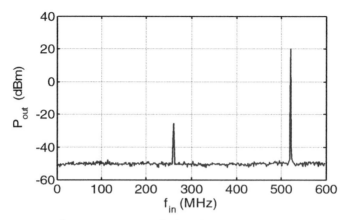

Figure 6.23 Frequency division experimentally found in distributed amplifier for f_{in} = 521 MHz, P_{in} = 16 dBm, and Γ_L = 0.75 $\angle$ 180°.

tuner measurements. The picture of the distributed amplifier board is similar to the one shown later in Figure 6.25, but the stabilization circuit is not included. As an example, Figure 6.23 shows that a measured frequency division by two is obtained for f_{in} = 521 MHz, P_{in} = 16 dBm, and Γ_L = 0.75 $\angle$ 180°.

6.4.2   Circuit Stabilization Technique

As discussed in Section 4.5, the design methodology (i.e., from device selection, synthesizing gate/drain-line elements from device packaged values, and layout optimizations) is applied in this section. The basic design goal of the work is to achieve high efficiency for SDR driver PA applications, and the power operation is ~27 dBm. Therefore, a medium-power device, for example, the LDMOS device (RD01MUS1),[2] is suitable. Low dc supply operation is required for the device, and it is typically about 7.5V. The drain loading effect of the device is not significant, but $X_{opt}(\omega)$ is important since it determines the drain-line cutoff frequency ω_c. A C_{opt} of 8.2 pF is extracted by means of device modeling (with inclusion of packaged properties). Therefore, effective drain-line elements L_i are synthesized according to (4.58), to form the desired ω_c (~0.8 GHz). Dummy drain termination is eliminated to improve the efficiency performance [61]. A simplified design schematic for a three-section LDMOS distributed amplifier applying a nonuniform drain line is shown in Figure 6.24.

The m-derived section is implemented at both terminations of the gate line. Each device is fed with a 5V drain supply voltage. Bias voltage of 2.1V is applied to each gate, resulting in Class AB operation, with quiescent current I_{dq} of ~110 mA. Power performance due to load termination R for n = 3 is investigated. Virtual impedance is seen by the transistor in both directions depending on selection, as illustrated in Section 5.2 and computed in (5.16) through (5.23). Broadband impedance matching employing lowpass LC elements is designed to transform impedance from 13Ω to

---

[2] An LDMOS n-type MOSFET packaged device from Mitsubishi (part number RD01MUS1) was used.

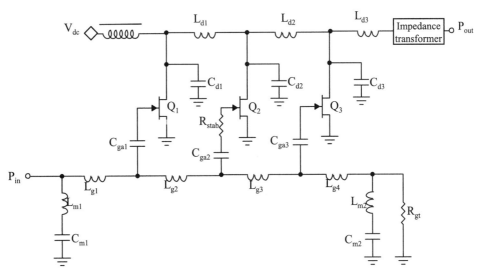

Figure 6.24 Simplified design schematic for a high-efficiency LDMOS distributed amplifier. The second version of the circuit includes the stabilization resistor R_{stab} at the second section of the distributed amplifier.

50Ω over the bandwidth of 100 to 700 MHz. A photograph of the high-efficiency LDMOS DPA is shown in Figure 6.25. The size of the board is 27 mm × 13 mm.

To find the best compromise between stability and circuit performance, it is essential to combine the large-signal stability analysis with a judicious stabilization strategy. Because the parametric oscillation is associated with a gain increase, circuit stability should be improved by introducing series resistors at the gate of any of the three transistors. Since the inclusion of series resistors will impact circuit performance, an evaluation to obtain the best placement and value is needed. Three stabilization configurations are compared here. Each case corresponds to the inclusion of a unique 5Ω series resistor at the gate of one of the transistors. The evolution of the critical poles versus P_{in} (with f_{in} = 200 MHz) for the three cases is plotted in Figure 6.21. As deduced from Figure 6.21, any of the three possibilities is sufficient to maintain the real part of the critical poles at negative values. However, the best stability margin is obtained by placing the series resistor at the second section, while the lowest is achieved when the series resistor is located at the first section.

PAE is the goal performance in this design, the simulated PAE (at P_{in} = 19 dBm) for the three configurations is shown in Figure 6.26, where they are compared to the original design. Results show that placing the resistor at the second or third section has a low impact on PAE. However, the resistor located at the first section offers the strongest degradation of PAE, especially at high frequencies. Therefore, location of the series resistor at the second section was eventually selected because it provides the best stability margin with low PAE degradation. Once the choice is made and before circuit fabrication, an exhaustive stability analysis (varying load and input drive) of the circuit with the series resistor connected at the second section is performed. This analysis serves to confirm circuit stability for any power and frequency of the input drive and for any circuit load. As an example, results of the small-signal stability analysis for the circuit with the stabilization resistor are

Figure 6.25 Photograph of a high-efficiency LDMOS distributed amplifier. The size of the board is 27 mm × 13 mm.

shown in Figure 6.20. We can observe how the stability margin has been increased compared to the original circuit (critical poles shifted leftward).

The three-section LDMOS DPA with stabilization resistor in the second section has been fabricated and characterized. Note that same board as shown in Figure 6.25 is used but the series resistor at gate R_{stab} of the second section is inserted. Measured PAE and gain results are superimposed in Figure 6.26. The rest of the measured performances include P_{out} ~30 dBm and a gain of 10 dB over the frequency range of interest (100 to 700 MHz). Contrary to the original circuit, no parametric oscillations are reported at any frequency and input drive. Figure 6.27 shows an example of measured spectra at 700 MHz, indicating that no oscillation is reported in the measured level.

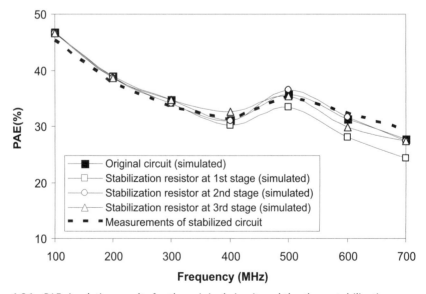

Figure 6.26 PAE simulation results for the original circuit and the three stabilization configurations for P_{in} = 19 dBm. The PAE measurement results are for a stabilized amplifier for P_{in} = 19 dBm.

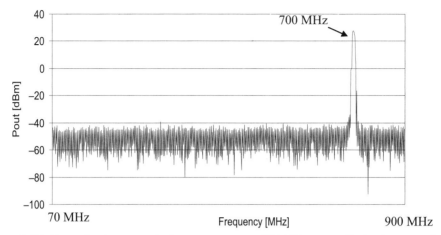

Figure 6.27   Example of measured of power spectra at 700 MHz; no oscillation is reported.

References

[1]   F. Bonani and M. Gilli, "Analysis of Stability and Bifurcations of Limit Cycles in Chuas's Circuit Through the Harmonic-Balance Approach," *IEEE Trans. Circuits and Systems*, Vol. 46, No. 8, pp. 881–890, Aug.1999.

[2]   V. Rizzoli and A. Neri, "State of the Art and Present Trends in Nonlinear Microwave CAD Techniques," *IEEE Trans. Microwave Theory and Techniques*, Vol. 36, No. 2, pp. 343–365, Feb. 1998.

[3]   J. Hale and H. Koçak, *Dynamics and Bifurcations*, New York: Springer-Verlag, 1991.

[4]   R. D. Martinez and R. C. Compton, "A General Approach for the S-Parameter Design of Oscillators with 1 and 2-Port Active Devices," *IEEE Trans. Microwave Theory and Techniques*, Vol. 40, No. 3, pp. 569–574, Mar. 1992.

[5]   R. W. Jackson, "Criteria for the Onset of Oscillation in Microwave Circuits," *IEEE Trans. Microwave Theory and Techniques*, Vol. 40, No. 3, pp. 566–569, Mar. 1992.

[6]   M. Ohtomo, "Stability Analysis and Numerical Simulation of Multidevice Amplifiers," *IEEE Trans. Microwave Theory and Techniques*, Vol. 41, No. 6/7, pp. 983–991, June/July 1993.

[7]   M. Odyniec, "Stability Criteria via S-Parameters," *Proc. 25th European Microwave Conf.*, Vol. 2, Sep. 1995.

[8]   M. Ohtomo, "Stability Analysis and Numerical Simulation of Multidevice Amplifiers," *IEEE Trans. Microwave Theory and Techniques*, Vol. 41, No. 6/7, pp. 983–991, June/July 1993.

[9]   S. Ramberger and T. Merkle, "A Symmetry Device to Speed Up Circuit Simulation and Stability Tests," *IEEE MTT-S Int. Microwave Symp. Digest*, Vol. 2, pp. 967–970, June 2002.

[10]   A. Costantini et al., "Stability Analysis of Multi-Transistor Microwave Power Amplifiers," *Gallium Arsenide Applications Symp. Technical Digest*, Paris, 2000.

[11]   V. Rizzoli and A. Lipparini, "General Stability Analysis of Periodic Steady State Regimes in Nonlinear Microwave Circuits," *IEEE Trans. Microwave Theory and Techniques*, Vol. MTT-33, No. 1, pp. 30–37, Jan. 1985.

[12]   A. Anakabe et al., "Efficient Nonlinear Stability Analysis of Microwave Circuits Using Commercially Available Tools," *32nd European Microwave Conf.*, Milan, Italy, pp. 1017–1020, Sep. 2002.

[13] A. I. Mees and L. O. Chua, "The Hopf Bifurcation Theorem and Its Applications to Nonlinear Oscillations in Circuits and Systems," *IEEE Trans. Circuits and Systems*, Vol. CAS-26, No. 4, pp. 235–254, Apr. 1979.

[14] A. Suárez et al., "Nonlinear Stability Analysis of Microwave Circuits Using Commercial Software," *IEE Electronics Lett.*, Vol. 34, No. 13, pp. 1333–1335, June 1998.

[15] G. E. Bodway, "Two Port Power Flow Analysis Using Generalized Scattering Parameters," *Microwave J.*, Vol. 10, No. 6, pp. 61–69, May 1967.

[16] E. R. Hauri, "Overall Stability Factor of Linear Two-Ports in Terms of Scattering Parameters," *IEEE J. Solid-State Circuits*, Vol. 6, No. 12, pp. 413–415, Dec. 1971.

[17] D. M. Pozar, *Microwave Engineering*, New York: Addison-Wesley,1990.

[18] A. Suárez and R. Quéré, *Stability Analysis of Nonlinear Microwave Circuits*, Norwood, MA: Artech House, 2003.

[19] P. Bianco, G. Ghione, and M. Pirola, "New Simple Proofs of the Two-Port Stability Criterion in Terms of the Single Stability Parameters µ1 (µ2)," *IEEE Trans. Microwave Theory and Techniques*, Vol. 36, No. 2, pp. 343–365, June 2001.

[20] F. Centurelli et al., "A Synthesis-Oriented Approach to Design Microwave Multidevice Amplifiers with a Prefixed Stability Margin," *IEEE Trans. Microwave and Guided Wave Lett.*, Vol. 49, No. 6, pp. 1073–1076, Mar. 2000.

[21] F. Centurelli et al., "A Synthesis-Oriented Conditional Stability Criterion for Microwave Multidevice Circuits with Complex Termination Impedance," *IEEE Trans. Microwave and Guided Wave Lett.*, Vol. 10, No. 11, pp. 460–462, Nov. 2000.

[22] J. M. Rollett, "Stability and Power-Gain Invariants of Linear two-ports," *IRE Trans. Circuit Theory*, Vol. 9, No. 1, pp. 29–32, Mar. 1962.

[23] D. Woods, "Reappraisal of the Unconditional Stability Criteria for Active 2-Port Networks in Terms of S Parameters," *IEEE Trans. Circuits and Systems*, Vol. 23, No. 2, pp. 73–81, Feb. 1976.

[24] A. Platzker, W. Struble, and K. T. Hetzler, "Instabilities Diagnosis and the Role of K in Microwave Circuits," *IEEE MTT-S Int. Microwave Symp. Dig.*, Vol. 3, pp. 1185–1188, June 1993.

[25] L. Samoska et al., "On the Stability of Millimeter-Wave Power Amplifiers," *IEEE MTT-S Int. Microwave Symp. Dig.*, Vol. 1, pp. 429–432, June 2002.

[26] R. G. Freitag et al., "Stability and Improved Circuit Modeling Considerations for High Power MMIC Amplifiers," *IEEE Microwave and Millimeter-Wave Monolithic Circuits Symp. Dig.*, pp. 125–128, May 1998.

[27] R. G. Freitag, "A Unified Analysis of MMIC Power Amplifier Stability," *IEEE MTT-S Int. Microwave Symp.*, Vol. 1, pp. 297–300, June 1992.

[28] M. Ohtomo, "Proviso on the Unconditional Stability Criteria for Linear Two Port," *IEEE Trans. Microwave Theory and Techniques*, Vol. 43, No. 5, pp. 1197–1200, May 1995.

[29] J. G. Kassakian and D. Lau, "An Analysis and Experimental Verification of Parasitic Oscillations in Paralleled Power MOSFET's," *IEEE Trans. Electron Devices*, Vol. ED-31, No. 7, pp. 959–963, July 1984.

[30] H. W. Bode, *Network Analysis and Feedback Amplifier Design*, New York: Van Nostrand, 1945.

[31] S. Ramberger and T. Merkle, "A Symmetry Device to Speed Up Circuit Simulation and Stability Tests," *IEEE MTT-S Int. Microwave Symp.*, Vol. 2, pp. 967–970, June 2002.

[32] A. Costantini et al., "Stability Analysis of Multi-Transistor Microwave Power Amplifiers," *Gallium Arsenide Applications Symp. Technical Digest*, Paris, 2000.

[33] M. L. Edwards and J. H. Sinsky, "A New Criterion for Linear 2-Port Stability Using a Single Geometrically Derived Parameter," *IEEE Trans. Microwave Theory and Techniques*, Vol. 40, No. 12, pp. 2303–2311, Dec. 1992.

[34]   W. Struble and A. Platzker, "A Rigorous Yet Simple Method for Determining Stability of Linear N-Port Networks," *15th Gallium Arsenide Integrated Circuit (GaAs IC) Symp. Tech. Dig.*, pp. 251–254, Oct. 1993.

[35]   G. Gonzalez, *Microwave Transistor Amplifiers—Analysis and Design*, Upper Saddle River, NJ: Prentice Hall, 1997.

[36]   S. Mons et al., "A Unified Approach for the Linear and Nonlinear Stability Analysis of Microwave Circuits Using Commercially Available Tools," *IEEE Trans. Microwave Theory and Techniques*, Vol. 47, No. 12, pp. 2403–2409, Dec. 1999.

[37]   A. Anakabe et al., "Efficient Nonlinear Stability Analysis of Microwave Circuits Using Commercially Available Tools," *32nd European Microwave Conf.*, Milan, Italy, pp. 1017–1020, Sep. 2002.

[38]   H. Nyquist, "Regeneration Theory," *Bell Syst. Tech. J.*, Vol. 11, pp.126–147, 1932.

[39]   J. M. Collantes, "Large-Signal Stability Analysis through Pole-Zero Identification," *Practical Analysis, Stabilization, and Exploitation of Nonlinear Dynamics in RF, Microwave and Optical Circuits*, Workshop at MS2007, IEEE, 2007.

[40]   A. Anakabe et al., "Detecting and Avoiding Odd-Mode Parametric Oscillations in Microwave Power Amplifiers," *Int. Journal on RF and Microwave Computer-Aided Engineering*, Vol. 15, No. 5, pp. 469–478, Sep. 2005.

[41]   A. Anakabe et al., "Analysis and Elimination of Parametric Oscillation in Monolithic Power Amplifiers," *IEEE MTT-S Int. Microwave Symp. Dig.*, Vol. 3, pp. 2181–2184, June 2002.

[42]   A. Anakabe et al., "Harmonic Balance Analysis of Digital Frequency," *IEEE Microwave Wireless and Component Lett.*, Vol. 12, No. 8, pp. 287–289, 2002.

[43]   J. M. Collantes and A. Suárez, "Period-Doubling Analysis and Chaos Detection Using Commercial Harmonic Balance Simulators," *IEEE Trans. Microwave Theory and Techniques*, Vol. 48, No. 4, pp. 574–581, Apr. 2000.

[44]   A. Suárez and R. Quéré, *Stability Analysis of Nonlinear Microwave Circuits*, Norwood, MA: Artech House, 2003.

[45]   K. B. Niclas et al., "On Theory Performance of Solid-State Microwave Distributed Amplifiers," *IEEE Trans. Microwave Theory Techniques*, Vol. MTT-31, No. 6, pp. 447–456, June 1983.

[46]   P. Gamand, "Analysis of the Oscillation Conditions in Distributed Amplifiers," *IEEE Trans. Microwave Theory and Techniques*, Vol. MTT-37, No. 3, pp. 637–640, Mar. 1989.

[47]   A. Anakabe et al., "Analysis of Odd Mode Parametric Oscillations in HBT Multistage Power Amplifiers," *11th European Gallium Arsenide and Other Compound Semiconductors Application Symp. (GaAs 2003)*, Munich, Germany, Oct. 2003.

[48]   F. Bonani and M. Gilli, "Analysis of Stability and Bifurcations of Limit Cycles in Chuas's Circuit Through the Harmonic-Balance Approach," *IEEE Trans. Circuits and Systems*, Vol. 46, No. 8, pp. 881–890, Aug. 1999.

[49]   S. A. Maas, *Nonlinear Microwave Circuits*, New York: IEEE Press, 1997.

[50]   P. Bolcato et al., "A Unified Approach of PM Noise Calculation in Large RF Multitone Autonomous Circuits," *2000 IEEE MTT-S Int. Microwave Symp. Digest*, Vol. 1, pp. 417–420, June 2000.

[51]   J. P. Fraysse et al., "2W Ku-Band Coplanar MMIC HPA using HBT for Flip-Chip Assembly," *2002 IEEE MTT-S Int. Microwave Symp. Digest*, Vol. 1, pp. 441–444, June 2002.

[52]   J. Portilla, H. García, and E. Artal, "High Power-Added Efficiency MMIC Amplifier for 2.4 GHz Wireless Communications," *IEEE Journal of Solid-State Circuits*, Vol. 34, No. 1, pp. 120–123, Jan. 1999.

[53]   R. Pintelom and J. Schoukens, *System Identification: A Frequency Domain Approach*, New York, IEEE Press, 2001.

[54] A. Mallet et al., "STAN: An Efficient Tool for Nonlineal Stability Analysis," *RF and Hyper Europe 2004, Microwave Power Amplifier Workshop*, Paris, Mar. 2004.

[55] A. Suárez et al., "Nonlinear Stability Analysis of Microwave Circuits Using Commercial Software," *IEE Electronics Lett.*, Vol. 34, No. 13, pp. 1333–1335, June 1998.

[56] K. Narendra et al., "Parametric Oscillations in Distributed Power Amplifiers," *Electronics Lett.*, Vol. 45, No. 25, pp. 1325–1326, Dec. 2009.

[57] A. Anakabe et al., "Automatic Pole-Zero Identification for Multivariable Large-Signal Analysis of RF and Microwave Circuits," *32nd European Microwave Conf.*, Paris, France, pp. 477–480, Sep. 2010.

[58] J. Jugo et al., "Closed-Loop Stability Analysis of Microwave Amplifiers," *Electronics Lett.*, Vol. 37, No. 4, pp. 226–228, Feb. 2001.

[59] S. Basu, S. A. Maas, and T. Itoh, "Stability Analysis for Large Signal Design of a Microwave Frequency Doubler," *IEEE Trans. Microwave Theory and Techniques*, Vol. 43, No. 12, pp. 2890–2898, Dec. 1995.

[60] R. Quéré et al., "Large Signal Design of Broadband Monolithic Microwave Frequency Dividers and Phase-Locked Oscillators," *IEEE Trans. Microwave Theory and Techniques*, Vol. 41, No. 11, pp. 1928–1938, Nov. 1993.

[61] K. Narendra et al., "pHEMT Distributed Power Amplifier Adopting Broadband Impedance Transformer," *Microwave Journal*, Vol. 51, pp.76–83, June 2013.

CHAPTER 7

Implementation of Distributed Amplifiers

The potential of traveling wave or distributed amplification for obtaining power gains over wide frequency bands was recognized as early as the mid-1930s when it was found that the gain-bandwidth performance is greatly affected by the capacitance and transconductance of the conventional vacuum tube [1]. However, the first theoretical analysis and its practical verification were obtained for very broadband vacuum-tube amplifiers more than a decade later [2, 3]. The basic concept was based on the idea of combining the interelectrode capacitances of the amplifying vacuum tubes with series wire inductors to form two lumped-element artificial transmission lines coupled by the tube transconductances. As a result, the distributed amplifier overcomes the difficulty of a conventional amplifier, whose frequency limit is determined by a factor that is proportional to the ratio of the transconductance of the tube to the square root of the product of its input grid-cathode and output anode-cathode capacitances. It overcomes the difficulty by paralleling the tubes in a special way that allows the capacitances of the tubes to be separated while the transconductances can be added almost without limit and not affect the input and output of the device. Since the grid-cathode and anode-cathode capacitances form part of lowpass filters that can be made to have a substantially uniform response up to filter cutoff frequencies, whose value can be conveniently set within a wide range by a suitable choice of the values of the external inductor coils, it became possible to provide amplification over much wider bandwidths than was achievable with conventional amplifiers.

7.1 Vacuum-Tube Distributed Amplifier

Figure 7.1 shows the basic circuit structure of a vacuum-tube distributed amplifier [2]. Here, an artificial transmission line consisting of the grid-cathode capacitances C_g and inductances between tubes L_g is connected between input terminals 1–1 and 2–2, with the characteristic impedance of the grid line defined as $Z_{01} = \sqrt{L_g/C_g}$. If the proper terminating impedance is connected to terminals 2–2 and if this transmission line is assumed to be lossless, then it can be shown that the driving-point impedance at terminals 1–1 is independent of the number of tubes so connected. In a similar fashion, a second transmission line is formed by making use of the anode-cathode capacitances C_p to shunt another set of coil inductances L_p, resulting in the similar characteristic impedance of the anode (or plate) line independent of the number of tubes as $Z_{02} = \sqrt{L_p/C_p}$. Impedances connected to terminals 3–3 and 4–4 are intended to be equal to the characteristic impedance of the anode line. The impedance connected to terminals 2–2 is called the grid termination, the impedance connected to terminals 3–3 is called the reverse termination, and the impedance connected to output terminals 4–4 is called the anode termination. These two artificial transmission lines are made to have identical velocities of propagation. The bandwidth of a distributed amplifier is determined by the cutoff frequency of the artificial line. In general, the higher this cutoff frequency, the lower the characteristic impedance of the line, and hence the less the voltage gain.

A signal generator connected to input terminals 1–1 will cause a wave to travel along the grid line. As this wave reaches the grids of the distributed tubes, currents will flow in the anode circuits of the tubes. Each tube will then send waves in the anode line in both directions. If the reverse termination is perfect, the waves that travel to the left in the anode line will be completely absorbed, and will not contribute to the output signal. The waves that travel to the right in the anode line are added all in phase, and the output voltage is thus directly proportional to the number of tubes. Hence, the effective transconductance of such a distributed stage may be increased to any desired limit, no matter how low the gain of each tube (or section) is (even if it less than unity). As long as the gain per section is greater than the transmission-line loss of the section, the signal in the anode line will increase and can be made enough large by using a sufficient number of tubes.

The total voltage gain A of the distributed amplifier consisting of n sections is written as

$$A = \frac{ng_m}{2}\sqrt{Z_{01}Z_{02}} \tag{7.1}$$

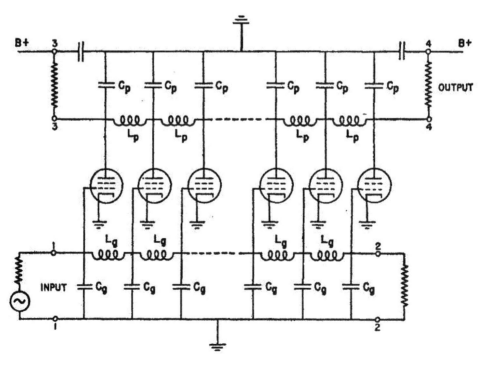

Figure 7.1   Basic structure of vacuum-tube distributed amplifier.

where g_m is the tube transconductance, Z_{01} is the characteristic impedance of the grid line, and Z_{02} is the characteristic impedance of the anode line. Assuming that both transmission lines are identical,

$$Z_{0\pi} = Z_{01} = Z_{02} = \frac{R}{\sqrt{1 - x_k^2}} \qquad (7.2)$$

where $x_k = f/f_c$, $R = 1/\pi f_c C$ ($C = C_g = C_p$), f is the frequency, and f_c is the cutoff frequency of the transmission line. Equation (7.1) can be rewritten under conditions given by (7.2) as

$$A = \frac{g_m R}{2} \frac{n}{\sqrt{1 - x_k^2}} \qquad (7.3)$$

where the second factor shows that the gain of the simple structure of the distributed amplifier shown in Figure 7.1 will be a function of frequency. This is due to the fact that the midshunt characteristic impedance of a lowpass constant-k filter section rises rapidly as the cutoff frequency is approached. This, in turn, causes the gain of the amplifier to increase sharply near cutoff, producing a large undesired peak.

Several methods can be used to eliminate this undesired peak and improve the frequency response, and one of them is to use adjacent coils, which are wound

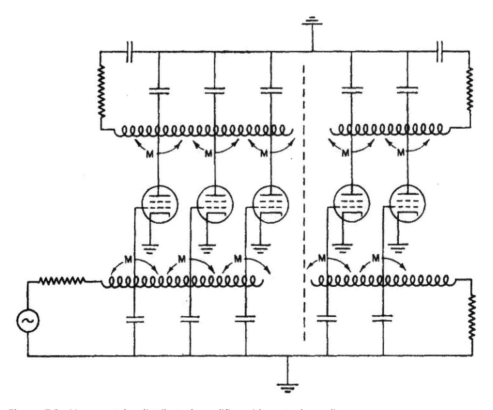

Figure 7.2 Vacuum-tube distributed amplifier with mutual coupling.

on the same form and in the same direction with large coupling coefficient M, as shown in Figure 7.2. In this case, each coupling section can be equated to the usual m-derived filter section. As a result, the total voltage gain A and phase shift ϕ for a distributed amplifier with n tubes can be written, respectively, as

$$A = \frac{g_m R_0}{2} \frac{nm^3}{\left[m^2 - (1-m^2)x_k^2\right]\sqrt{m^2 - x_k^2}} \tag{7.4}$$

$$\phi = 2n\tan^{-1}\frac{mx_k}{\sqrt{m^2 - x_k^2}} \tag{7.5}$$

where $R_0 = 1/\pi f_0 C_g$; $x_k = f/f_0$; $f_0 = g_m/\pi\sqrt{C_g C_p}$, which is Wheeler's bandwidth-index frequency; and m is the design parameter selected for desired tolerance [2]. Note that the presence of the parasitic capacitance distributed throughout the transmission-line coil windings results in lowering the amplifier cutoff frequency and in altering the impedance of the transmission lines, thus making it difficult to terminate properly [3]. An improvement of the gain/frequency response near the cutoff frequency can be achieved by the insertion of extra sections into the grid or anode line, by the use of a network whose image (or terminating) impedance (at a shunt-capacitance point) falls to zero at the cutoff frequency, or by the use of lowpass networks containing resistive elements [4, 5]. In addition, the rise of gain can be eliminated by having different propagation functions for the sections in a distributed amplifier [6].

To improve both the gain/frequency characteristic by making it flatter and the phase-shift/frequency characteristic by making it more linear, a staggering principle can be applied when the lumped lines in the distributed amplifier are arranged such that the anode-line traveling wave and the grid-line traveling wave are not in phase at corresponding points along the lines [7]. In a distributed amplifier embodying the constant-k LC filter network as the elements of the lumped lines, the stagger is introduced by making the cutoff frequency of the grid line a little higher than that of the anode line. At a given frequency, a line with a higher cutoff frequency produces a smaller phase than one with a lower cutoff frequency. The difference between the phase shifts produced by the two lines increases continuously as the frequency is increased.

The overall gain characteristic and phase shift of a staggered n-tube distributed amplifier with constant-k lowpass LC filter network is given, respectively, by

$$A = g_m \frac{Z_{0\pi}}{2} \frac{\sin n\psi/2}{\sin \psi/2} \tag{7.6}$$

$$\phi = (n-1)\left(\sin^{-1}x_k - \sin^{-1}qx_k\right) \tag{7.7}$$

where
$\psi = \theta_p - \theta_g$
$\theta_p = 2\sin^{-1}x_k$ = phase shift introduced by the individual section of the anode line

$\theta_g = 2 \sin^{-1} q x_k$ = phase shift introduced by the individual section of the grid line

$x_k = f/f_{cp}$

$q = f_{cp}/f_{cg}$

f_{cp} = anode-line cutoff frequency

f_{cg} = grid-line cutoff frequency.

In order for the amplitude to be able to fall to zero just below the cutoff frequency of the anode line, the last factor in (7.6) has to vanish at f_{cp}, resulting in $(n\psi/2) = \pi$ at $x_k = 1$. Hence, the ratio between grid-line and anode-line cutoff frequencies can be obtained as

$$q = \sin\left(\frac{\pi}{2} - \frac{\pi}{n}\right) \tag{7.8}$$

Similarly, the gain and phase-shift characteristics of a distributed amplifier with m-derived filter sections can be improved by staggering with different m-values of the anode and grid lines when the grid line has the larger m [7]. In practical implementation of a vacuum-tube distributed amplifier, the distributed cascode circuit can be used to minimize degeneration at higher frequencies caused by the common lead inductances and feedback capacitances of the tubes [8]. However, if the attenuation in the grid line due to grid loading is substantial, the staggering of the lines may not necessarily provide much improvement in the gain characteristic, although the phase-shift characteristic is approximately the same as for the lossless case [9]. Generally, the effects of staggering the lines in distributed amplifiers based on vacuum tubes and transistors are different because their electrical behavior is characterized by different equivalent circuit representations [10].

If the attenuation in the grid line is neglected in order to obtain a general analysis of distributed amplifiers operating as large-signal devices, all of the tubes' currents add in phase as they progress toward the output end of the amplifier. Consequently, the output current is just n times the current that one tube produces in the output resistance and the output power varies as the square of the number of tubes. This is in contrast to the dc power input, which varies directly with the number of tubes. As a result, it is desirable to use as large a number of tubes as possible to increase efficiency, which increases directly with the number of tubes.

In a distributed amplifier designed for flat frequency response, the output power is constant with frequency as long as the driving power remains constant. Hence, the output power calculated for low frequencies should apply for any frequency in the operating region. Figure 7.3 shows the instantaneous values of anode voltage and anode current for idealized tube voltage-ampere characteristics in a Class AB operation [11]. In this case, the instantaneous anode voltage $v_a(\omega t)$ varies according to

$$v_a(\omega t) = \begin{cases} V_m(1 - \cos \omega t) & -\theta \le \omega t < \theta \\ V_m(1 - \cos \theta) & \theta \le \omega t < 2\pi - \theta \end{cases} \tag{7.9}$$

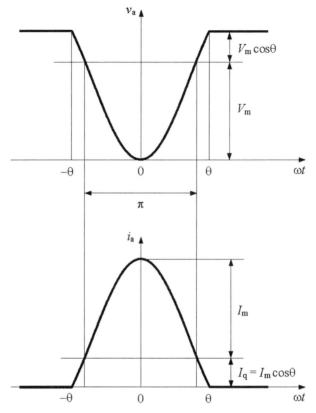

Figure 7.3 Instantaneous values of anode voltage and anode current in Class AB operation.

where the conduction angle 2θ indicates the part of the RF current cycle during which a device conduction occurs.

The instantaneous anode current $i_a(\omega t)$ can be written as

$$
i_a(\omega t) = \begin{cases} I_m(\cos \omega t - \cos \theta) & -\theta \leq \omega t < \theta \\ 0 & \theta \leq \omega t < 2\pi - \theta \end{cases} \tag{7.10}
$$

with a quiescent current $I_q = I_m \cos\theta$.

The dc value of the anode voltage must be equal to the anode supply voltage V_a as

$$
\begin{aligned}
V_a &= \frac{V_m}{2\pi} \int_{-\theta}^{\theta} (1 - \cos \omega t) d\omega t + (2\pi - \theta)(1 - \cos\theta) \\
&= \frac{V_m}{\pi} (\pi + \theta \cos\theta) - \pi \cos\theta - \sin\theta
\end{aligned} \tag{7.11}
$$

The dc anode current I_0 is the average value of the instantaneous anode current given by

$$
I_0 = \frac{I_m}{2\pi} \int_{-\theta}^{\theta} (\cos \omega t - \cos\theta) d\omega t = \frac{I_m}{2\pi} (\sin\theta - \theta\cos\theta) \tag{7.12}
$$

The peak value of the fundamental component of the anode voltage is written as

$$V_1 = \frac{V_m}{\pi}\left[2\int_0^\theta (1 - \cos\omega t)\cos\omega t\, d\omega t + \int_\theta^{2\pi-\theta} (1 - \cos\theta)\cos\omega t\, d\omega t \right]$$

$$= \frac{V_m}{\pi}(\sin\theta - \theta) \tag{7.13}$$

The peak value of the fundamental component of the anode current is obtained by

$$I_1 = \frac{I_m}{2\pi}\int_{-\theta}^{\theta}(\cos\omega t - \cos\theta)\cos\omega t\, d\omega t = \frac{I_m}{2\pi}(\theta - \sin\theta) \tag{7.14}$$

The efficiency η is calculated as the ratio of the output fundamental power to the input dc power. At low frequencies the anode circuit of the distributed amplifier with resistive terminations is only 50% efficient since half the output power from tubes is dissipated in the reverse termination. Consequently, the output power is one-half the product of the fundamental current and fundamental voltage. As a result,

$$\eta = \frac{1}{2}\left(\frac{V_1 I_1}{2}\right) \Big/ \left(V_a I_0\right)$$

$$= \frac{1}{8}\frac{(\theta - \sin\theta)^2}{(\pi + \theta\cos\theta - \pi\cos\theta - \sin\theta)(2\sin\theta - \theta\cos\theta)} \tag{7.15}$$

From (7.15), it follows that the maximum theoretical efficiency of the distributed amplifier is derived to be about 30% and occurs with an anode current conduction angle of about 225° [11]. However, since the average value of the dc current drawn from the anode supply will decrease with frequency supply, the distributed amplifier will tend to become more efficient as the operating frequency increases. The change in efficiency is related to the change in the average plate current from no-signal conditions to full-signal conditions.

7.2 Microwave GaAs FET Distributed Amplifiers

The distributed amplifier using hybrid technology with lumped elements was first investigated based on silicon bipolar transistors in 1959 [12], MOSFETs in 1965 [13], and MESFETs in 1968–1969 [14, 15]. In circuits that employ FETs as active elements, the gate and drain loading plays a very significant role in the operation and high-frequency performance of the distributed amplifier. Therefore, it is only recently, with the availability of good-quality microwave GaAs FETs, that distributed amplifiers have again become popular at microwave frequencies [16]. In this case, GaAs FETs are used as the active devices, and the input and output lines can represent the periodically loaded microstrip transmission lines. With such an arrangement, the factors degrading the expected performance such as device input and output resistances and capacitances are either completely eliminated or their effect is included in the design. The resultant distributed amplifiers exhibit very low sensitivities to process variations and are relatively easy to design and simulate. In the early 1980s, the technology of distributed amplification was further improved by implementing the silicon and semi-insulating GaAs MMICs, which provide low loss, small size, high reliability, circuit design flexibility, and a high level of integration. The first 0.5- to 14-GHz monolithic GaAs FET distributed (or traveling-wave) amplifier was designed in 1981 [17].

7.2.1 Basic Configuration with Microstrip Lines

The simplified schematic representation of a four-section GaAs FET distributed amplifier is shown in Figure 7.4, where the microstrip lines are periodically loaded with the complex gate and drain impedances of the devices, thus forming lossy transmission-line structures of different characteristic impedance and propagation constant [18]. An RF signal applied at the input end of the gate line travels down the line to the other end, where it is absorbed by the terminating impedance connecting at the end of the gate line, which includes the gate dummy resistor R_1. However, a significant portion of the signal is proportionally dissipated by the gate circuits of the individual FETs along the way. The input signal sampled by the gate circuits at different phases (and generally at different amplitudes) is transferred to the drain line through the FET transconductances. If the phase velocity of the signal at the drain line is identical to the phase velocity of the gate line, then the signals on the drain line add, forming a traveling wave. The addition will be in phase only for the forward-traveling signal. Any signal that travels backward, and is not fully canceled by the out-of-phase additions, will be absorbed by the terminating impedance connecting at the end of the drain line, which includes the drain dummy resistor R_2. The gate and drain capacitances of the FET effectively become part of the gate and drain transmission lines, while the gate and drain resistances introduce loss on these lines.

In conventional power amplifiers, it is impossible to increase the gain-bandwidth product by just paralleling the FETs because the resulting increase in transconductance g_m is compensated for by the corresponding increase in the input and output capacitances. The distributed power amplifier overcomes this problem by adding the

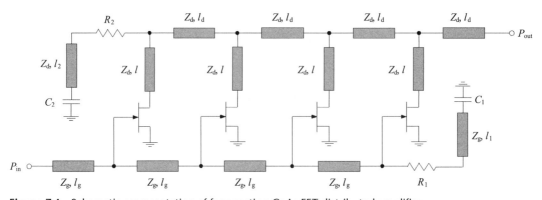

Figure 7.4 Schematic representation of four-section GaAs FET distributed amplifier.

individual device transconductances without adding their input and output capacitances, which are now the parts of the artificial gate and drain transmission lines, respectively. If the spacing between FETs is small compared to the wavelength, the characteristic impedances of the gate and drain lines shown in Figures 7.5(a) and (b), respectively, can be approximated as

$$Z_g = \sqrt{\frac{L_g}{C_g + \dfrac{C_{gs}}{l_g}}} \tag{7.16}$$

$$Z_d = \sqrt{\frac{L_d}{C_d + \dfrac{C_{ds}}{l_d}}} \tag{7.17}$$

where

C_{gs} = gate-source capacitance

C_{ds} = drain-source capacitances of the unit FET cell

l_g and l_d = lengths of the unit gate-line and drain-line sections, respectively

L_g, C_g and L_d, C_d = per-unit-length inductance and capacitance of the gate and drain lines, respectively.

Here, the effects of the gate resistance R_{gs} and drain resistance R_{ds} are neglected. Note that the characteristic impedance expressions in (7.16) and (7.17) are clearly independent of the number of FETs used in the circuit.

As a result, the amplifier available gain G for an n-section circuit found by approximating the gate and drain lines as continuous structures can be written as

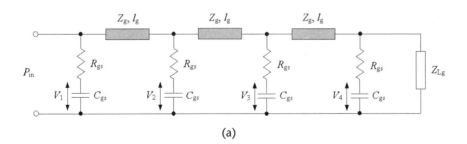

(a)

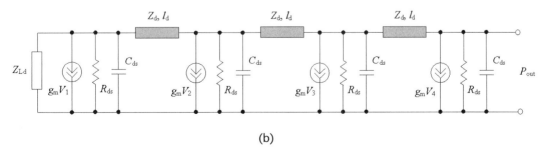

(b)

Figure 7.5 Simplified equivalent-circuit diagram of FET distributed amplifier.

$$G = \frac{g_m^2 Z_g Z_d}{4} \left| \frac{\exp(-n\gamma_g l_g) - \exp(-n\gamma_d l_d)}{\exp(-\gamma_g l_g) - \exp(-\gamma_d l_d)} \right|^2 \tag{7.18}$$

where the propagation constants γ_g and γ_d are simplified using small-loss approximation as

$$\gamma_g = \frac{\omega^2 C_{gs}^2 R_{gs}}{2 l_g} Z_g + j\omega \sqrt{L_g \left(C_g + \frac{C_{gs}}{l_g} \right)} \equiv \alpha_g + j\beta_g \tag{7.19}$$

$$\gamma_d = \frac{1}{2 R_{ds} l_d} Z_d + j\omega \sqrt{L_d \left(C_d + \frac{C_{ds}}{l_d} \right)} \equiv \alpha_d + j\beta_d \tag{7.20}$$

Under normal operating conditions, the signals in the gate and drain lines are near synchronism when $\beta_g l_g \cong \beta_d l_d$, and (7.18) can be simplified for $Z_g \cong Z_d \equiv Z_0$ and small losses to

$$G = \frac{g_m^2 Z_0^2}{4} \frac{\left[\exp(-n\alpha_g l_g) - \exp(-n\alpha_d l_d) \right]^2}{\left(\alpha_g l_g - \alpha_d l_d \right)^2} \tag{7.21}$$

from which it follows that, as the number of unit cells or sections n is increased, the available gain G does not increase monotonically and approaches zero in limit as n increases.

For values of $n\alpha_g l_g \leq 1$ and when the drain-line losses are negligible compared to the gate-line losses, (7.21) can be rewritten as

$$G \cong \frac{n^2 g_m^2 Z_0^2}{4} \left(1 - \frac{n\alpha_g l_g}{2} - \frac{n^2 \alpha_g^2 l_g^2}{6} \right)^2 \tag{7.22}$$

which means that, in this operating condition, the available gain G can be made proportional to n^2. In this case, by using the expression for the gate-line attenuation constant α_g given in (7.19), one can find that

$$n Z_0 R_{gs} \omega^2 C_{gs}^2 \leq 2 \tag{7.23}$$

which defines the upper limit to the total gate periphery that can be used in a practical distributed amplifier or the maximum number of unit sections for a given FET device. Similar results can be obtained by applying a theoretical analysis based on matrix technique by employing a finite number of active and passive circuit elements [19].

In view of the gate-line and drain-line losses, from (7.21) it follows that the power gain of a distributed amplifier approaches zero as $n \to \infty$. This happens due to the fact that the input voltage on the gate line decays exponentially, so the input signal does not reach FETs at the end of the amplifier gate line and, similarly, the amplified signals from FETs near the beginning are attenuated along the drain line. This implies that, for a given set of FET parameters, there will be an optimum value of n that maximizes the power gain of a distributed amplifier. Hence, by differentiating (7.21) with respect to n and setting the result to zero, the optimum number of sections n_{opt} can be determined from

$$n_{opt} = \frac{1}{\alpha_g l_g - \alpha_d l_d} \ln\left(\frac{\alpha_g l_g}{\alpha_d l_d}\right) \tag{7.24}$$

which depends on the device parameters, line lengths, and frequency through the attenuation constants given in (7.19) and (7.20). For example, it is necessary to calculate the power gain of a distributed amplifier operated from 1 to 12 GHz, with a maximum gain at 10 GHz. Assuming that $\omega R_{gs} C_{gs} = 0.5$, $Z_0/R_{gs} = 4$, and $Z_0/R_{ds} = 0.2$ for the specified device parameters and $Z_g = Z_d = Z_0 = 50\Omega$, from (7.19) and (7.20) it follows that $\alpha_g l_g = 0.5$ and $\alpha_d l_d = 0.1$, resulting in $n_{opt} = \ln(0.5/0.1)/(0.5 - 0.1) = 4.0$ or four sections.

Figure 7.6 shows the frequency dependence of a power gain for different numbers of FET unit cells or sections [18]. The figure clearly shows that there is an optimum number n_{opt} that provides a maximum frequency bandwidth with minimum gain variations and reasonable power gain. For example, a power gain of 9±1 dB over a bandwidth of 1 to 13 GHz was obtained for a four-cell distributed amplifier with a total GaAs FET gate periphery of 4 × 300 μm. Note that the resistive part of the gate loading typically results in a 3-dB gain reduction. The effect of the drain loading is not as significant; however, the power gain can be increased by about 1 dB for increased values of the drain loading resistance.

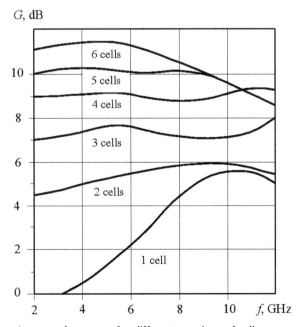

Figure 7.6 Power gain versus frequency for different numbers of cells.

Replacing the conventional GaAs MESFETs with high-performance HEMT devices in a five-section monolithic distributed amplifier will result in significant improvements in the power gain and noise figure. For example, use of 0.35-μm-gate-length HEMTs provides a low-noise figure that is 2 dB lower and a power gain that is 2.5 dB higher than achieved using 0.5-μm-gate-length MESFET devices in a frequency range from 2 to 20 GHz [20].

7.2.2 Basic Configuration with Lumped Elements

The equivalent gate and drain artificial transmission lines based on lumped inductors and capacitors are shown in Figures 7.7(a) and (b), respectively [21, 22]. For a constant-k type transmission line, the phase velocity is a well-known function of the cutoff frequency f_c of the line. By requiring the phase shift between each gate-line and drain-line section to be equal, the cutoff frequency for the gate transmission line $f_{cg} = 1/2\pi R_{gs}C_{gs}$ and the cutoff frequency for the drain transmission line $f_{cd} = 1/2\pi R_{ds}C_{ds}$ must also be equal. As a result, the available gain G of the lumped distributed amplifier can be written as

$$G = \frac{g_m^2 Z_{01} Z_{02} \sinh^2\left[\frac{n}{2}\left(\alpha_d - \alpha_g\right)\right]\exp\left[-n\left(\alpha_d + \alpha_g\right)\right]}{4\left[1 + \left(\frac{f}{f_{cg}}\right)^2\right]\left[1 - \left(\frac{f}{f_c}\right)^2\right]\sinh^2\left[\frac{1}{2}\left(\alpha_d - \alpha_g\right)\right]} \tag{7.25}$$

where α_g and α_d are the attenuations on the gate and drain lines per section, and $Z_{01} = \sqrt{L_g/C_{gs}}$ and $Z_{02} = \sqrt{L_d/C_{ds}}$ are the characteristic impedances of the gate and drain line, respectively.

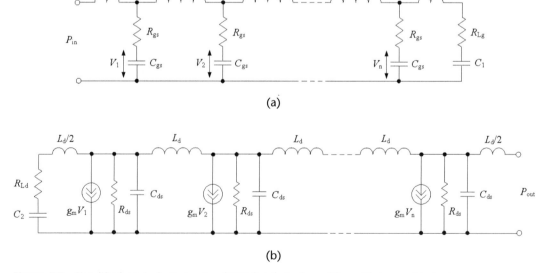

(a)

(b)

Figure 7.7 Simplified equivalent circuits of FET distributed amplifier with lumped inductors.

The attenuation on the gate and drain lines is the critical factor controlling the frequency response of a distributed amplifier. When attenuation per section is sufficiently small, the corresponding attenuations on the gate and drain lines can be given by

$$\alpha_g = \frac{x_k^2}{\sqrt{1 - \left[1 - \left(\frac{f_c}{f_{cg}}\right)^2\right]x_k^2}} \frac{f_c}{f_{cg}} \qquad (7.26)$$

$$\alpha_d = \frac{1}{\sqrt{1 - x_k^2}} \frac{f_{cd}}{f_c} \qquad (7.27)$$

where $x_k = f/f_c$ is the normalized frequency and $f_c = 1/\pi\sqrt{L_g C_{gs}} = 1/\pi\sqrt{L_d C_{ds}}$ [22]. From (7.26) and (7.27), it follows that the gate-line attenuation is more sensitive to frequency than the drain-line attenuation, and the drain-line attenuation does not vanish in the low-frequency limit, unlike attenuation in the gate line. Therefore, the frequency response of the distributed amplifier can be expected to be predominantly controlled by the attenuation on the gate line. Generally, the attenuation on the gate and drain lines can be decreased by making f_c/f_{cg} and f_{cd}/f_c small when the transistor having high f_{cg} and low f_{cd} has to be chosen for a given f_c.

The maximum gain-bandwidth product of the distributed amplifier can be estimated by

$$\sqrt{G_0}f_{1dB} \approx 0.8f_{max} \qquad (7.28)$$

where G_0 is the low-frequency available gain of the amplifier, f_{1dB} is the frequency at which the power gain of the amplifier falls below G_0 by 1 dB, and

$$f_{max} = \frac{g_m}{4\pi C_{gs}}\sqrt{\frac{R_{ds}}{R_{gs}}} \qquad (7.29)$$

is the frequency at which the maximum available gain (MAG) of the FET becomes unity [22].

7.2.3   Capacitive Coupling

The attenuation of gate line α_g increases rapidly with frequency, as shown in (7.19), resulting in a lower power gain at high bandwidth frequencies. In this case, if the gate-line attenuation can be made very small, the input signal is nearly evenly applied to all FETs in the amplifier and the power gain will remain constant over a wide frequency band. However, since the gate-line attenuation is directly proportional to the gate-source capacitance C_{gs}, then it is possible to reduce its effect by connecting a series capacitor C to each gate, as shown in Figure 7.8 [23, 24]. As a result, since the effective gate capacitance is reduced by a factor of $q/(1 + q)$ when $C = qC_{gs}$, the gate-line attenuation α_g decreases by a factor of $q/(1 + q)$ and the gate-circuit

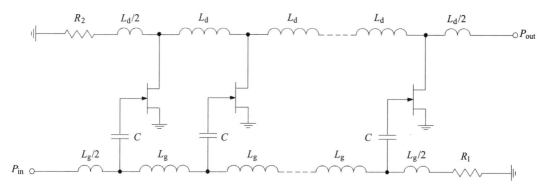

Figure 7.8 Schematic of a distributed amplifier with series capacitors at the FET gates.

cutoff frequency f_g is increased by a factor of $(1 + q)/q$ at a fixed frequency. The gate voltage, however, divides between C and C_{gs}, and the FET can now be considered as a modified device having an effective gate-source capacitance of $C'_g = qC_{gs}/(1 + q)$ and an effective transconductance of $g'_m = qg_m/(1 + q)$ [25]. The series capacitor also reduces the gain per device, but the overall amplifier gain cannot be reduced if more devices are connected or a larger gate periphery is used. Moreover, the series capacitance and gate-source capacitance form a voltage divider, allowing for an increased signal level along the gate line, resulting in significantly higher output power and efficiency for a distributed amplifier.

With a much larger total FET periphery, drain-line loading begins to limit the output power, particularly at the upper end of the operating band, resulting in a low or even negative real part of the impedance at the drains closer to the output. In this case, a capacitor can be inserted between the drain line and the drain of any FET with a low real part of the impedance, thus decreasing the drain-line loading and increasing the impedance at the drains [26]. As a result, a higher total FET periphery can be accommodated and higher output power can be achieved. Figure 7.9 shows an example of a three-cell GaAs FET distributed amplifier that uses capacitive drain coupling. This circuit with a drain coupling capacitor connected to FET_3, which operates from 14 to 37 GHz, also features varying gate periphery and capacitive gate coupling. Inserting the 0.25-pF capacitor between FET_3 and the drain line substantially increases the real part of the impedances at the drains of FET_2 and FET_3 over the frequency range from 18 to 38 GHz. This, in turn, results

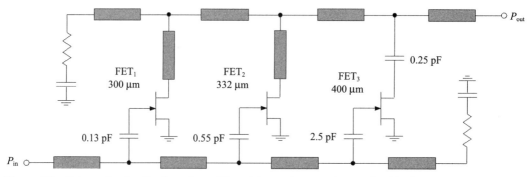

Figure 7.9 Schematic of a distributed amplifier with a series capacitor at the FET drain.

in higher and flatter output power performance up to much higher frequencies, with a power increase of 1.5 dB at 18 GHz and 5 dB at 27 GHz compared to the circuit without drain capacitive coupling.

7.3   Tapered Distributed Amplifier

As an alternative, to compensate for the attenuation that results from the gate finite input resistance so that the FETs in the distributed network are not driven equally, equal drive to each transistor can be restored by increasing the characteristic impedance of the gate line in a systematically tapered manner from the input of the gate line to its end toward the gate load resistor [27]. In a manner analogous to gate voltage equalization, the voltage at the drains of all of the FETs can be made the same by tapering the drain-line impedance, but by systematically decreasing the impedance along the line toward the output load [2]. Improved performance in terms of the smaller gain flatness and wider frequency bandwidth can also be achieved by using a concept of the declining drain-line lengths when the lengths of the drain-line elements between the FET drains become shorter with optimized values the closer the drain line is located toward the output terminal [28].

The effect of a tapered drain line in terms of current distribution for a two-section distributed amplifier is shown in Figure 7.10, where the first FET device operates into a section of the drain line with a characteristic impedance Z_0, and the entire drain current i_d flows to the next section [2]. If the next section has a lower characteristic impedance of $Z_0/2$, one-third of the incident drain current from the second FET device will cancel the reflected current from the first FET device at the junction of the second FET device. The remaining two-thirds of the drain current from the second FET device and four-thirds of the drain current from the first FET device add and propagate toward the end of the second section of the drain line into the new third section. At the next junction, the third section should have a characteristic impedance equal to $Z_0/3$. This process continues where each successive transmission-line section has a characteristic impedance of Z_0/n, where n is the

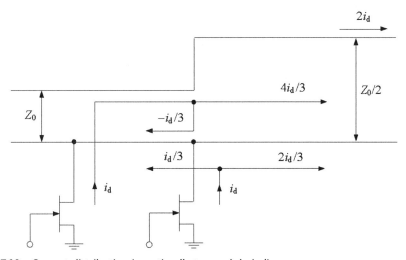

Figure 7.10   Current distribution in optimally tapered drain line.

number of sections. The entire current of the FET devices may thus be effectively used in the load without the necessity of half the drain current flowing into the load and half the drain current flowing into the reverse termination. In this case, Note that current equalization is difficult to achieve in practice due to unequal drive voltages on the gate line and FET process variation, and there exists a small range of useful realizable impedances for microstrip-line practical implementation.

Figure 7.11(a) shows the general structure of a distributed FET amplifier, where $Z_{g(i)}$ and $Z_{d(i)}$ are the optimum characteristic impedances of the gate- and drain-line

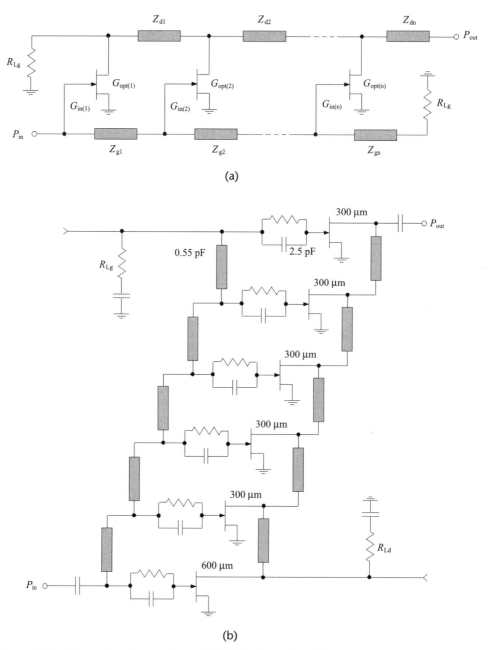

(a)

(b)

Figure 7.11   Schematics of nonuniform pHEMT distributed amplifiers.

ith sections, while R_{Lg} and R_{Ld} are the gate and drain dumping resistors, respectively [29]. In this case, the optimum input and output capacitances of each FET device are absorbed into the artificial gate and drain lines to synthesize the optimum characteristic impedances $Z_{\text{g(i)}}$ and $Z_{\text{d(i)}}$. In the particular case of uniform distributed amplifiers, assuming an identical gate voltage amplitude on each transistor, the characteristic admittances of the drain-line sections can be given as

$$Y_{\text{d(1)}} = G_{\text{opt}} \tag{7.30}$$

$$Y_{\text{d(i)}} = G_{\text{opt}}\left(\frac{G_{\text{opt}}}{G_{\text{opt}} + G_{\text{Ld}}} + i - 1\right) \qquad 2 \le i \le n \tag{7.31}$$

where

  G_{opt} = optimum output conductance of each transistor

  $Y_{\text{d(1)}} = 1/Z_{\text{d(1)}}$ = optimum characteristic admittance of the first drain-line
    section

  $Y_{\text{d(i)}} = 1/Z_{\text{d(i)}}$ = optimum characteristic admittance of the ith drain-line section

  $G_{\text{Ld}} = 1/R_{\text{Ld}}$ = drain dummy conductance.

The resulting optimum output power P_{out} of the uniform distributed amplifier is defined as

$$P_{\text{out}} = \left(\frac{G_{\text{opt}}}{G_{\text{opt}} + G_{\text{Ld}}} + n - 1\right)P_{\text{max}} \tag{7.32}$$

where P_{max} is the maximum power at the 1-dB gain compression point, and n is the total number of transistors within the amplifier.

In the case of nonuniform distributed amplifiers, the generalized optimum power-matching structure can be analytically determined to add the individual power contribution in the direction of the output power as

$$Y_{\text{d(1)}} = G_{\text{opt(1)}} \tag{7.33}$$

$$Y_{\text{d(i}\geq 2)} = \left(\frac{G_{\text{opt(1)}}^2}{G_{\text{opt(1)}} + G_{\text{Ld}}} + \sum_{k=2}^{i} G_{\text{opt(k)}}\right) \qquad 2 \le i \le n \tag{7.34}$$

$$P_{\text{out}} = \frac{G_{\text{opt(1)}}}{G_{\text{opt(1)}} + G_{\text{Ld}}} P_{\text{max(1)}} + \sum_{k=2}^{n} P_{\text{max(k)}} \tag{7.35}$$

where $P_{\text{max(1)}}$ is the maximum power of the first device, P_{out} is the amplifier output power, and $G_{\text{opt(k)}}$ and $P_{\text{max(k)}}$ are the optimum output conductance and output power of the kth transistor, respectively [29].

Note that, in the case of moderate frequency bandwidth applications ($f_{\text{max}}/f_{\text{min}}$ < 3), the drain dumping load R_{Ld} can be removed so that each transistor could be ideally matched and yield its maximum output power, resulting in

$$Y_{d(i)} = \sum_{k=1}^{i} G_{opt(k)} \qquad 1 \le i \le n \qquad (7.36)$$

$$P_{out} = \sum_{k=2}^{n} P_{max(k)} \qquad (7.37)$$

To achieve the equal-gate voltage distribution, the characteristic impedances of the gate-line sections $Z_{g(i)} = 1/Y_{d(i)}$ are defined as

$$Y_{g(i)} = \sum_{k=i}^{n} G_{in(k)} \qquad 1 \le i \le n \qquad (7.38)$$

$$Z_{Lg} = 1/G_{in(n)} \qquad (7.39)$$

and the electrical lengths $\theta_{g(i)}$ and $\theta_{d(i)}$ of the corresponding gate- and drain-line sections must always verify $\theta_{g(i)} = \theta_{d(i)}$.

Figure 7.11(b) shows the simplified circuit schematic of a monolithic nonuniform distributed amplifier composed of six amplifying cells and implemented in a 0.25-μm power pHEMT process, where the first transistor represents a 600-μm HEMT and the other transistors represent 300-μm HEMTs [29]. Here, discrete series capacitors couple each transistor to the gate line and act as voltage dividers to ensure equal drive levels on the transistor gates. Implanted GaAs resistors shunt the series MIM capacitors to supply gate bias. As a result, an output power of 30 dBm with a power gain of 7 dB and a PAE of greater than 20% was achieved across the frequency band from 4 to 19 GHz at a drain supply voltage of 8V. By optimizing the nonuniform nature of the gate and drain lines and using the series capacitors at the device gates, the average 5.5 W output power and 25% PAE were achieved over 2 to 15 GHz for a five-cell monolithic distributed amplifier using a high-voltage 0.25-μm AlGaN/GaN HEMT on SiC technology with a total gate device periphery of 2 mm at a 20V drain bias [30].

The efficiency can be further increased for the same multiple-octave frequency bandwidth when the resistive termination is neglected and an optimum load can be presented to each active device by corresponding tapering of the drain-line characteristic impedance [31]. In addition, the transmission-line lengths can be adjusted such that the transistor currents add in-phase and the FET output capacitances can be absorbed into the transmission line. If the optimum load resistance $R_{opt(k)}$ for the kth transistor is a known quantity for the process and is extracted from load-pull data, the unknown drain-line characteristic impedances $Z_{d(k)}$ can be calculated from

$$Z_{d(k)} = \frac{R_{opt}\,(\Omega\text{-mm})}{\sum\limits_{i=1}^{k} W_{Q_i}} \qquad (7.40)$$

where $R_{opt} = R_{opt(k)} W_{Q_k}$ and W_{Q_k} is the gate width of the kth transistor [32]. At low frequency, the individual FET optimum load resistances $R_{opt(k)}$ will combine in parallel and this parallel combination should be equal to the load impedance R_L to maximize the output power of the amplifier, where $Z_{d(k)} = R_L$.

Table 7.1   Parameters of a Ten-Cell Nonuniform Distributed Amplifier

Parameters	FET number	Equal FET cells		Unequal FET cells	
		W_Q (mm)	Z_d (mm)	W_Q (mm)	Z_d (mm)
FET R_{opt} (Ω-mm) = 120	1	0.24	500	0.60	200
R_L (Ω) = 50	2	0.24	250	0.20	150
Total FET width (mm) = 2.4	3	0.24	167	0.20	120
Number of cells = 10	4	0.24	125	0.20	100
Supply voltage (V) = 30	5	0.24	100	0.20	86
Max RF power (W) = 9.0	6	0.24	83	0.20	75
	7	0.24	71	0.20	67
	8	0.24	63	0.20	60
	9	0.24	56	0.20	55
	10	0.24	50	0.20	50

Table 7.1 shows the device and transmission-line parameters corresponding to the 10-cell nonuniform distributed amplifiers with equal and unequal device gate sizes. In the case of equal gate widths for each cell, the maximum characteristic impedance of the transmission line is equal to 500W, which is difficult to realize with microstrip lines using normal GaN on SiC process. However, by making the first FET cell larger by three times than that of the others, the maximum characteristic impedance of the first transmission line can be significantly reduced to an acceptable value. The estimated maximum RF output power shown in Table 7.1 is calculated assuming a sinusoidal output voltage across the load as $V_{dd}^2/2R_L$, where V_{dd} is the drain supply voltage.

A circuit schematic of a monolithic 10-cell nonuniform distributed amplifier designed to operate over a 10:1 bandwidth including as much of the Ku-band as possible is shown in Figure 7.12(a) [32]. In this case, by using (7.40) and assuming a 35Ω load impedance, the first FET cell was sized at 520 μm with the remaining nine cells sized at 320 μm each for a total periphery of 3.4 mm. To transform the standard 50Ω load to 35Ω impedance, a quarterwave microstrip line centered near the upper band edge was used, as shown in Figure 7.12(b). As a result, the saturated output power greater than 8W with a peak value of 13W and PAE greater than 20% with a peak value of 38% were achieved over a frequency bandwidth of 1.5 to 17.0 GHz at a drain supply voltage of 30V for an input power of 32 dBm using a 0.25-μm GaN HEMT on SiC technology. With some process modifications resulting in higher current handling capability for passive elements and increased gain and voltage handling at the device unit cell, a minimum output power of 11W with a minimum PAE of 28% was achieved across 2 to 18 GHz at a drain supply voltage of 35V [33].

Greater efficiency can be provided at lower frequencies when the power-added efficiencies of 30% to 60% with an output power of around 40 dBm over the frequency bandwidth of 100 MHz to 2.2 GHz have been achieved for a monolithic four-cell distributed amplifier with a tapered drain line using a 0.5-μm GaN HEMT on Si process [34]. A low-temperature cofired ceramic (LTCC) technology is an attractive choice for fabricating power amplifiers because it provides a sufficiently high circuit density integration, low RF loss, and good thermal performance. In

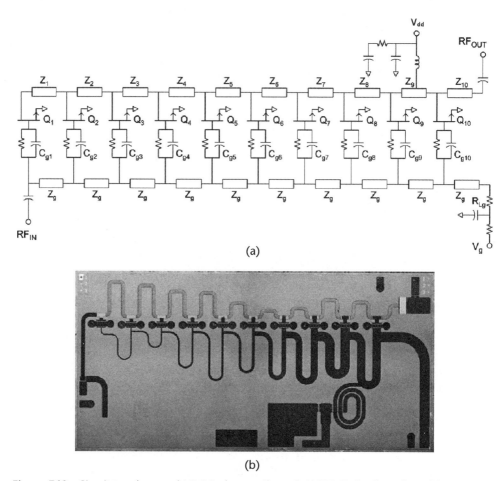

(a)

(b)

Figure 7.12 Circuit topology and MMIC of nonuniform GaN HEMT distributed amplifier.

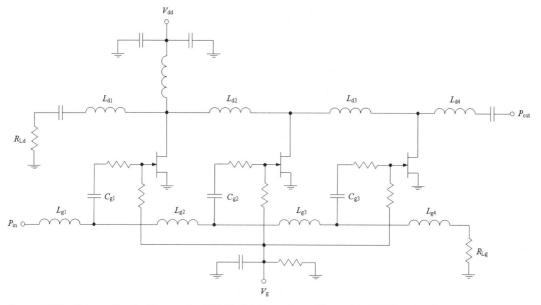

Figure 7.13 Schematic of a three-cell pHEMT distributed amplifier using LTCC technology.

this case, the discrete active devices can be placed directly on top of 400-μm silver-filled vias, which provide a good thermal dissipation to ground [35]. As a result, a five-cell distributed amplifier using pHEMT devices with a 2.1-mm gate periphery each provides a 1W output power over the frequency bandwidth from 800 MHz to 2.1 GHz with a 3.2V supply voltage. Here, a broadband multisection output impedance transformer (incorporating two 4:1 coupled-coil transformers in series with additional matching elements) is required to transform the optimum output impedance of 3.3Ω to a standard 50Ω load throughout the band.

A three-cell design for a distributed amplifier using pHEMT devices with a 1.9-mm gate periphery each, whose circuit schematic is shown in Figure 7.13, can provide a 2W output power over the frequency range from 0.6 to 2.2 GHz with a PAE of greater than 30% at a 12V supply voltage [36]. In this case, the drain-line termination was set well above 50Ω, since the impedance at that point in the circuit is approximately 150Ω, and it could be made even larger at the expense of gain flatness and stability with only 1% to 2% efficiency drop due to the extensive drain-line impedance tapering. A 6W distributed amplifier based on five LMOSFET devices having a 5-mm gate width each can achieve a PAE of greater than 30% over the frequency bandwidth from 100 MHz to 1.8 GHz at a 28V supply voltage [36]. With a hybrid implementation using a Rogers RT5880 substrate and discrete transistors, a higher than 30% PAE over a frequency bandwidth of 20 MHz to 2.5 GHz was achieved for a 5W three-cell distributed amplifier with an optimized tapered drain line using a 0.35-μm GaN HEMT on SiC process at a supply voltage of 28 V [37].

7.4   Power Combining

Because there is a strong demand for solid-state power amplifiers to provide high output power, high efficiency, and a wide bandwidth in different microwave and millimeter-wave radar and communication systems, distributed amplifiers can be considered a good candidate for ultrawideband operation because their bandwidth performance is dominated by a high cutoff frequency for artificial input and output transmission lines. However, output power is generally limited by the drain-line termination and maximum total gate periphery that can be included in a single-stage design. In this case, it is necessary to use power-combining schemes that employ a transmission-line Wilkinson power divider or coupled-line Lange-type power dividers and combiners [38].

Figure 7.14 shows the circuit schematic of a monolithic two-stage distributed amplifier, where the input signal is equally divided into the gate lines, each employing four 150-μm FETs and using a Wilkinson power divider without the isolation resistor [39]. Such a configuration with a single-section Wilkinson divider had contributed to obtaining a decade bandwidth performance because of the good input matching characteristics of the individual amplifiers. In this circuit, the FETs excited from the two separate gate lines are combined on a single drain line, effectively giving a 4 × 300-μm drain periphery, thus doubling the output power over a frequency range of 2 to 20 GHz. Figure 7.15 shows the circuit diagram of a monolithic balanced nonuniform distributed amplifier for which an input power divider and an output power combiner represent the monolithic Lange couplers, which were designed to

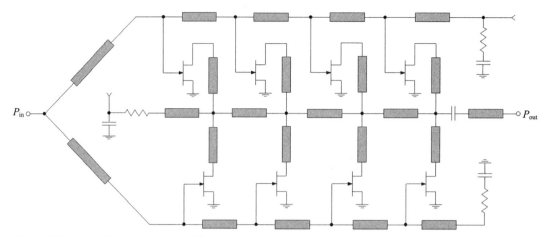

Figure 7.14 Distributed power amplifier with eight FET cells.

achieve operation over 6 to 18 GHz using highly overcoupled lines [40]. In this case, some output power degradation at upper operating frequencies because of the loss in artificial gate and drain lines due to the parasitic resistors in transistors was compensated for by using a shunt short-circuited quarterwave microstrip line at the output of each distributed amplifier. As a result, the fabricated monolithic balanced nonuniform distributed amplifier using a 0.25-μm AlGaN/GaN HEMT

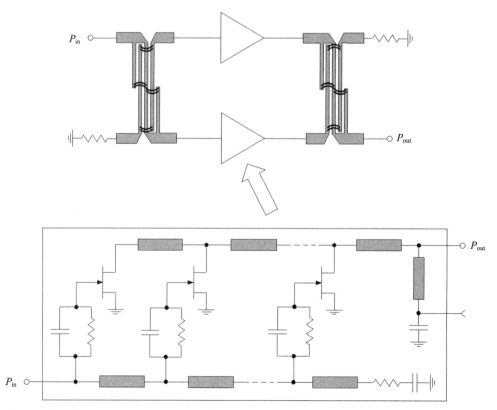

Figure 7.15 Schematic of a balanced distributed amplifier with Lange couplers.

technology was able to deliver an output power of more than 10W over 6 to 18 GHz at a drain supply voltage of 40V.

7.5   Bandpass Configuration

Similar to the conventional distributed amplifiers representing a lowpass configuration, it is possible to achieve sufficient power gain over extreme bandwidths with the bandpass distributed amplifiers limited only by the high-frequency effects of the amplifying devices [8]. In this case, additional inductive elements are placed in parallel with the shunt capacitance of the FETs, and this allows the gate capacitance to be effectively reduced, thus increasing the upper operating frequency for a given FET [41, 42]. However, there has to be some drop in fractional bandwidth owing to the introduction of a lower cutoff frequency. Besides, better noise performance and linear phase response can be achieved with the bandpass distributed amplifiers. However, the basic bandpass distributed amplifier has an amplitude response that is inherently nonflat and that may require an additional parameter optimization procedure or insertion of additional compensating circuits.

The gate and drain lines in such a bandpass distributed amplifier have capacitors in the series arms of the gate and drain lines, thus preventing direct biasing of the gate and drain terminals in the usual manner by means of two common dc power supplies connected to both lines. An alternative structure for the three-cell bandpass distributed amplifier with the series inductors and shunt series LC circuits in the gate and drain lines is shown in Figure 7.16 [43]. The values of the corresponding inductances in the gate and drain circuits are calculated from

$$L_{g1} = L_{d1} = \left(\frac{1-m}{1+m} \right) \frac{CR_L^2}{2} \tag{7.41}$$

$$L_{g2} = L_{d2} = \left(\frac{2}{m} \right) \frac{1}{\omega_0^2 C} \tag{7.42}$$

where
   $R_L = R_{Lg} = R_{Ld}$ = load impedance
   $\omega_0 = 2\pi\sqrt{f_1 f_2}$ = center bandwidth frequency
   f_1 = lower bandwidth frequency
   f_2 = upper bandwidth frequency
   $m = f_1/f_2$
   $C = C_g = C_d$ = effective capacitance.

The effective capacitance is determined for particular device input and output capacitances, load impedance, and boundary frequencies. For example, $C = 3.18$ pF for $f_1 = 1$ GHz, $f_2 = 3$ GHz, $R_L = 50\,\Omega$, and GaAs FET NE72218 with $C_{in} = 0.96$ pF and $C_{out} = 0.64$ pF. In this case, the inductor values are calculated from (7.41) and (7.42) as $L_{g1} = L_{d1} = 1.99$ nH and $L_{g2} = L_{d2} = 7.97$ nH. As a result, the passband of this bidirectional and symmetric distributed amplifier with surface-mounted

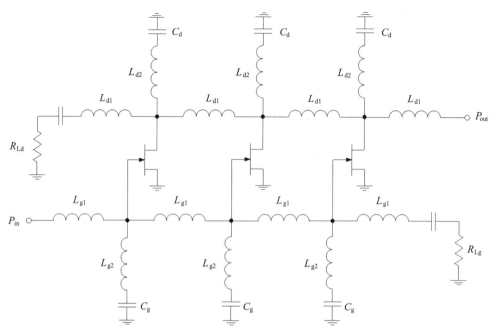

Figure 7.16 Schematic of a three-cell bandpass distributed amplifier.

inductors covers a frequency bandwidth from 842 MHz to 3.17 GHz with a power gain of around 10 dB and linear phase response [43].

7.6 Parallel and Series Feedback

Among the various factors affecting the amplifier gain are the input and output loss factors of the active device. The input loss factor determines the rate at which the input signal decays along the gate line, and the output loss factor affects the growth rate of the output signal along the drain line. As a result, the optimum number of transistors to be used in the distributed amplifier is determined by the input and output loss factors, which can be changed to achieve higher gain values through the use of feedback provided by the active element.

Figure 7.17 shows the small-signal equivalent circuit of a common-source FET device, where $Y_f = G_f + jB_f$ is a parallel feedback admittance and $Z_f = R_f + jX_f$ is a series feedback impedance. For studying the effects of various feedback elements on the maximum available gain (MAG) of the two-port network derived in [44], a perturbation method can be used where the feedback is assumed small and only the first-order correction term ΔMAG is considered. As a result, for a series resistive feedback,

$$\Delta\text{MAG} = -R_f g_m \left[\frac{2\omega^2 R_{gs} R_{ds} C_{gs} C_{ds} \left(g_m R_{ds} C_{ds} - 2C_{gs} \right)}{8\omega^2 C_{gs}^3 R_{gs}^2} \right.$$

$$\left. + \frac{2 g_m C_{gs} \left(R_{gs} + R_{ds} \right) - \left(g_m R_{ds} \right)^2 C_{ds}}{8\omega^2 C_{gs}^3 R_{gs}^2} \right] \qquad (7.43)$$

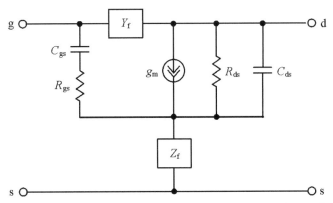

Figure 7.17   General FET model to evaluate the effect of series and parallel feedback.

and for a series reactive feedback,

$$\Delta \text{MAG} = X_f g_m \frac{4\omega^2 C_{gs}^2 R_{gs} - g_m^2 R_{ds}}{8\omega^3 C_{gs}^3 R_{gs}^2} \tag{7.44}$$

where $X_j = \omega L_f$ if the feedback element is an inductance, and $X_f = -1/\omega C_f$ if it is a capacitance [45]. From (7.44), it follows that capacitive series feedback increases MAG at low frequencies, acting as positive feedback. In contrast, inductive series feedback acts as a negative feedback at low frequencies. At the same time, at higher frequencies when the numerator in (7.44) becomes positive, capacitive feedback is negative and inductive feedback is positive. The role of the resistor as a positive or negative feedback element is not only frequency dependent but also dependent on various FET parameters at a given frequency. Specifically, series source resistance acts as a negative feedback element only if the numerator in (7.43) is positive. At low frequencies, the increase in gain due to positive capacitive feedback and the reduction in gain due to negative resistance feedback do not have the same frequency dependence. Therefore, a series RC feedback cannot be used for broadband loss compensation, unlike a parallel RC feedback, which can lead to the possibility of broadband loss compensation.

Similarly, for a parallel resistive feedback,

$$\Delta \text{MAG} = G_f g_m \left[\frac{\left(g_m R_{ds} \right)^2 + 2 g_m R_{ds} \left(\omega^2 C_{gs}^2 R_{gs}^2 + \omega^2 C_{gs}^2 R_{gs} R_{ds} + 1 \right)}{8 \left(\omega^2 C_{gs}^2 R_{gs} \right)^2} \right.$$
$$\left. + \frac{4 R_{gs} R_{ds} \omega^2 C_{gs}^2}{8 \left(\omega^2 C_{gs}^2 R_{gs} \right)^2} \right] \tag{7.45}$$

and for a parallel reactive feedback,

$$\Delta \text{MAG} = B_f g_m \frac{4\omega^2 C_{gs}^2 R_{gs} - g_m^2 R_{ds}}{8\omega^3 C_{gs}^3 R_{gs}^2} \tag{7.46}$$

where $B_j = \omega C_f$ if the feedback element is a capacitance, and $B_f = -1/\omega L_f$ for inductive feedback [45]. It can be seen from (7.45) that a parallel resistive element always gives negative feedback regardless of frequency or FET parameters. However, according to (7.46), the role of capacitive and inductive elements in giving positive or negative parallel feedback is reversed compared to that for series feedback.

Note that the maximum level of positive feedback usable in a given distributed amplifier is limited by the requirement of unconditional stability of the amplifier over the entire frequency range. Generally, the risk of oscillations increases with greater device transconductance g_m, feedback gate-drain capacitance C_{gs}, or transmission-line cutoff frequency $f_c = 1/\pi\sqrt{L_g C_{gs}}$; however, the parasitic resistances R_{gs} and R_{ds} tend to moderate the oscillation phenomena. The detailed analysis of a two-cell distributed structure has shown that the oscillation conditions can be satisfied at high frequencies and are due to an internal loop formed by the transconductance and the feedback capacitance of the active devices, combined with the transmission lines [46]. The oscillations cannot occur due to reflections at the terminations (Z_{Lg} and Z_{Ld}) because the oscillation frequency is lower than the cutoff frequency of the transmission lines and an active feedback is necessary to provide gain in the loop to generate oscillations. The feedback gate-drain capacitance C_{gd} of the active devices strongly modifies the behavior of distributed amplifiers, but mainly when the frequency is high and transistors with high transconductance are used.

Figure 7.18 shows the circuit schematic of a monolithic three-cell distributed driving amplifier, in which each active unit cell represents a cascade pHEMT amplifier with a self-biasing circuit through the parallel and series feedback resistors [47]. The feedback resistor values were determined for high stability and high gain in a distributed amplifier using 200-μm pHEMTs as input devices and 480-μm pHEMTs as output devices (187Ω for parallel feedback resistors and 53Ω for series feedback resistors). The conventional drain termination resistor was eliminated for increasing output power and efficiency in this amplifier. As a result, a maximum available gain

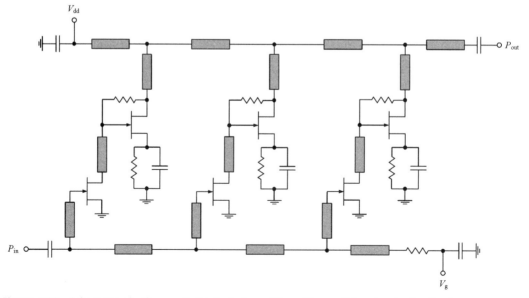

Figure 7.18 Schematic of a three-cell distributed amplifier with parallel and series feedback.

of more than 25 dB across 5 to 20 GHz and a stability factor of larger than unity over all frequencies were simulated, and a small-signal gain of 16±0.6 dB over 5 to 21 GHz and an output power of more than 22 dB at 1-dB gain compression point over 6 to 18 GHz were achieved.

7.7   Cascode Distributed Amplifiers

In practical implementations of a vacuum-tube distributed amplifier, the distributed cascode circuit was used to minimize the degeneration at higher frequencies caused by the common lead inductances and feedback capacitances of the tubes [8]. By using a cascode connection of transistors, it is possible to significantly improve the isolation between the input and output artificial transmission lines. The cascode MESFET cell is characterized by a higher output shunt resistance, which reduces the loading of the drain line, and by a lower gate-drain capacitance than the common-source cell, which reduces negative feedback at the high end and provides for the possibility of automatic gain control. As a result, a higher MAG can be achieved for the cascode cell over a very wide frequency range. For example, a cascode monolithic distributed amplifier based on 0.25-μm InP HEMT technology was able to produce a gain as high as 12±1 dB from 5 to 60 GHz and a noise figure as low as 2.4 to 4.0 dB in the Ka-band [48].

Figure 7.19 shows the circuit schematic of a three-cell nonuniform cascode AlGaN/GaN HEMT distributed amplifier exhibiting an output power of 5W to 7.5W and a PAE of 20% to 33% over the operating frequency range from dc to 8 GHz [49]. In this case, three cascode-connected AlGaN/GaN HEMT cells each having a 0.3-μm gate length and a 1-mm gate periphery were employed to design and fabricate a high-power monolithic distributed amplifier. The drain-line dummy load was removed, and the gate- and drain-line sections were optimized to maximize the output power and efficiency and provide a flat gain throughout the operating frequency range. Note that the lines nearer the output are of lower characteristic impedance and, hence, are wider and more able to supply higher dc current in a Class A bias mode. The corresponding center conductor widths of CPW drain lines

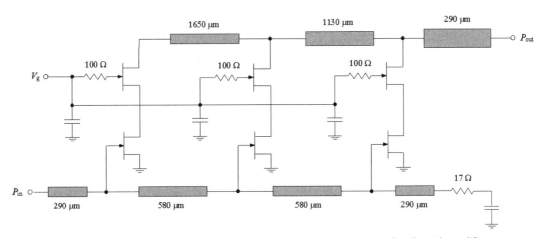

Figure 7.19   Schematic of a three-cell nonuniform cascode GaN HEMT distributed amplifier.

are 27 μm (60Ω), 40 μm (50Ω), and 66 μm (30Ω), with ground-to-ground spacing of 80 μm. The nine-cell monolithic cascode distributed amplifier using a 0.2-μm AlGaN/GaN low-noise GaN HEMT technology with an f_T of ~75 GHz was able to achieve 1W to 4W from 100 MHz to over 20 GHz with a noise figure of around 3 dB at a drain supply voltage of 30V [50].

Figure 7.20(a) shows the MMIC of a five-cell cascode GaN HEMT distributed amplifier (Cree CMPA0060025F), which operates between 20 MHz and 6.0 GHz

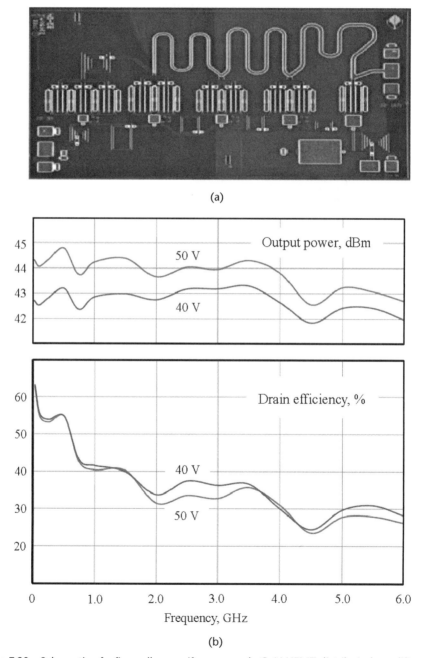

(a)

(b)

Figure 7.20 Schematic of a five-cell nonuniform cascode GaN HEMT distributed amplifier.

[51]. The amplifier typically provides 17 dB of small-signal gain and an average 30W (from 42 to 45 dBm) of saturated output power with a drain efficiency of better than 23% (better than 30% up to 4.0 GHz), as shown in Figure 7.20(b) for an input power of 32 dBm and two different drain supply voltages of 40V and 50V. To achieve high efficiency, a nonuniform approach was used in the design of the drain line where the characteristic impedances change cell by cell and the output reverse termination was eliminated. Proper design of the gate and drain lines and resizing of the individual cells provide a reasonable load-line impedance for each cell.

The dual-gate GaAs FET device, which is equivalent to a cascode-connected single-gate device, has an input impedance that is comparable to that of a single-gate device, but a much higher isolation and output impedance. Note that high reverse isolation in the device is necessary for high amplifier isolation to achieve better operation stability, and high device output impedance improves gain flatness and output VSWR. Figure 7.21 shows the circuit schematic of a monolithic four-cell dual-gate MESFET distributed amplifier, which provides a power gain of 6.5±0.5 dB with greater than 25-dB isolation across the frequency range of 2 to 18 GHz [52]. With dual-gate AlGaN/GaN HEMT technology, broadband performance can be improved when a small-signal gain of 12±1 dB over a bandwidth of 2 to 32 GHz with a peak PAE of 16% and a peak output power of about 30 dBm is achieved for a five-cell monolithic dual-gate distributed amplifier [53]. Using a two-stage distributed amplifier when the first stage consists of six 4 × 50-μm dual-gate GaN HEMTs and the second stage consists of six 4 × 100-μm dual-gate GaN HEMTs allows the small-signal gain to be increased to more than 20 dB with a peak output power of 33 dBm over a bandwidth of 2 to 18 GHz [54].

When using bipolar technology, superior gain-bandwidth performance of the HBT cascode cell compared to a conventional common-emitter HBT shows that the cascode offers as much as 7 dB more maximum available gain, especially at higher bandwidth frequencies [55, 56]. Figure 7.22(a) shows the circuit schematic of a three-cell cascode distributed amplifier based on SiGe HBT devices [57]. Here,

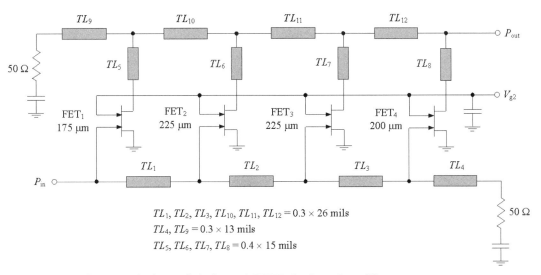

$TL_1, TL_2, TL_3, TL_{10}, TL_{11}, TL_{12} = 0.3 \times 26$ mils
$TL_4, TL_9 = 0.3 \times 13$ mils
$TL_5, TL_6, TL_7, TL_8 = 0.4 \times 15$ mils

Figure 7.21   Schematic of a four-cell dual-gate MESFET distributed amplifier.

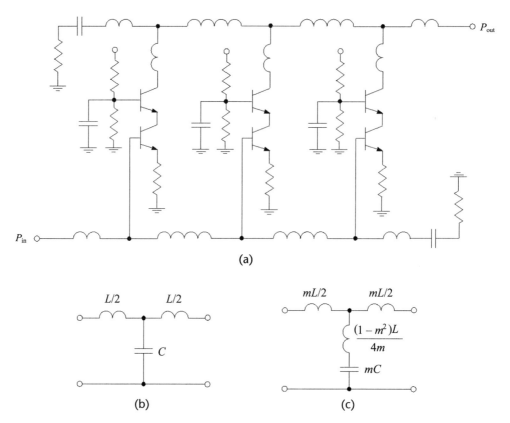

Figure 7.22 Schematic of a three-cell bipolar cascode distributed amplifier.

the emitter degeneration resistor increases the device input impedance and helps to reduce the output distortion. Instead of a constant-k T-section consisting of a series lumped inductor L and a shunt capacitor C with the frequency-independent characteristic impedance $Z_0 = \sqrt{L/C}$ shown in Figure 7.22(b), the m-derived T-section shown in Figure 7.22(c) can be used by adding a parallel inductor to provide an additional degree of freedom [58]. Both sections still maintain the same input and output impedances, but the m-derived T-section has an LC series resonance in its shunt arm. This resonance provides the ability to modify the passband attenuation. As a result, the m-derived T-section has a flatter passband and a better input reflection coefficient than its corresponding T-section with constant Z_0. The choice of a filter section is limited by the available die area, complexity of design, and the technology used. Being implemented in SiGe BiCMOS or HBT technology, such a cascode distributed power amplifier can achieve a measured passband from 100 MHz to 50 GHz with a 1-dB compression power gain varying from 6 to 8.5 dB with an output power of 4.2±2 dBm [57]. In this case, the m-derived filter sections were used for the drain line, and the constant-k T-sections were used for the gate line.

To further improve the gain-bandwidth characteristics required for high-bit-rate telecommunication systems, the design approach based on the attenuation compensation technique when a common-collector stage is followed by a cascode transistor pair in each distributed cell can be used. In this case, a gain of 12.7 dB over a bandwidth of 50 MHz to 27.5 GHz was achieved for the fabricated monolithic

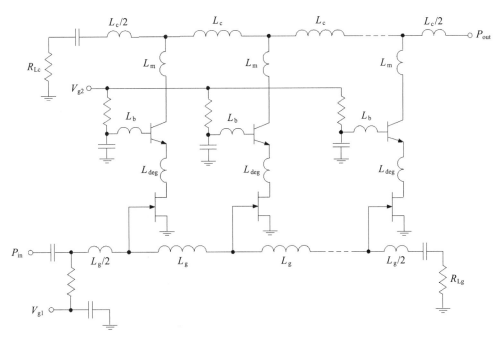

Figure 7.23   Schematic of a six-cell HEMT-HBT cascode distributed amplifier.

three-cell "common-collector cascode" HBT distributed amplifier [59]. The measured midband saturate output power of 17.5 dBm with a peak PAE of 13.2% and the 3-dB output power bandwidth greater than 77 GHz were achieved with a monolithic eight-cell nonuniform cell-scaled cascode distributed amplifier using a 0.13-μm SiGe BiCMOS technology [60].

Monolithic integration of HEMT and HBT devices in a single chip can combine the advantages of both processes, such as a low-noise figure and high input impedance from HEMTs and low $1/f$ noise, higher linearity, and high current driving capability from HBTs. In this case, the monolithic distributed amplifier can be designed using the stacked 2-μm InGaP/GaAs HBT and 0.5-μm AlGaAs/GaAs HEMT process for each cascode-connected cell [61]. To simplify the design procedure, the modified m-derived lowpass T-section can be used where the capacitance is fixed and only the inductance varies with m. Figure 7.23 shows the circuit schematic of a monolithic six-cell HEMT-HBT cascode distributed amplifier where the peaking inductances L_{deg} and L_{b} are utilized to resonate with the parasitic capacitances of the common-source and common-base transistors, resulting in a gain peaking at the cutoff frequency. For a supply voltage of 4.5V with a total dc current consumption of 50 mA, the distributed amplifier achieves an average small-signal gain of 8.5 dB and a 3-dB bandwidth of wider than 43.5 GHz.

7.8   Extended Resonance Technique

An extended-resonance power-combining technique can be used to space transistors in a distributed manner to the proper distance from each other to form a resonant power combining/dividing structure where quarter- or half-wavelength spacing between transistors can be avoided [62]. The spacing is such that the input/output admittance of one transistor is converted to its conjugate value at the input/output of the next transistor. In this case, both gate- and drain-line termination resistors used in a traveling-wave structure are eliminated, resulting in a lower frequency bandwidth but higher efficiency. Since the magnitude of the voltage at each transistor is the same, the gain of the extended-resonance power-combining amplifier with equal-size n transistors is equal to the gain of a single-device amplifier, while its power capability is increased by n times. This approach enables a compact circuit to be designed that is particularly suitable for MMICs.

Figure 7.24(a) shows the circuit schematic of a MESFET distributed amplifier using an extended-resonance technique to combine powers from n devices, where the gates and drains are sequentially linked with transmission lines [63]. Here, θ_{gk} and θ_{dk} are the electrical lengths of the transmission lines connecting each device for the gate and drain lines, respectively, where $1 \leq k \leq n - 1$. The quarterwave transmission-line transformers are used at the input and output of the amplifier to match with the respective 50Ω source and load impedances. The gate and drain extended resonance circuits can be designed separately after calculating the simultaneous conjugate-match admittances. It is assumed that each device has the same gate admittance $Y_g = G_g + jB_g$ and drain admittance $Y_d = G_d + jB_d$. The extra susceptance jB_g connected to the device input is provided by shunt capacitors or inductors,

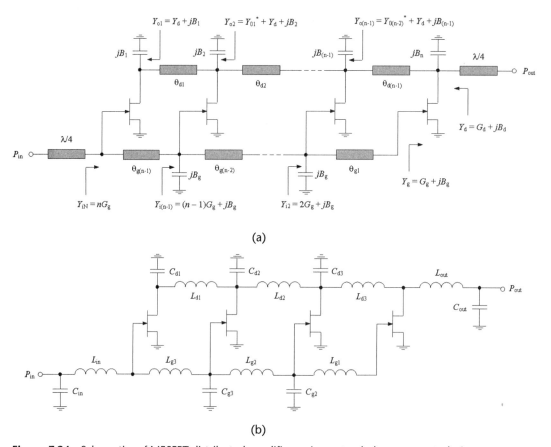

(a)

(b)

Figure 7.24 Schematics of MESFET distributed amplifiers using extended resonance technique.

which can be realized in the form of open- or short-circuit stubs. To provide proper power combining, the voltage phase difference between successive drains should be equal to the voltage phase shift between successive gates. An example of a lumped four-device version of a distributed extended-resonance amplifier is shown in Figure 7.24(b) [64].

As shown in Figure 7.24(a), the gate-line length $\theta_{g(n-1)}$ transforms the admittance $Y_g = G_g + jB_g$ from the Nth device to its conjugate value $Y_g^* = G_g - jB_g$ at the location of the next device. Adding the gate admittance of the next device, the imaginary components cancel, resulting in the gate admittance at the $(n-1)$th device being equal to $2G_g$. Then, a shunt susceptive element B_g is placed at this device so that line length θ_{g2} transforms the resulting admittance $Y_{i2} = 2G_g + jB_g$ from the second device to its conjugate value $Y_{i2}^* = 2G_g - jB_g$ at the next device. This process continues all the way to the gate of the first device where the admittance $Y_{in} = nG_g$ can be matched to a given source impedance using a quarterwave transformer. Similarly, but only in the reverse direction, the drain-line length θ_{d1} transforms the admittance $Y_{o1} = Y_d + jB_1$, where the shunt susceptance jB_1 is connected to the device drain, from the first device to its conjugate value $Y_{o1}^* = G_d - j(B_d + B_1)$ at the location of the second device. Then, the drain-line length θ_{d2} transforms the resulting admittance $Y_{o2} = 2G_d + j(B_2 - B_1)$ at the second device to $Y_{o3} = 2G_d + j(B_2 - B_2 + B_1)$ at the third device. This continues to the drain of the nth device, where the resulting susceptance is canceled by jB_n. Consequently, an input signal applied to the gate of the first device will be divided equally among all devices. Then, each device amplifies $1/n$ of the input power and delivers it to the output combining circuit where the power from each device is recombined at the load. Assuming lossless transmission lines, it can be shown that the total output power is equal to n times the power generated by each device.

Note that, due to the resonant nature of the dividing and combining circuits, their 3-dB frequency bandwidth is limited to about 5%. However, to maximize the bandwidth performance, the optimized lowpass ladder-type networks can be used in the dividing and combining circuits. Besides, the broadband matching circuit is required at the input to match the low impedance presented by the transistors at the gate to the 50Ω source. As a result, measured output power of around 32 dBm with 1-dB flatness and a PAE of 20% to 40% from 4 to 9 GHz were achieved for a hybrid four-device distributed extended-resonance amplifier using AlGaAs/InGaAs pHEMT transistors [65]. To increase the overall output power capability, a 40-device distributed multicell multistage amplifier based on the extended-resonance technique was designed in which each multicell includes four 0.25-μm AlGaAs/InGaAs pHEMT devices with a 600-μm gate width and a total MMIC gate width of 24 mm, resulting in a small-signal gain of 15 dB and an output power of around 32 dBm at 1-dB compression point within the frequency bandwidths of 25 to 30 GHz and 28 to 34 GHz [66].

7.9   Cascaded Distributed Amplifiers

High gain over a wide frequency range can be achieved by cascading several stages of a single-stage amplifier, thus creating an artificial active transmission line that

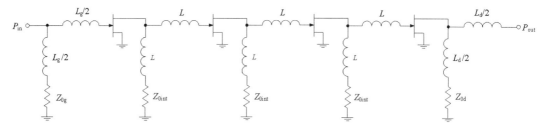

Figure 7.25 Schematic of a four-cascaded single-stage distributed amplifier.

includes active-device parameters. Figure 7.25 shows the schematic representation of a four-cascaded single-stage distributed amplifier (4-CSSDA), where each stage is based on equal-size transistors and equal-value characteristic impedances for all cascades. A feature of this arrangement is that the need to maintain the characteristic impedance $Z_{0\text{int}}$ of the intermediate stages at the standard impedance of 50Ω is eliminated. An increase in this intermediate impedance results in a higher available gain since the gate voltage at each intermediate stage is correspondingly increased.

The available gain for an n-cascaded single-stage distributed amplifier (CSSDA) can be calculated as

$$G_{\text{CSSDA}} = \frac{g_m^{2n} Z_{0\text{int}}^{2(n-1)} Z_{0g} Z_{0d}}{4} \qquad (7.47)$$

where g_m is the device transconductance, Z_{0g} is the characteristic impedance of the input gate line, and Z_{0d} is the characteristic impedance of the output drain line [67]. An RF signal from a matched generator will be coupled by the transconductance of the active device at each stage, and finally terminated by the matched output load port. The amplified signal is valid only up to the cutoff frequency, which is controlled by the gate circuits. Unlike the conventional distributed amplifiers, it is only necessary to equalize the characteristic impedances of the input gate and output drain ports of the active device involved at each stage, and this can be done by adding an extra capacitance with the output capacitance C_{ds} to equalize with the input capacitance C_{gs}. In this case, the CSSDA is characterized by two main features when the input and output stages are the only stages to match with the 50Ω source and load, respectively, and the intermediate characteristic impedance $Z_{0\text{int}}$ can be optimized to boost the overall available gain according to (7.47).

In order for the CSSDA to produce available gain equal to or higher than the forward available gain for an ideal lossless n-stage conventional distributed amplifier (CDA) derived from (7.1) as

$$G_{\text{CDA}} = \frac{n^2 g_m^2 Z_{0g} Z_{0d}}{4} \qquad (7.48)$$

from (7.47) and (7.48), it follows that

$$g_m Z_{0\text{int}} > \sqrt[(n-1)]{n} \qquad (7.49)$$

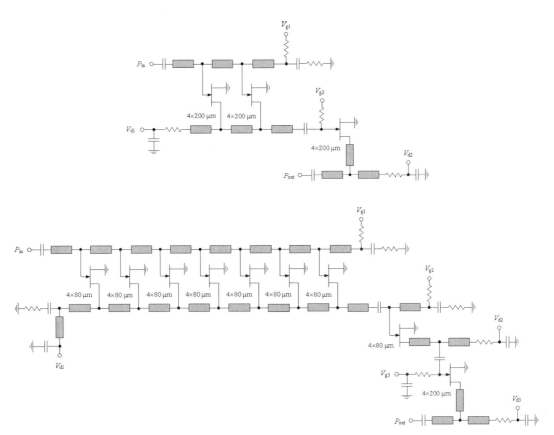

Figure 7.26 Schematics of cascaded conventional and single-stage distributed amplifiers.

which shows that the required interstage characteristic impedance $Z_{0\text{int}}$ decreases for the same devices as the stage number n increases [68]. Hence, a cascade of four single-stage FETs with intermediate characteristic impedance of 86.6Ω and device transconductance $g_m = 28$ mS should yield a 50Ω matched distributed amplifier configuration with a 20-dB available gain over a wide frequency band, compared to only a 9-dB available gain for a conventional distributed amplifier according to (7.48). In a practical case of the four-CSSDA using low-noise MESFET devices with $g_m = 55$ mS, a power gain of 39±2 dB over a frequency bandwidth of 0.8 to 10.8 GHz was measured [68].

Due to the typical second-order lowpass filter configurations, the bandwidth of the two-CSSDA is band limited compared to the CDA. As the number of stages of the amplifier increases, the low-frequency gain also increases so it is not easy to design a flat gain performance for a multistage CSSDA. Therefore, to provide a wider bandwidth with high-gain performance, the broadband distributed amplifier can combine the CDA and CSSDA with a different number of stages [69]. As a result, the forward available gain of the distributed amplifier combining the n-stage CDA and n-stage CSSDA is given by

$$G = \frac{n^2 g_m^2 Z_{0g} Z_{0d}}{4} \frac{g_m^{2n} Z_{0\text{int}}^{2(n-1)} Z_{0g} Z_{0d}}{4} \tag{7.50}$$

Figure 7.26(a) shows the circuit schematic of a monolithic two-stage CDA cascaded with a single-stage CSSDA using 0.15-μm pHEMT technology, which provides a small-signal gain of 19±1 dB over the frequency range of 0.5 to 27 GHz. To extend the amplifier gain-bandwidth performance for millimeter-wave applications, a monolithic seven-stage CDA cascaded with a two-stage CSSDA using the same technology with a die size of 1.5 × 2 mm^2 was designed, as shown in Figure 7.26(b), achieving a small-signal gain of 22±1 dB over the frequency range of 0.1 to 40 GHz with a total dc consumption of 484 mW [69]. The group delay of 30±10 ps is sufficiently flat over the whole bandwidth, which is very important for digital optical communications.

The gain response and efficiency of the CSSDA are increased if the intermediate impedance $Z_{0int} = R_{var} + j\omega L_{var}$ at the drain terminal of each active device is included, as shown in Figure 7.27 for a lumped three-stage cascaded reactively terminated single-stage distributed amplifier (CRTSSDA) [70]. Although the bandwidth of this amplifier is also limited by the gate and drain inductances L, it can be substantially improved by the inclusion of the inductance L_{var} and resistance R_{var}. The effect of the reactive termination is to enhance the voltage swing across the input gate-source

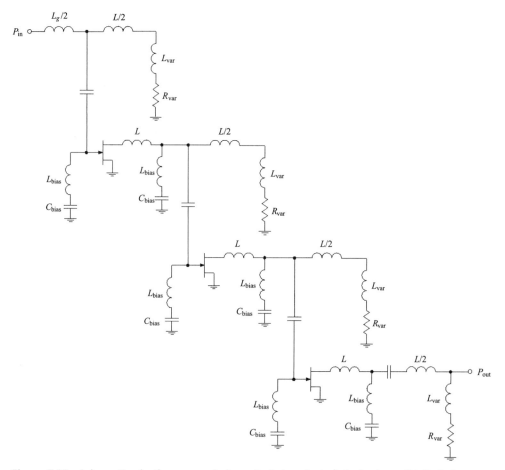

Figure 7.27   Schematic of a three-cascaded reactively terminated single-stage distributed amplifier.

capacitance C_{gs} of each active device, which results in an increased output drain current from each device. This consequently improves the amplifier overall gain performance over the multioctave bandwidth. In this case, to provide a flat gain response over the desired bandwidth, it is simply necessary to adjust the impedance Z_{0int}, because the effect of the inductance L_{var} is negligible at lower frequencies (in the range of 10 kHz to 1 GHz) when the intermediate impedance can be written as $Z_{0int} = R_{var}$. The selection of the bias components L_{bias} and C_{bias} also plays a critical role in optimizing the bandwidth and must have minimum intrinsic parasitics.

The initial value of the resistance R_{var} can be calculated from (7.49), which is dependent on the device transconductance g_m and the number of stages n constituting the CRTSSDA. However, the calculated value of R_{var} will have to be optimized in order to achieve the required small-signal response. The inductive component L_{var} will have an effect on the small-signal gain at higher frequencies (over 2 GHz). The primary effect of this component is to alter the magnitude of the small-signal level at the input port of the respective device of the CRTSSDA chain to be amplified by its device transconductance. The initial value of the inductance can be calculated from

$$L_{var} \geq \frac{\sqrt[(n-1)]{n}}{g_m \omega} \tag{7.51}$$

The fabricated three-stage CRTSSDA based on a 0.25-μm double pHEMT technology, with a gate periphery of 360 μm for each transistor with a self-biased mode of operation (gates are directly grounded through the inductances L_{bias}) providing a dc current of 120 mA, achieved a gain of 26±1.5 dB, an input and output return loss of better than 9.6 dB (VSWR of better than 2:1), and a PAE of greater than 12.6% across the frequency bandwidth of 2 to 18 GHz [71]. The output power of greater than 24.5 dBm with a PAE of greater than 27% across 2 to 18 GHz was achieved when a pHEMT device with a gate width of 720 μm was used in a final stage and a pHEMT device with a gate width of 200 μm was used in a first stage [72].

7.10   Matrix Distributed Amplifiers

The concept of the matrix amplifier combines the processes of additive and multiplicative amplification in one and the same module. Its purpose, therefore, is to combine the characteristic features of both principles, namely, to increase the gain of the additive amplifier concept and the bandwidth of the multiplicative amplifier concept. This can be accomplished in a module whose size is significantly reduced when compared with the traditional amplifier types of similar gain and bandwidth performance. In its most general form, the matrix amplifier consists of an array of m rows and n columns of active devices. Each column is linked to the next by inductors or transmission-line elements connected at the input and output terminals of each transistor, composing a lattice of circuit elements. For m active tiers, there are $2m$ idle ports that are terminated into power-dissipating loads. By adding the vertical dimension to the horizontal dimension of the distributed amplifier in the form of the $n \times m$ rectangular array, the multiplicative and additive process in one and the same module is achieved. The advantages of the matrix amplifier include

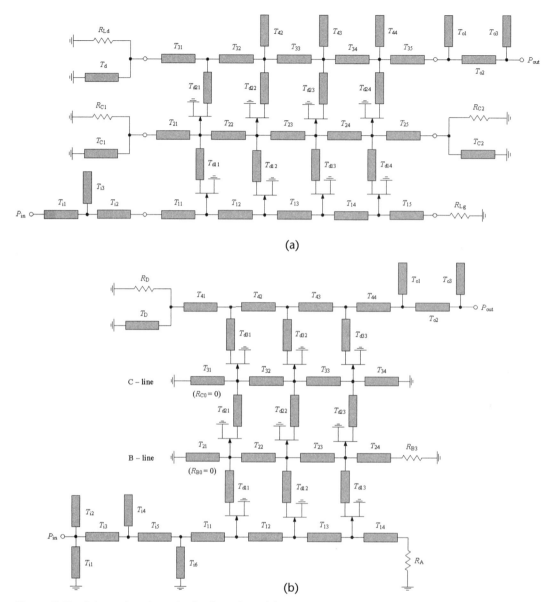

(a)

(b)

Figure 7.28 Schematics of matrix distributed amplifiers.

significantly higher gain and reverse isolation over wide bandwidths at considerably reduced size.

   Figure 7.28(a) shows the circuit schematic of a distributed matrix MESFET
amplifier in the form of a 2×4 array representing a six-port flanked by the input
and output four-ports [73, 74]. The active six-port incorporates the transistor
characterized by its set of Y-parameters, the network of transmission-line elements
represented by their respective characteristic impedances and electrical lengths,
and the open-circuit shunt stubs capacitively loading the drain line. In contrast,
the input and output four-ports contain only passive circuit elements, that is, the
terminations of the amplifier idle networks and simple input and output matching

networks. Each idle port is terminated into either a resistor or an impedance consisting of a resistor shunted by a short transmission line that allows biasing of the active devices without any power dissipation in the termination resistors. The choice for the termination elements is critical for gain flatness, noise figure, gain slope, and operational stability. Based on a rigorous solution for voltages and currents involving GaAs MESFETs with 0.25×200-μm gate dimensions, the 2×4 matrix amplifier was fabricated with an overall size of 0.5×0.24 in. using a 10-mil-thick quartz substrate, achieving a large-signal gain of 11.6 ± 1.5 dB from 2 to 21 GHz with an output power of 100 mW [73]. In addition, note that the matrix amplifier can offer a most desirable compromise between its broadband maximum noise figure on one hand and its gain and VSWR performance on the other. As a result, a computer-optimized two-tier (2×4) GaAs MESFET matrix amplifier could provide a noise figure of $F = 3.5\pm0.7$ dB with an associated gain of 17.8 ± 1.6 dB across the frequency band of 2 to 18 GHz [75]. The monolithic 2×3 matrix amplifier using 0.2-μm pseudomorphic InGaAs HEMT technology achieved a 20-dB gain and a 5.5-dB noise figure over the frequency band of 6 to 21 GHz [76].

The circuit schematic of a 3×3 matrix MESFET amplifier is shown in Figure 7.28(b), where the left port of artificial transmission line B and both ports of artificial transmission line C are terminated into short circuits [77]. The input and output matching circuits are necessary to improve the reflection coefficients of the amplifier, and biasing of the active devices is easily provided through the short-circuited idle ports. The theoretical analysis of the amplifier circuit shows that a low-frequency gain of the matrix distributed amplifier with three tiers ($m = 3$) can be estimated by

$$G_{3\times n} \cong \left[\frac{2(ng_m)^3 Y_0}{(G_A + Y_0)(G_{B0} + G_{Bn} + nG_{ds})(G_{C0} + G_{Cn} + nG_{ds})(G_{D0} + Y_0 + nG_{ds})} \right]^2$$

(7.52)

where $G_{ds} = 1/R_{ds}$ is the device drain-source conductance, $Y_0 = 1/Z_0$ is the characteristic admittance of the artificial transmission lines, and n is the number of MESFETs per tier. If one terminal of the artificial transmission line is short-circuited, the stability of the amplifier can only be maintained if the other terminal of the same line is terminated into a finite impedance or a short. By using GaAs MESFETs with 0.35×200-μm gate dimensions and termination resistors $R_A = 29\Omega$, $R_{B3} = 49\Omega$, and $R_D = 212\Omega$ ($R_{B0} = R_{C0} = R_{C3} = 0$), a noise figure of 5.2 ± 1.2 dB and a gain of 27.7 ± 0.9 dB were achieved across the frequency band of 6 to 18 GHz. The low-frequency gain of the matrix distributed amplifier with two tiers ($m = 2$) can be calculated from

$$G_{2\times n} \cong \frac{1}{2} g_{m1} g_{m2} n Z_0 \frac{Z_{0c} R_{ds1}}{n Z_{0c} + 2R_{ds1}} \frac{\sinh(b)\exp(-b)}{\sinh(b/n)}$$

(7.53)

where Z_{0c} is the characteristic impedance of the central line and $b = (n/4)(Z_0/R_{ds2})$ [78].

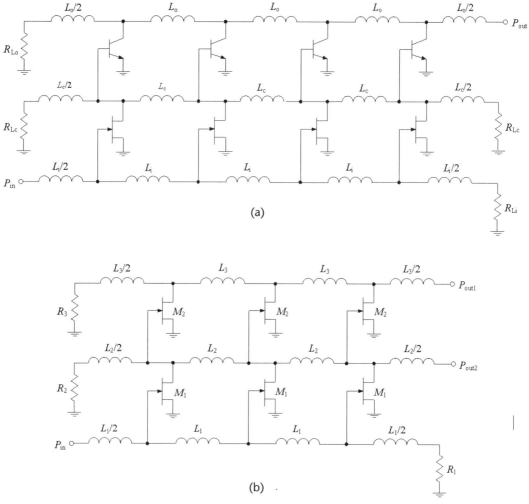

Figure 7.29   Schematics of (a) a lumped matrix amplifier and (b) an active balun.

The integration on the same chip of the active devices based on different technologies can merge the advantages inherent to these technologies. For example, using simultaneously the HEMT and HBT devices on a single chip allows high-gain performance in a multioctave frequency band and achievement of low dc power consumption and noise figure. Figure 7.29(a) shows the simplified circuit schematic of a 2×4 HEMT-HBT matrix amplifier, where the first tier consists of HEMTs and the second tier is replaced by HBT devices [79]. Here, the higher input capacitance of each HBT is absorbed in the central line, whose characteristic impedance can be different from 50Ω without degradation of the input and output matching. As a result, a flat gain of 18 dB, which is only by 1 dB less than that for the HEMT matrix amplifier and 2 dB higher compared to the HBT matrix amplifier; a noise figure of 5 to 6 dB, which is close to that for the HBT matrix amplifier and more than 2 dB better than the noise figure of the HEMT matrix amplifier; and a more than 40% reduction in the dc-power consumption compared to the HEMT matrix amplifier were achieved across the frequency band up to 30 GHz.

The matrix balun, which is based on the matrix amplifier concept, can provide a decade bandwidth and a high gain, while having small size, compared to the conventional active and passive baluns. Figure 7.29(b) shows the simplified circuit schematic of a 2×3 HEMT matrix balun, where the phase balance is achieved by utilizing the fact that the phase difference between two rows in a matrix amplifier with common-source transistors is 180° [80]. The analytical expression for the common-mode rejection ratio (CMMR) for this matrix balun with finite output conductances for zero normalized frequency Ω is defined as

$$\text{CMRR} = \left| \frac{S_{21} - S_{31}}{S_{21} + S_{31}} \right| = \left| \frac{3g_{m2}Z_0R_3 + R_3 + \left(\dfrac{3R_3}{R_{ds2}} + 1 \right)Z_0}{3g_{m2}Z_0R_3 - R_3 - \left(\dfrac{3R_3}{R_{ds2}} + 1 \right)Z_0} \right| \qquad (7.54)$$

where

    Z_0 = characteristic impedance of the artificial transmission lines

    R_3 = idle-port termination resistance of the output line

    g_{m2} = transconductance

    R_{ds2} = drain-source resistance of the transistors connected to the output line.

As a result, a 2×3 matrix balun implemented in a 0.15-μm GaAs mHEMT technology with a chip size of 0.9×1.1 mm^2 ($R_1 = R_2 = 39\Omega$ and $R_3 = 29\Omega$) achieved more than a decade bandwidth of 4 to 42 GHz with a CMRR of greater than 15 dB, a gain of 2±1 dB, and a maximum phase imbalance of 20° with a power consumption of 20 mW. The same matrix balun circuit may also be biased for amplification and used as a matrix amplifier. In this case, the circuit exhibited a 10.5-dB gain up to 63 GHz with a 1-dB ripple above 5.5 GHz and a power consumption of 67 W.

7.11 CMOS Distributed Amplifiers

Unlike the semi-insulating GaAs process, which provides high quality lumped inductors and transmission lines, a CMOS-based implementation is advantageous in that it results in lower costs and a higher level of integration. One of the first designs of a four-cell CMOS distributed amplifier was based on a 0.6-μm CMOS process with a three-layer Al-metal interconnect, for which a flat gain of 6.5±1.2 dB over a bandwidth from 500 MHz to 4 GHz with approximately linear phase over the passband was achieved [81]. Figure 7.30(a) shows the basic circuit schematic of a four-cell lumped CMOS distributed amplifier. In this case, if the gate- and drain-line inductors are matched, and the drain capacitance is made equal to the gate capacitance for each transistor, then the input and output currents are phase synchronized. Another modification to the basic circuit relates to the proper gate- and drain-line termination. The impedance seen looking into the LC artificial transmission lines will exhibit a strong deviation from the nominal impedance near the cutoff frequency of the lines. Ideally, all four ports would be image-impedance matched to the lines to eliminate reflections. However, it is not practical to realize direct image-impedance matching. Thus, the method used will be to insert the

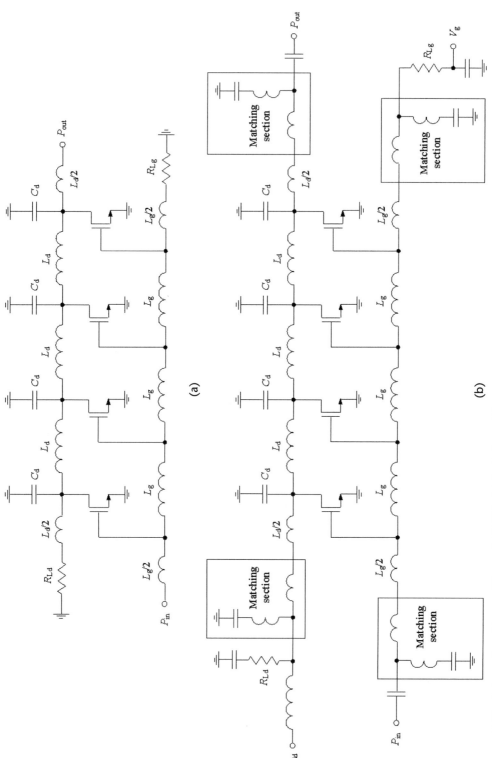

Figure 7.30 Schematics of lumped four-cell CMOS distributed amplifiers.

m-derived half-sections between the lines and the input port, output port, and terminations. These half-sections will greatly improve the impedance matching, while also allowing simple resistive terminations to be used. The modified circuit of a lumped four-cell CMOS distributed amplifier with matching sections is shown in Figure 7.30(b) [81]. Based on 0.18-μm SiGe BiCMOS technology with six-layer metal interconnects and a final two thick-copper layers to realize low-loss inductors and using only nMOS transistors, a frequency bandwidth was extended from 500 MHz to 22 GHz with a flat gain of 7±0.7 dB and input return loss of better than 10 dB over most of the bandwidth [82].

Silicon-on-insulator (SOI) CMOS technologies can provide high-gain performance at the millimeter-wave frequencies required for low-power broadband microwave and optical systems. For example, a 0.12-μm SOI CMOS process offers a low-parasitic nMOS transistor with a peak f_T in excess of 150 GHz for a gate length of less than 60 nm. Here, since the integration of the low-loss 50Ω microstrip lines is difficult, the coplanar waveguide (CPW) structures were implemented on the last 1.2-μm-thick metal layers to reduce the parasitic capacitances to the substrate [83]. As a result, a gain of 4±1.2 dB over the bandwidth of 4 to 91 GHz and an 18-GHz output 1-dB compression point of 10 dBm were measured for the cascode five-cell distributed amplifier with a power consumption of 90 mW. The cascode three-cell distributed amplifier implemented in a 45-nm SOI CMOS process with a peak f_T in excess of 230 GHz, whose value strongly depends on the layout parasitics and may reach 380 GHz, achieved a 3-dB bandwidth of 92 GHz and a peak gain of 9 dB with a gain ripple of 1.5 dB and an input return loss of better than 10 dB [84]. Note that the noise behavior of a distributed amplifier over entire frequency band depends significantly on the number of cells n. For example, the best low-frequency noise performance is achieved for larger values of n, whereas the lowest noise figures are reached at high frequencies for smaller values of n [85]. This noise behavior is attributed to the fact that the drain noise is inversely proportional to n, whereas the gate noise is proportional to n. Besides, for the same number of cells, a cascode CMOS distributed amplifier demonstrates better noise performance over the most of the frequency bandwidth, especially at higher frequencies, compared to the conventional CMOS distributed amplifier with the transistors in a common-source configuration.

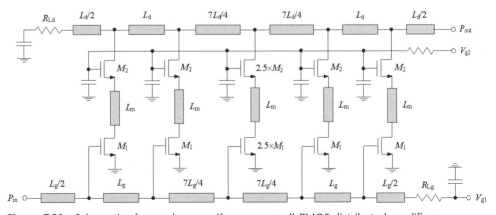

Figure 7.31 Schematic of cascode nonuniform seven-cell CMOS distributed amplifier.

By employing a nonuniform architecture for the artificial input and output transmission lines, the CMOS distributed amplifier exhibits enhanced performance in terms of gain and bandwidth. Figure 7.31 shows the circuit schematic of a cascode nonuniform seven-cell distributed amplifier using standard 0.18-μm CMOS technology where the transistor sizes and inductance values of the center cell are 2.5 times as large as those of the other cells [86]. In this design, the parameters of the common-source transistors are designed in consideration of the cutoff frequency of the input line and the transconductance of the gain stages. On the other hand, the common-gate transistors are designed to provide an output capacitance equivalent to the input capacitance of the gain stages such that matched input and output lines can be utilized to optimize the phase response of the distributed amplifier. The values of the inductive elements L_m between cascode transistors need to be adjusted for maximum bandwidth extension and better noise performance [87]. Finally, the high-impedance CPW structures were employed to realize the required gate and drain inductances, resulting in a passband gain of 9.5 dB and a 3-dB bandwidth of 32 GHz for this distributed amplifier.

The main drawback of an integrated CMOS implementation of the single-ended common-source amplifiers including system-on-chip solutions is that parasitic interconnects, bondwires, and package inductors degenerately degrade their gain-bandwidth performance. Specifically, for a packaged single-ended distributed amplifier that exhibits a unity-gain bandwidth of 4 GHz, there is a bandwidth degradation of 27% compared to its unpackaged performance [81]. Figure 7.32 shows the circuit schematic of a fully differential four-cell CMOS distributed amplifier with the ideal passive components [88]. The characteristic impedances of the gate

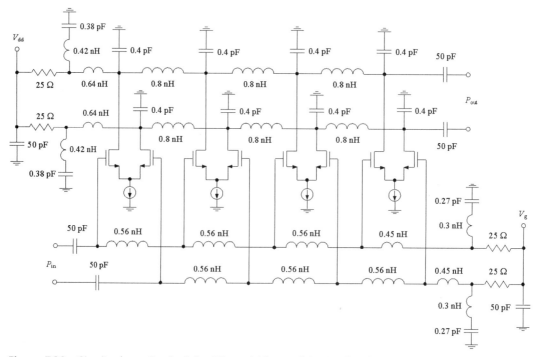

Figure 7.32 Circuit schematic of a fully differential four-cell CMOS distributed amplifier.

and drain lines are designed to be 25Ω each to provide a load impedance of 50Ω for fully differential signals. The highest achievable line cutoff frequency for the 0.6-µm CMOS process is about 10 GHz and is limited by the smallest practical values of the circuit inductances and capacitances. Since high-quality tail current sources are essential to achieving high common-mode and power-supply rejection ratios, a regulated cascode current source is employed in each stage. Such an architecture also minimizes undesirable signal coupling between stages through the common substrate. To improve impedance matching between the termination resistors and the artificial transmission line over a wide range of frequencies, two pairs of *m*-derived half-sections are used, and the drain and gate bias voltages are supplied through the termination ports. As a result, the measured results for this fully differential CMOS distributed amplifier with transistor gate widths of 400 µm demonstrated a bandwidth of 1.5 to 7.5 GHz, which is about 50% greater than for a single-ended counterpart, but obtained at the expense of increased power consumption, die size, and noise figure.

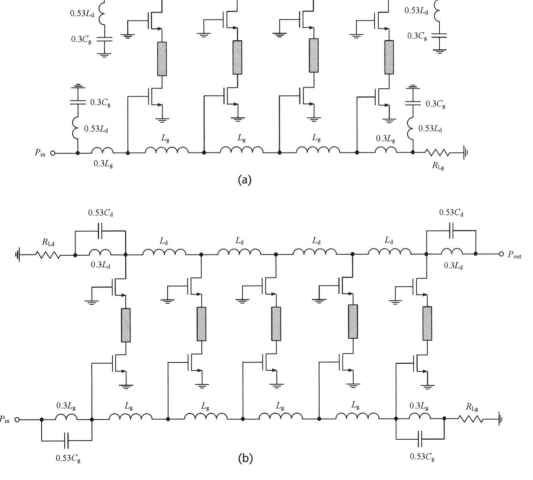

Figure 7.33   Schematics of CMOS distributed amplifiers with (a) bisected-*T*–type and (b) π-type *m*-sections.

Bisected-T m-derived filter sections at the input and output of the distributed amplifiers, as shown in Figure 7.33(a) for $n = 4$, are widely used to improve matching and gain flatness near the amplifier cutoff frequency. In this case, a bisected-T–type m-section matches a T-type k-section (or cascade of them) on one side and matches the constant real impedance on the other side. Thus, the bisected-T m-section can couple power from a real source into a cascade of T-type k-sections over the full frequency range from dc to cutoff frequency. Similarly, the bisected-π m-section can couple power from a real source into a cascade of π-type k-sections over the same frequency range [89]. Here, the shunt $0.3C$ capacitance in the matching section is connected in parallel with the adjacent capacitance from the first k-section. This can be added together into a single transistor or gain cell that is 80% of the size of a full gain cell, thus resulting in a higher voltage gain by

$$\frac{A_\pi}{A_T} = 1 + \frac{0.6}{n} \tag{7.55}$$

than its T-type equivalent. The second factor $0.6/n$ comes from an extra transistor area in the matching sections at both the beginning and end of the artificial transmission line. It follows from (7.55) that the gain boost is higher for smaller n, but gives appreciable improvement over the typical range of n, specifically of 1 dB for $n = 5$. In addition, because inductors are generally lossy and difficult to accurately model at microwave frequencies, the π-type topology reduces both the number and the size of inductors. Figure 7.33(b) shows the circuit schematic of a cascode five-cell CMOS distributed amplifier with bisected π-type m-sections. In a practical implementation, an overall area reduction of 17% (excluding pads) was achieved for a π-type topology [89]. To extend the flat bandwidth and improve the input matching of a cascode distributed amplifier, the gate artificial transmission line based on coupled inductors in conjunction with series-peaking inductors in cascode gain stages can be used [90].

The cascaded CMOS distributed structure is an alternative configuration to exhibit simultaneous high gain and wide operating bandwidth. When compared with the matrix CMOS amplifier topology, which has the same low-frequency gain characteristic, the cascaded structure offers robustness to high-frequency mismatches in the signal delays along the individual paths, thus compromising the overall gain [91]. It also has no loading effect at interstage artificial transmission lines, yielding a larger operating bandwidth, and the omission of the idle drain terminations at intermediate cascaded stages or interstages results in a significant gain improvement. The basic structure of a generalized cascaded CMOS distributed amplifier includes artificial transmission lines in the form of the constant-k LC network with a simultaneous match to both T- and π-sections [92]. It consists of m cascading stages of n-cell distributed amplifiers with matched idle drain terminations. To achieve a gain improvement, the idle terminations are omitted, except for the output mth stage, so that the current and voltage waves along the interstage drain/gate lines are enhanced, and hence the total gain is increased. In this case, the tapered structure offers an additional bandwidth improvement as compared to the case of the uniform-drain cascaded distributed amplifier, with only a small sensitivity-to-impedance ratio variation. Figure 7.34 shows the circuit schematic of

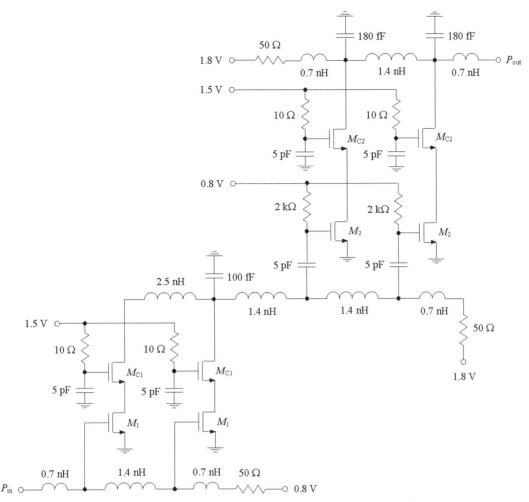

Figure 7.34 Schematic of a cascaded double-cell cascode CMOS distributed amplifier.

a cascaded tapered-drain double-stage cascode CMOS distributed amplifier along with the component parameters [92]. The size of the amplifier transistors M_1 and M_2 was selected to be $W/L = 100\ \mu m/0.18\ \mu m$, whereas that of the cascoded transistors M_{C1} and M_{C2} was selected to be twice the smaller gate periphery as $W/L = 50\ \mu m/0.18\ \mu m$, so that the drain/gate capacitance ratio was equal to 0.25 for $n = 2$. The measured results demonstrated an output driving capability of -2.5 dBm and a gain of 14 dB with a noise figure of 5.5 to 7.5 dB over the bandwidth from 1.0 to 13.8 GHz.

References

[1]   W. S. Percival, "Improvements In and Relating to Thermionic Valve Circuits," British Patent 460,562, Jan. 1937.

[2] E. L. Ginzton et al., "Distributed Amplification," *Proc. IRE*, Vol. 36, pp. 956–969, Aug. 1948.

[3] W. H. Horton, J. H. Jasberg, and J. D. Noe, "Distributed Amplifiers: Practical Considerations and Experimental Results," *Proc. IRE*, Vol. 38, pp. 748–753, July 1950.

[4] H. G. Bassett and L. C. Kelly, "Distributed Amplifiers: Some New Methods for Controlling Gain/Frequency and Transient Responses of Amplifiers Having Moderate Bandwidths," *Proc. IEE, Part III: Radio and Commun. Eng.*, Vol. 101, No. 69, pp. 5–14, 1954.

[5] W. K. Chen, "Distributed Amplifiers: Survey of the Effects of Lumped-Transmission-Line Design on Performance," *Proc. IEE*, Vol. 114, pp. 1065–1074, Aug. 1967.

[6] W. K. Chen, "Distributed Amplification: A New Approach," *IEEE Trans. Electron Devices*, Vol. ED-14, pp. 215–221, Apr. 1967.

[7] D. G. Sarma, "On Distributed Amplification," *Proc. IEE, Part B: Radio and Electronic Eng.*, Vol. 102, pp. 689–697, Nov. 1955.

[8] F. C. Thompson, "Broad-Band UHF Distributed Amplifiers Using Band-Pass Filter Techniques," *IRE Trans. Circuit Theory*, Vol. CT-7, pp. 8–17, Oct. 1960.

[9] W. K. Chen, "The Effects of Grid Loading on the Gain and Phase-Shift Characteristics of a Distributed Amplifier," *IEEE Trans. Circuit Theory*, Vol. CT-16, pp. 134–137, Feb. 1969.

[10] W. K. Chen, "Theory and Design of Transistor Distributed Amplifiers," *IEEE J. Solid-State Circuits*, Vol. SC-3, pp. 165–179, June 1968.

[11] J. A. Gallagher, "High-Power Wide-Band RF Amplifiers," *IRE Trans. Aerospace and Electronic Syst.*, Vol. AES-1, pp. 141–151, Oct. 1965.

[12] L. H. Enloe and P. H. Rogers, "Wideband Transistor Distributed Amplifiers," *1959 IEEE Int., Solid-State Circuits Symp. Dig.*, pp. 44–45.

[13] G. W. McIver, "A Travelling-Wave Transistor," *Proc. IEEE*, Vol. 53, pp. 1747–1748, Nov. 1965.

[14] G. Kohn and R. W. Landauer, "Distributed Field-Effect Amplifiers," *Proc. IEEE*, Vol. 56, pp. 1136–1137, June 1968.

[15] W. Jutzi, "A MESFET Distributed Amplifier with 2 GHz Bandwidth," *Proc. IEEE*, Vol. 57, pp. 1195–1196, June 1969.

[16] J. A. Archer, F. A. Petz, and H. P. Weidlich, "GaAs FET Distributed Amplifier," *Electronics Lett.*, Vol. 17, p. 433, June 1981.

[17] Y. Ayasli et al., "Monolithic GaAs Travelling-Wave Amplifier," *Electronics Lett.*, Vol. 17, pp. 413–414, June 1981.

[18] Y. Ayasli et al., "A Monolithic GaAs 1–13 GHz Traveling-Wave Amplifier," *IEEE Trans. Microwave Theory Tech.*, Vol. MTT-30, pp. 976–981, July 1982.

[19] K. B. Niclas et al., "On Theory and Performance of Solid-State Microwave Distributed Amplifiers," *IEEE Trans. Microwave Theory Tech.*, Vol. MTT-31, pp. 447–456, June 1983.

[20] S. G. Bandy et al., "A 2–20 GHz High-Gain Monolithic HEMT Distributed Amplifiers," *IEEE Trans. Microwave Theory Tech.*, Vol. MTT-35, pp. 1494–1500, Dec. 1987.

[21] E. W. Strid and K. R. Gleeson, "A DC-12 GHz Monolithic GaAs FET Distributed Amplifier," *IEEE Trans. Microwave Theory Tech.*, Vol. MTT-30, pp. 969–975, July 1982.

[22] J. B. Beyer et al., "MESFET Distributed Amplifier Design Guidelines," *IEEE Trans. Microwave Theory Tech.*, Vol. MTT-32, pp. 268–275, Mar. 1984.

[23] B. Kim and H. Q. Tserng, "0.5 W 2–21 GHz Monolithic GaAs Distributed Amplifier," *Electronics Lett.*, Vol. 20, pp. 288–289, Mar. 1984.

[24] Y. Ayasli et al., "Capacitively Coupled Traveling-Wave Power Amplifier," *IEEE Trans. Microwave Theory Tech.*, Vol. MTT-32, pp. 1704–1709, Dec. 1984.

[25] S. N. Prasad, J. B. Beyer, and I. K. Chang, "Power-Bandwidth Considerations in the Design of MESFET Distributed Amplifiers," *IEEE Trans. Microwave Theory Tech.*, Vol. MTT-36, pp. 1117–1123, July 1988.

[26] M. J. Schindler et al., "A *K/Ka*-Band Distributed Power Amplifier with Capacitive Drain Coupling," *IEEE Trans. Microwave Theory Tech.*, Vol. MTT-36, pp. 1902–1907, Dec. 1988.

[27] C. Z. Den Brinker and M. Parkyn, "Amplifiers," British Patent 1,235,472, June 1971.

[28] K. B. Niclas et al., "The Declining Drain Line Lengths Circuit—A Computer Derived Design Concept Applied to a 2–28.5-GHz Distributed Amplifier," *IEEE Trans. Microwave Theory Tech.*, Vol. MTT-34, pp. 427–435, Apr. 1986.

[29] C. Duperrier et al., "New Design Method of Uniform and Nonuniform Distributed Power Amplifiers," *IEEE Trans. Microwave Theory Tech.*, Vol. MTT-49, pp. 2494–2500, Dec. 2001.

[30] J. Gassmann et al., "Wideband, High-Efficiency GaN Power Amplifiers Utilizing a Non-Uniform Distributed Topology," *2007 IEEE MTT-S Int. Microwave Symp. Dig.*, pp. 615–618.

[31] B. M. Green et al., "High Efficiency Monolithic Gallium Nitride Distributed Amplifier," *IEEE Microwave and Guided Wave Lett.*, Vol. 10, pp. 270–272, July 2000.

[32] C. Campbell et al., "A Wideband Power Amplifier MMIC Utilizing GaN on SiC HEMT Technology," *IEEE J. Solid-State Circuits*, Vol. SC-44, pp. 2640–2647, Oct. 2009.

[33] E. Reese et al., "Wideband Power Amplifier MMICs Utilizing GaN on SiC," *2010 IEEE MTT-S Int. Microwave Symp. Dig.*, pp. 1230–1233.

[34] C. Xie and J. Pavio, "A High Efficiency Broadband Monolithic Gallium Nitride Distributed Power Amplifier," *2008 IEEE MTT-S Int. Microwave Symp. Dig.*, pp. 307–310.

[35] L. Zhao, A. Pavio, and W. Thompson, "A 1 Watt, 3.2 VDC, High Efficiency Distributed Power PHEMT Amplifier Fabricated Using LTCC Technology," *2003 IEEE MTT-S Int. Microwave Symp. Dig.*, Vol. 3, pp. 2201–2204.

[36] L. Zhao et al., "A 6 Watt LDMOS Broadband High Efficiency Distributed Power Amplifier Fabricated Using LTCC Technology," *2002 IEEE MTT-S Int. Microwave Symp. Dig.*, pp. 897–900.

[37] S. Lin, M. Eron, and A. E. Fathy, "Development of Ultra Wideband, High Efficiency, Distributed Power Amplifiers Using Discrete GaN HEMTs," *IET Circuits Devices Syst.*, Vol. 3, pp. 135–142, May 2009.

[38] D. E. Meharry et al., "Multi-Watt Wideband MMICs in GaN and GaAs," *2007 IEEE MTT-S Int. Microwave Symp. Dig.*, pp. 631–634.

[39] Y. Ayasli et al., "2–20-GHz GaAs Traveling-Wave Power Amplifier," *IEEE Trans. Microwave Theory Tech.*, Vol. MTT-32, pp. 290–295, Mar. 1984.

[40] S. Masuda et al., "Over 10W C-Ku Band GaN MMIC Nonuniform Distributed Power Amplifier with Broadband Couplers," *2010 IEEE MTT-S Int. Microwave Symp. Dig.*, pp. 1388–1391.

[41] P. N. Shastry and J. B. Beyer, "Bandpass Distributed Amplifiers," *Microwave and Optical Technology Lett.*, Vol. 2, pp. 349–354, Oct. 1989.

[42] P. N. Shastry, A Kajjam, and Z. M. Li, "Bandpass Distributed Amplifier Design Guidelines," *Microwave and Optical Technology Lett.*, Vol. 10, pp. 215–218, Nov. 1995.

[43] N. P. Mehta and P. N. Shastry, "Design Guidelines for a Novel Bandpass Distributed Amplifier," *Proc. 35th Europ. Microwave Conf. Dig.*, Vol. 1, pp. 1–4, 2005.

[44] J. M. Rollett, "Stability and Power-Gain Invariants of Linear Twoports," *IRE Trans. Circuit Theory*, Vol. CT-9, pp. 29–32, Mar. 1962.

[45] M. Riaziat et al., "Feedback in Distributed Amplifiers," *IEEE Trans. Microwave Theory Tech.*, Vol. MTT-38, pp. 212–215, Feb. 1990.

[46] P. Gamand, "Analysis of the Oscillation Conditions Distributed Amplifiers," *IEEE Trans. Microwave Theory Tech.*, Vol. MTT-37, pp. 637–640, Mar. 1989.

[47] H. T. Kim et al., "6–18 GHz MMIC Drive and Power Amplifiers," *J. Semiconductor Technology and Science*, Vol. 2, pp. 125–131, June 2002.

[48] C. Yuen, Y. C. Pao, and N. G. Bechtel, "5-60-GHz High-Gain Distributed Amplifier Utilizing InP Cascode HEMT's," *IEEE J. Solid-State Circuits*, Vol. SC-27, pp. 1434–1438, Oct. 1992.

[49] B. M. Green et al., "High-Power Broad-Band AlGaN/GaN HEMT MMICs on SiC Substrates," *IEEE Trans. Microwave Theory Tech.*, Vol. MTT-49, pp. 2486–2493, Dec. 2001.

[50] K. W. Kobayashi et al., "Multi-Decade GaN HEMT Cascode-Distributed Power Amplifier with Baseband Performance," *2009 IEEE RFIC Symp. Dig.*, pp. 369–372.

[51] R. S. Pengelly et al., "A Review of GaN on SiC High Electron Mobility Power Transistors and MMICs," *IEEE Trans. Microwave Theory Tech.*, Vol. MTT-60, pp. 1764–1783, June 2012.

[52] W. Kennan, T. Andrade, and C. C. Huang, "A 2–18-GHz Monolithic Distributed Amplifier Using Dual-Gate GaAs FETs," *IEEE Trans. Microwave Theory Tech.*, Vol. MTT-32, pp. 1693–1697, Dec. 1984.

[53] R. Santhakumar et al., "Monolithic Millimeter-Wave Distributed Amplifiers Using AlGaN/GaN HEMTs," *2008 IEEE MTT-S Int. Microwave Symp. Dig.*, pp. 1063–1066.

[54] R. Santhakumar et al., "Two-Stage High-Power High-Power Distributed Amplifier using Dual-Gate GaN HEMTs," *IEEE Trans. Microwave Theory Tech.*, Vol. MTT-59, pp. 2059–2063, Aug. 2011.

[55] K. W. Kobayashi et al., "A 2–32 GHz Coplanar Waveguide InAlAs/InGaAs-InP HBT Cascode Wave Distributed Amplifier," *1995 IEEE MTT-S Int. Microwave Symp. Dig.*, pp. 215–218.

[56] J. P. Fraysse et al., "A 2W, High Efficiency, 2-8GHz, Cascode HBT MMIC Power Distributed Amplifier," *2000 IEEE MTT-S Int. Microwave Symp. Dig.*, pp. 529–532.

[57] J. Aguirre and C. Plett, "50-GHz SiGe HBT Distributed Amplifiers Employing Constant-*k* and *m*-Derived Filter Sections," *IEEE Trans. Microwave Theory Tech.*, Vol. MTT-52, pp. 1573–1579, May 2004.

[58] Y. Chen et al., "A 11GHz Hybrid Paraphase Amplifier," *1986 IEEE Int. Solid-State Circuits Conf. Dig.*, pp. 236–237.

[59] S. Mohammadi et al., "Design Optimization and Characterization of High-Gain GaInP/GaAs HBT Distributed Amplifiers for High-Bit-Rate Telecommunication," *IEEE Trans. Microwave Theory Tech.*, Vol. MTT–48, pp. 1038–1044, June 2000.

[60] J. Chen, and A. M. Niknejad, "Design and Analysis of a Stage-Scaled Distributed Power Amplifier," *IEEE Trans. Microwave Theory Tech.*, Vol. MTT-59, pp. 1274–1283, May 2011.

[61] H. Y. Chang et al., "Design and Analysis of a DC–43.5-GHz Fully Integrated Distributed Amplifier Using GaAs HEMT-HBT Cascode Gain Stages," *IEEE Trans. Microwave Theory Tech.*, Vol. MTT-59, pp. 443–455, Feb. 2011.

[62] A. Martin, A. Mortazawi, and B. C. De Loach, "A Power Amplifier Based on an Extended Resonance Technique," *IEEE Microwave and Guided Wave Lett.*, Vol. 5, pp. 329–331, Oct. 1995.

[63] A. Martin, A. Mortazawi, and B. C. De Loach, "An Eight-Device Extended-Resonance Power-Combining Amplifier," *IEEE Trans. Microwave Theory Tech.*, Vol. MTT-46, pp. 844–850, June 1998.

[64] A. Martin and A. Mortazawi, "A New Lumped-Elements Power-Combining Amplifier Based on an Extended Resonance Technique," *IEEE Trans. Microwave Theory Tech.*, Vol. MTT-48, pp. 1505–1515, Sep. 2000.

[65]   X. Jiang and A. Mortazawi, "A Broadband Power Amplifier Design Based on the Extended Resonance Power Combining Technique," *2005 IEEE MTT-S Int. Microwave Symp. Dig.*, pp. 835–838.

[66]   R. Lohrman, H. Gill, and S. Koch, "A Novel Distributed Multicell Multistage Amplifier Structure," *Proc. 33rd Europ. Microwave Conf.*, pp. 379–382, 2003.

[67]   J. Y. Liang and C. S. Aitchison, "Gain Performance of Cascade of Single-Stage Distributed Amplifier," *Electronics Lett.*, Vol. 31, pp. 1260–1261, July 1995.

[68]   B. Y. Banyamin and M. Berwick, "Analysis of the Performance of Four-Cascaded Single-Stage Distributed Amplifiers," *IEEE Trans. Microwave Theory Tech.*, Vol. MTT-48, pp. 2657–2663, Dec. 2000.

[69]   K. L. Deng, T. W. Huang, and H. Wang, "Design and Analysis of Novel High-Gain and Broad-Band GaAs pHEMT MMIC Distributed Amplifiers with Travelling-Wave Gain Stages," *IEEE Trans. Microwave Theory Tech.*, Vol. MTT-51, pp. 2188–2196, Nov. 2003.

[70]   A. S. Virdee and B. S. Virdee, "2–18GHz Ultra-Broadband Amplifier Design Using a Cascaded Reactively Terminated Single Stage Distributed Concept," *Electronics Lett.*, Vol. 35, pp. 2122–2123, Nov. 1999.

[71]   A. S. Virdee and B. S. Virdee, "Experimental Performance of Ultra-Broadband Amplifier Design Concept Employing Cascaded Reactively Terminated Single-Stage Distributed Amplifier Configuration," *Electronics Lett.*, Vol. 36, pp. 1554–1556, Aug. 2000.

[72]   A. S. Virdee and B. S. Virdee, "A Novel High Efficiency Multioctave Amplifier Using Cascaded Reactively Terminated Single-Stage Distributed Amplifier Configuration," *2001 IEEE MTT-S Int. Microwave Symp. Dig.*, Vol. 1, pp. 519–522.

[73]   K. B. Niclas and R. R. Pereira, "The Matrix Amplifier: A High-Gain Module for Multioctave Frequency Bands," *IEEE Trans. Microwave Theory Tech.*, Vol. MTT-35, pp. 296–306, Mar. 1987.

[74]   K. B. Niclas, R. R. Pereira, and A. P. Chang, "On Power Distribution in Additive Amplifiers," *IEEE Trans. Microwave Theory Tech.*, Vol. MTT-38, pp. 1692–1700, Nov. 1990.

[75]   K. B. Niclas, R. R. Pereira, and A. P. Chang "A 2–18 GHz Low-Noise/High-Gain Amplifier Module," *IEEE Trans. Microwave Theory Tech.*, Vol. MTT-37, pp. 198–207, Jan. 1989.

[76]   K. W. Kobayashi et al., "A 6–21-GHz Monolithic HEMT 2 × 3 Matrix Distributed Amplifier," *IEEE Microwave and Guided Wave Lett.*, Vol. 3, pp. 11–13, Jan. 1993.

[77]   K. B. Niclas and R. R. Pereira, "On the Design and Performance of a 6–18 GHz Three-Tier Matrix Amplifier," *IEEE Trans. Microwave Theory Tech.*, Vol. MTT-37, pp. 1069–1077, July 1989.

[78]   C. Paoloni and S. D'Agostino, "A Design Procedure for Monolithic Matrix Amplifier," *IEEE Trans. Microwave Theory Tech.*, Vol. MTT-45, pp. 135–139, Jan. 1997.

[79]   C. Paoloni, "HEMT-HBT Matrix Amplifier," *IEEE Trans. Microwave Theory Tech.*, Vol. MTT-48, pp. 1308–1312, Aug. 2000.

[80]   M. Ferndahl and H. O. Vickes, "The Matrix Balun—A Transistor-Based Module for Broadband Applications," *IEEE Trans. Microwave Theory Tech.*, Vol. MTT-57, pp. 53–60, Jan. 2009.

[81]   B. M. Ballweber, R. Gupta, and D. J. Allstot, "A Fully Integrated 0.5–5.5-GHz CMOS Distributed Amplifier," *IEEE J. Solid-State Circuits*, Vol. SC-35, pp. 231–239, Feb. 2000.

[82]   G. A. Lee, H. Ko, and F. De Flaviis, "Advanced Design of Broadband Distributed Amplifier Using a SiGe BiCMOS Technology," *2003 IEEE RFIC Symp. Dig.*, pp. 703–706.

[83]   J. O. Plouchart et al., "A 4–91-GHz Travelling-Wave Amplifier in a Standard 0.12-μm SOI CMOS Microprocessor Technology," *IEEE J. Solid-State Circuits*, Vol. SC-39, pp. 1455–1461, Sep. 2004.

[84] J. Kim and J. F. Buckwalter, "A 92 GHz Bandwidth Distributed Amplifier in a 45-nm SOI CMOS Technology," *IEEE Microwave and Wireless Comp. Lett.*, Vol. 21, pp. 329–331, June 2011.

[85] F. Ellinger, "60-GHz SOI CMOS Traveling-Wave Amplifier with NF below 3.8 dB from 0.1 to 40 GHz," *IEEE J. Solid-State Circuits*, Vol. SC-40, pp. 553–558, Feb. 2005.

[86] L. H. Lu, T. Y. Chen, and Y. J. Lin, "A 32-GHz Non-Uniform Distributed Amplifier in a 0.18-μm CMOS," *IEEE Microwave and Wireless Comp. Lett.*, Vol. 15, pp. 745–747, Nov. 2005.

[87] P. Heydari, "Design and Analysis of a Performance-Optimized CMOS UWB Distributed LNA," *IEEE Trans. Solid-State Circuits*, Vol. SC-42, pp. 1892–1905, Sep. 2007.

[88] H. T. Ahn and D. J. Allstot, "A 0.5–8.5-GHz Fully Differential CMOS Distributed Amplifier," *IEEE J. Solid-State Circuits*, Vol. SC-37, pp. 985–993, Aug. 2002.

[89] A. Kopa and A. B. Apsel, "Alternative *m*-Derived Termination for Distributed Amplifiers," *2009 IEEE MTT-S Int. Microwave Symp. Dig.*, pp. 921–924.

[90] K. Entesari, A. R. Tavakoli, and A. Helmy, "CMOS Distributed Amplifiers with Extended Flat Bandwidth and Improved Input Matching Using Gate Line with Coupled Inductors," *IEEE Trans. Microwave Theory Tech.*, Vol. MTT-57, pp. 2862–2871, Dec. 2009.

[91] J. C. Chien and L. H. Lu, "40-Gb/s High-Gain Distributed Amplifiers with Cascaded Gain Stages in 0.18-μm CMOS," *IEEE J. Solid-State Circuits*, Vol. SC-42, pp. 2715–2725, Dec. 2007.

[92] A. Worapishet, I. Roopkom, and W. Surakampontorn, "Theory and Bandwidth Enhancement of Cascaded Double-Stage Distributed Amplifiers," *IEEE Trans. Circuits and Systems-I: Regular Papers*, Vol. CAS-57, pp. 759–772, Apr. 2010.

CHAPTER 8

Distributed Power Amplifiers

A distributed power amplifier is simply a distributed amplifier in which high-power devices are used instead of small-signal transistor devices [1–4]. Distributed amplifiers have already demonstrated high performance for small-signal broadband operation, but have limited output power performance [5–9]. They represent an attractive candidate for high-power SDR applications [10–16]. Each transistor demonstrates a strongly frequency-dependent power behavior so that the overall output power is only a small fraction of the combined power capabilities for all active devices [17–19]. The limitation to achieving high output power with a distributed power amplifier can be addressed on both the device technology level and circuit design level [19]. The realization of distributed power amplifiers has posed a significant challenge due to the electrical and thermal limitations of GaAs or HBT transistor technology. In recent years, AlGaN/GaN technology has established itself as a strong contender for such applications, because of its large electron velocity, bandgap, breakdown voltage V_{bk} for current-gain cutoff frequency f_τ, and sheet carrier concentration [15, 16, 19–21]. Few recent distributed power amplifier circuit design techniques, such as drain impedance tapering [12–14], nonuniform device periphery [2, 3], cascode [21, 22], cascaded nonidentical transistors [10], dual-fed [23], and extended resonance power combining [24], lead to high output power performance over the desired bandwidth frequency of operation.

8.1 Dual-Fed Distributed Power Amplifier

In a conventional distributed amplifier, current combining efficiency at the drain line is poor due to the fact that the current splitting on the drain line into two branches forms waves traveling both toward the load termination and toward the dummy termination. The tapered drain-line distributed amplifier (DA) [13,14,25,26] eliminates the drain-line reverse wave by suitable tapering of the drain-line impedance. The dual-fed distributed power amplifier topology proposed by Aitchison et al. [27–29] allows efficient power combining at the load termination. A similar technique that realizes a Lange coupler and Wilkinson combiner at the input and output of the DA to improve output power, gain, and PAE has been studied by D'Agostino and Paoloni [30–32]. An approach by Liang and Aitchison [29] showed efficiency improvement by reducing backward wave propagation at the drain transmission line. Eccleston's approach [33, 34] investigated when a microstrip line periodically loaded with short open-circuit stubs can be used in place of a transmission line to reduce the size.

This section discusses the high-power, dual-fed distributed power amplifier (DPA) with termination adjustment, which demonstrated remarkable bandwidth-efficiency improvement over conventional distributed amplifiers. Instead of a lumped transmission line, discrete approach LC components are utilized and followed by selection of an optimum resonance frequency of the splitter/combiner [35, 36].

The conventional distributed amplifier consists of an input port on one side of the gate line and an output port on the opposite side of the drain line. In a basic dual-fed DPA, the unused gate and drain ports, which are known as the dummy termination, will be terminated in appropriate characteristic impedances. The two ports of the gate line are simultaneously fed and both ends of the drain line are assumed to be output ports. In this work, the termination of the dual-fed DPA is modified, where remarkable bandwidth extension is possible with optimum impedance termination selection at both end ports of the gate line [23], as shown in Figure 8.1. In a similar manner, by adjusting the termination impedance at the drain line, efficiency is maximized over the entire bandwidth range.

Under linear conditions superposition will apply to its operation and it follows that the output due to forward gain from the left to right gate input signal will appear at the right-hand drain port and similarly the output due to forward gain from the right to the left gate input signal will appear at the left-hand gain port [29]. These two output signals can be combined to give the total output power. The reverse gain must also be taken into account and, if the phase of the splitter/combiner is appropriate, the currents due to reverse gain flow out of the same output port and are added vectorially to the forward gain [28].

Reference [31] explains that if both terminations' end ports are terminated with appropriate impedance values (i.e., higher than Z_0), gain improvement is achieved while maintaining the same device structure. D'Agostino and Paoloni have demonstrated significant output power, gain and efficiency improvement over a conventional DA by applying Wilkinson [31] and Lange couplers [30, 37]. Nevertheless, this work is focused on improving the bandwidth-efficiency response over conventional DA with modified dual-fed distributed power amplifier (DFDA). Beyer et al. have shown analytically that the gate-line attenuation α_g is more sensitive to frequency response than the drain-line attenuation α_d [5].

The proposed concept in this work provides optimum impedance at both ends of the DA distributed amplifier gate line to extend the bandwidth response. With reference to Figure 8.2, $Z_a(\omega) = R_a(\omega) + jX_a(\omega)$ and $Z_b(\omega) = R_b(\omega) + jX_b(\omega)$ can be defined as Thevenin impedances for the driving sources e_a and e_b, respectively, at

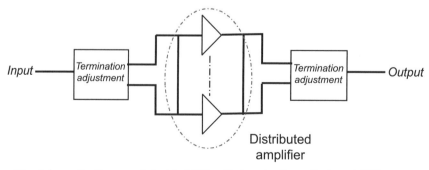

Figure 8.1 Schematic diagram of dual-fed DPA with termination adjustment [23].

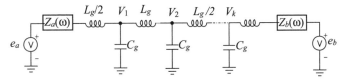

Figure 8.2 Termination adjustment is placed at both ends of the input of DPA gate line to provide efficient gate-line adjustment over a wide frequency range.

both ends of the distributed amplifier gate line. Figure 8.2 can be simplified to Figure 8.3(a). Neglecting e_b as shown in Figure 8.3(b), voltage V_1 can be expressed as

$$V_1 = \frac{e_a}{\dfrac{R_a(\omega) + jX_a(\omega)}{Z_{OT}(\omega/\omega_c)} + 1} \tag{8.1}$$

where $Z_{OT}(\omega/\omega_c)$ represents the T-section constant-k LC network terminated with matched characteristic impedance [28]. Note that $\omega_c = 1/\sqrt{L_g C_g}$ and is known as the line cutoff frequency.

Equation (8.1) can be investigated to gather the V_1 behavior over frequency. Assuming e_a is set to $V_1 = 1 \angle 0°\text{V}$, an optimum $X_a(\omega)$ can be numerically determined for various cases of $R_a(\omega)$. Figure 8.4 shows a plot of V_1 in magnitude; note that the magnitude can be increased over a wide frequency range with proper selection of $R_a(\omega) + jX_a(\omega)$.

Superposition can be now applied to Figure 8.3(a) considering also e_b (and Z_b). The total resulting voltage $V_1(\omega)$ is expressed by

$$V_1 = \frac{Z_t^L e_a}{Z_t^L + j\omega L_g/2 + Z_a(\omega)} + \frac{Z_t^L e_b}{Z_t^R + j\omega L_g/2 + Z_b(\omega)} \tag{8.2}$$

where

$$Z_t^L = \frac{j\omega L_g/2 + Z_b(\omega)}{1 - \omega^2/4\omega_c^2 + j\omega C_g Z_b(\omega)} \quad \text{and} \quad Z_t^R = \frac{j\omega L_g/2 + Z_a(\omega)}{1 - \omega^2/4\omega_c^2 + j\omega C_g Z_a(\omega)}$$

For equal injection at both ends of the gate line, $e_a = e_b$, the voltage response $V_1(\omega)$ from (8.2) can be widened over a wide frequency range in similar manner by terminating optimum $Z_a(\omega)$ and $Z_b(\omega)$. Termination impedance adjustments

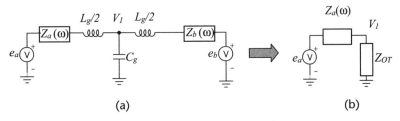

(a)   (b)

Figure 8.3 (a) Simplification of Figure 8.2 and (b) exclusion of e_b.

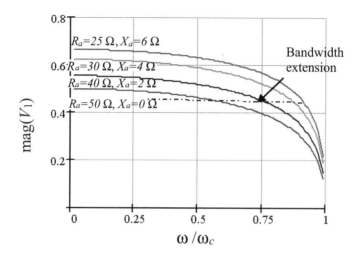

Figure 8.4 Voltage $V_1(\omega)$ over normalized frequency (ω / ω_c) with proper selection of $R_a(\omega) + jX_a(\omega)$.

of $Z_a(\omega)$ and $Z_b(\omega)$ are accomplished by tuning the resonance frequency f_o of the Wilkinson splitter at the gate-line input. A termination adjustment network is realized with the Wilkinson splitter approach. The splitter offers broad bandwidth and equal phase characteristics at each of its output ports [35]. The splitter employs $\lambda/4$ transmission-line sections at the design center frequency, which can have unrealistic dimensions at low RF frequencies, where the wavelength is large [36]. Due to size constraints, a lumped-element equivalent network that replaces the $\lambda/4$ transmission line would be preferable and is shown in Figure 8.5(a). This network is equivalent to the original only at the center frequency f_o. Consequently, the expected performance (insertion loss, return loss, isolation, and so on) should be similar to that exhibited by the distributed-form divider for a narrow bandwidth centered in f_o, wide enough for most applications. The "π" LC equivalent networks exhibit lowpass behavior [Figure 8.5(b)], rejecting high frequencies, while the response of the classical splitter repeats at odd multiples of the center frequency ($3f_o$ and $5f_o$, mainly) [36].

A 1-GHz microstrip Wilkinson splitter or combiner can occupy about 6 cm² on FR-4 PCB, whereas this lumped-element version occupies less than 1 cm². Lumped-element circuits with a higher Q than distributed circuits have the advantage of smaller size, low cost, and wide bandwidth characteristics [36]. The element values are given by the following equations:

$$C = \frac{1}{2\pi f_o Z_o} \tag{8.3}$$

$$L = \frac{Z_o}{2\pi f_o} \tag{8.4}$$

where Z_o and f_o are characteristic impedance and resonance frequency of the lumped elements, respectively, as shown in Figure 8.5(b).

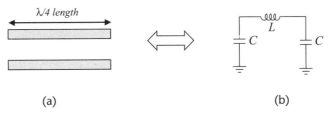

Figure 8.5 Lumped element π-section: (a) transmission line and (b) lumped elements.

To investigate the output drain line of the modified dual-fed DPA, consider Figure 8.6. Forward currents (I_{o1} and I_{o2}), summing in phase on the correspondent output ports, are considered [31]. For simplicity, the circuit is analyzed at low frequencies. By assuming that the input voltage of each gate line is $V_i/\sqrt{2}$, the output current toward each output port is expressed as

$$I_{oi} = \frac{nR_{ds}}{\left(nZ_{0o} + 2R_{ds}\right)} g_m \frac{V_i}{\sqrt{2}} \qquad 1 = 1,2,\dots,n \qquad (8.5)$$

where

n = number of FETs

R_{ds} = drain-source resistance of the active devices

g_m = device transconductance

Z_{0o} = load impedance of the output ports (Z_{0a} and Z_{0b}).

The total output power P_{out}, computed as the sum of the contribution from I_{o1} and I_{o2}, is given by

$$P_{\text{out}} = \frac{1}{2}\left|I_{o1}\right|^2 Z_{0a} + \frac{1}{2}\left|I_{o2}\right|^2 Z_{0b} \qquad (8.6)$$

and when $Z_{0a} = Z_{0b} = Z_0$, P_{out} can be simplified as follows:

$$P_{\text{out}} = \frac{n^2 R_{ds}^2 Z_0^2}{\left(nZ_0 + 2R_{ds}\right)^2} g_m^2 \frac{V_i^2}{2} \qquad (8.7)$$

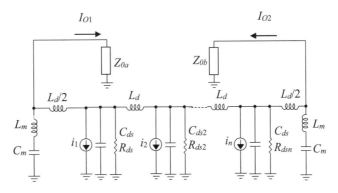

Figure 8.6 Theoretical circuit analysis of the drain of the modified dual-fed DPA [31].

The power delivered from input source P_{in} is

$$P_{\text{in}} = \frac{1}{2}\frac{V_i^2}{Z_o} \tag{8.8}$$

where V_i represents the supply voltages, and Z_o is the gate-line characteristic imped-ance. The overall gain of the topology, G, is given by

$$G = \frac{P_{\text{out}}}{P_{\text{in}}} = \frac{n^2 R_{ds}^2 Z_0^2}{\left(nZ_0 + 2R_{ds}\right)^2} g_m^2 \tag{8.9}$$

From (8.9), when increasing Z_0, higher gain and output power can be obtained. Unfortunately, the correspondent degradation of the output reflection is a critical limitation [31]. The gain as a function of the output loads Z_{0a} and Z_{0b} for different number of FETs ($n = 2, 3, 4$) and the correspondent return loss are discussed in [31]. One can conclude from [31] that the correct selection of Z_{0a} and Z_{0b} for a chosen n will lead to optimum power performance without trading off output return loss. In this work, it is worthwhile to select the f_o of the drain line, to improve the output matching over the entire bandwidth.

Therefore, a medium-power device, for example, an LDMOS n-type MOSFET packaged device[1] is suitable. Low dc supply operation is required for the device, which typically is about 7.5V. The breakdown voltage V_{bk} of the device is ~25V, which is close to the computation value from (4.47). The drain loading effect of the device is not significant, but $X_{\text{opt}}(\omega)$ is important because it determines the drain-line cutoff frequency ω_c. The effective device input and output capacitances are $C_{\text{in}} = 14$ pF and $C_{\text{out}} = 8$ pF, respectively. They are extracted by means of device modeling (with inclusion of packaged properties). Therefore, effective drain-line elements L_i are synthesized by means of (4.58) to form the desired ω_c (~800 MHz). The dummy drain termination is eliminated to improve device efficiency.

Design of the modified dual-fed DPA is very similar to that of a high-efficiency DA (as discussed in Chapter 5); additionally, a termination adjustment network is implemented. Hence, the method of synthesizing a DA, comprised of the steps of device selection, determining an appropriate topology, synthesizing gate/drain-line elements from device packaged values, and so on, is applied. High-Q discrete inductors and capacitors from Coilcraft, Inc., and Murata, Inc., are selected for their lower equivalent series resistor (ESR) value and minimum part tolerance. The gain of the amplifier increases with additional devices until the optimum number of devices at a given frequency is reached [36]. Any device added beyond this optimum number is not driven sufficiently to excite the signal in the drain line, which will induce attenuation in the extra section of the drain line. Based on (4.37), n_{opt} is 3.26 for an RD01MUS1 device. Due to this, three devices have been used in our design.

Inductances of gate and drain lines could be determined from the value of the line image impedance and ω_c. As for the RF standard, gate- and drain-line image impedance are set to 50Ω. In order for the currents on the drain line to interfere

---

[1] The device part number is RD01MUS1, from Mitsubishi Corp., Kanagawa, Japan.

constructively (add in phase), the phase shift per section on gate and drain lines must be equalized. Synchronization of the phase velocities between the gate and drain lines is achieved by adjusting the inductance value of the gate line (the capacitively coupled technique [49], in which discrete capacitors in series form to each gate of the transistor, is not implemented). The *m*-derived filter section serves wideband image impedance image termination and is placed at both ends of gate and drain lines. The *m*-derived half section is designed with *m* = 0.6 for best flatness across the bandwidth [39]. For the termination adjustment network, *L* and *C* are selected according to (8.3) and (8.4), where an optimum f_o and Z_o will be identified in simulation analysis. For reasons of convenience, Z_o is set to 70.7Ω, but f_o will be determined. The modified dual-fed DPA topology is shown in Figure 8.7.

Figure 8.8 shows that by applying a termination adjustment with proper selection of f_o at the gate line, an improvement in bandwidth is obtained, although a

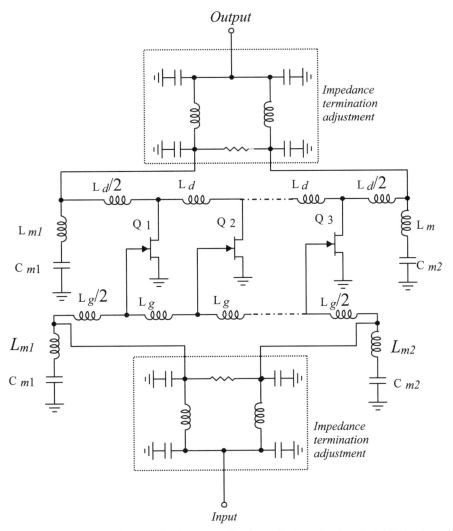

Figure 8.7 Modified dual-fed DPA having a termination adjustment network at both gate and drain lines. The *m*-derived filter section is terminated at both ends of the gate- and drain-line termination [23].

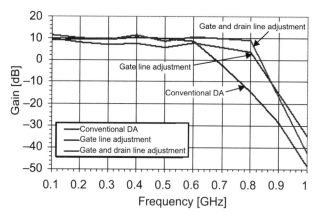

Figure 8.8 Gain response for a few cases: conventional distributed amplifier, applying termination adjustment at gate line, and termination adjustment at gate and drain line.

small degradation in gain is expected due to loss of the splitter. The measured loss of the splitter is characterized as approximately 3 dB over the entire bandwidth. Nevertheless, the gain of the amplifier from low frequency can be boosted in a similar way at the drain line. The f_o selection at the drain line does not improve bandwidth performance. As shown in Figure 8.9, the gain response of the modified DFDA is improved by 200 MHz compared to the conventional distributed amplifier. Simulation analysis (with harmonic balance) of the modified dual-fed distributed amplifier with a nonlinear device and passive models showed that optimum selection of f_o (i.e., $f_o \sim 600$ MHz) led to bandwidth extension with minimum gain peaking (see Figure 8.8).

Hence, as a guideline, tuning the f_o of the termination adjustment network in a range that is less than the line cutoff frequency f_c yet greater than the conventional distributed amplifier center frequency can lead to significant bandwidth extension and is a promising aspect over the conventional distributed amplifier.

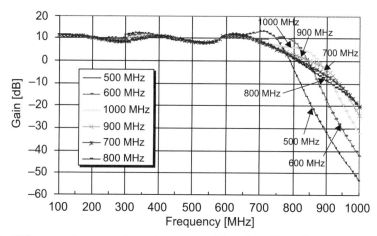

Figure 8.9 Gain versus frequency for various cases of f_o tuning. Fine selection of $f_o \sim 600$ MHz leads to bandwidth extension with minimum gain peaking.

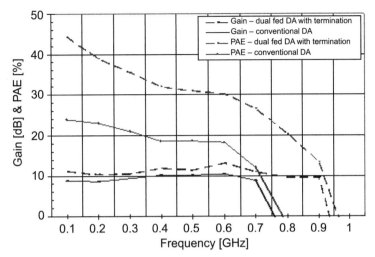

Figure 8.10 Gain and PAE comparison for modified dual-fed DPA and conventional distributed amplifier having same input and output artificial transmission line and a dc biasing scheme at the same condition (V_{GS} = 2.1V and V_{DS} = 7.8V) and same P_{in} = 17 dBm.

Performance of the three-section modified dual-fed DPA is compared to a conventional distributed amplifier. Both topologies used the same device, same input, and same output artificial transmission line, as well as a dc biasing scheme at the same condition (V_{GS} = 2.1V and V_{DS} = 7.8V) and same P_{in} = 17 dBm for both amplifiers. The resonance frequency f_o is selected to be 600 MHz. The gain of the conventional distributed amplifier is 9.5 ± 1 dB, whereas for the modified dual-fed DPA it increased by ~2 dB across the 100- to 900-MHz bandwidth (Figure 8.10). The PAE for a conventional distributed amplifier is lower than 20%, whereas the PAE for the modified dual-fed DPA is increased by ~10%, as shown in Figure 8.10. The remarkable achievement of the modified dual-fed DPA in the power-efficiency operation range is tremendously improved by 200 MHz over a

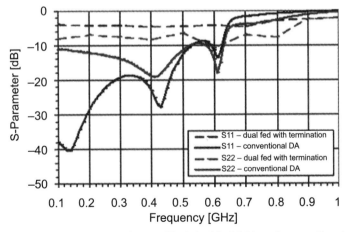

Figure 8.11 *S*-parameter comparison for modified dual-fed DPA and conventional distributed amplifier having the same input and output artificial transmission line and a dc biasing scheme at the same condition (V_{GS} = 2.1V and V_{DS} = 7.8V) and same P_{in} = 17 dBm.

conventional distributed amplifier. Refer to Figure 8.11 for the *S*-parameter data over the bandwidth of interest.

8.2   Tapered Termination Cascaded Distributed Power Amplifier

Several gain stage configurations that are widely used in distributed amplifiers are shown in Figure 8.12 [10, 13, 38]. Figure 8.12(a) is a common-source type that provides a decent gain and very large bandwidth. Figure 8.12(b) is a cascode structure used to enhance reverse isolation. This structure does not provide higher transconductance g_m than the common-source transistor and, thus, does not have considerable gain advantage over that in Figure 8.12(a). Figure 8.12(c) shows the cascade common-source gain structure, where two identical transistors are connected with each other through a peaking inductor and a series resistor [12]. Figure 8.12(d) shows a proposed topology for two cascaded, nonidentical transistors with interstage tapered impedance [10].

Due to the loading effect of the bigger size transistor, dc-RF energy conversion is not optimum as the frequency increases toward the cutoff frequency. These amplifiers, in fact, have never exhibited high power and high gain performance simultaneously. Beyer et al. have showed that the gain in a conventional distributed amplifier cannot be increased indefinitely by adding more sections [5]. However, few works have shown high output power with distributed amplifier topology while preserving reasonable gain [14, 25, 22, 40].

Achieving high output power with a distributed amplifier is very challenging. Conventional distributed amplifier design is different from a reactively matched

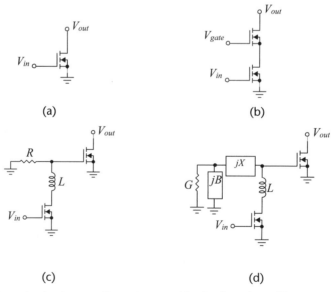

(a)   (b)

(c)   (d)

Figure 8.12   Several gain stage configurations used in distributed amplifiers: (a) common-source, (b) cascode [38], (c) two cascaded, identical transistors with common-source gain [39], and (d) cascaded nonidentical transistors and interstage tapered impedance [12].

design because the drain line is fixed to approximately 25Ω with constant-k ladder [41], and could be a higher value if the drain line is tapered. RF current swing is the main limitation to power increases due to the high impedance of the tapered drain line [14]. To increase power, the device gate periphery must be increased, but this will increase input capacitance C_{gs} as well. A capacitively coupled technique [49] allows for a power-handling increase with constant bandwidth and gain. However, this technique never demonstrated high gain and high output power simultaneously.

Many techniques have been reported to improve gate-line matching problems in distributed amplifiers such as tapering of device gate widths [42], tapering the capacitors connected in series with transistor inputs [43], tapering the gate line to gain equal voltages at the transistors by moving the characteristic impedance higher than 50Ω [50], and tapering the gate line to reduce gate-line mismatching [41]. Image impedance of the constant-k filter causes gain peaking (expansion) close to cutoff frequency ω_c. The first attempt to achieve a wideband matching solution using nonidentical high-f_τ transistors in a cascaded DPA is discussed in [13], where an adjustable interstage matching network between the two nonidentical high-f_τ transistors is introduced.

A constant-k network isolates the parasitic capacitances of the transistors to form a wide bandwidth response with lumped inductance. This frequency-dependent impedance implies that fixed impedance (e.g., 50Ω termination) cannot provide ideal matching transformation. A constant real termination causes ripples as the signal moves toward the cutoff frequency due to unmatched power that reflects back to the input since the imaginary part has not been canceled [41]. It is worthwhile to trade off the low-frequency matching to improve matching at high frequencies.

A high-f_τ transistor typically has a lower input parasitic capacitance C_{gs} and breakdown voltage V_{bk}. Examples include SiGe, HBT, and pHEMT devices coupled to the input of the power transistor (i.e., GaN HEMT device) with tapered imped-ance termination Z_T, as shown in Figure 8.13. Note that Z_T improves matching near the cutoff frequency ω_c with minimum trading off at low frequencies. Typically, the output capacitance C'_{ds} of the high-f_τ transistor is much lower than C''_{gs} of the power transistor. In this work we make the assumption that $C''_{gs} \approx 2 \times C'_{ds}$; however, the concept is applicable to any ratio of C''_{gs}/C'_{ds}. Figure 8.14 shows a general con-figuration for two nonidentical transistors (high-f_τ transistor and power transistor) terminated with interstage tapered impedance matching. The impedance Z_{in} seen by

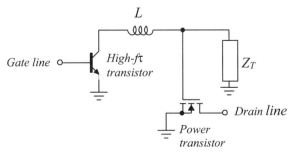

Figure 8.13   Schematic of a cascaded DPA with nonidentical transistors and controlled interstage tapered impedance for wideband solution.

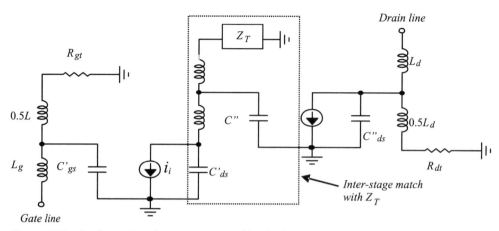

Figure 8.14 Configuration of a two-stage nonidentical transistor. The first transistor is high-f_τ and the second is a power transistor. An interstage matching configuration with tapered termination is included.

the current source i_i is illustrated in Figure 8.15. The Z_T could be proposed with any suitable broadband interstage impedance termination. In this work, a proposal for Z_T that consists of three elements (jX, jB, and G) is used, as shown in Figure 8.16.

To begin the explanation of the concept technique shown in Figure 8.15, the impedance Z_{in} seen by the current source i_i will be analyzed. Figure 8.16, which is a simplified schematic of Figure 8.15, will be used. The analysis will focus on achieving optimum matching up to ω_c in order to overcome reflection (poor matching) caused by a constant-k network near ω_c. Note that V_{gs} has proportional behavior to Z_{int}. The impedance Z_{int} is given by

$$Z_{int} = \frac{4 - 4XB - 2\omega LB + j2(\omega GL + 2GX)}{4G - 2\omega GXC - \omega^2 GLC + j(2\omega C + 4B - 2\omega XCB - \omega^2 LCB)} \qquad (8.10)$$

where L and C values will determine the cutoff frequency ω_c of the line. From Figure 8.16, $C/2$ is referred to C''_{gs}. The elements, for example, G, X, and B, are shown in Figure 8.12(d).

The real and imaginary parts of Z_{int} at ω_c are analyzed as a function of jX and jB. A numerical example is given for illustration purposes, where $L = 6.93$ nH and

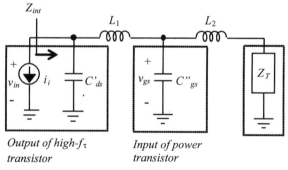

Figure 8.15 Schematic to illustrate the impedance Z_{int}, which is seen by the current source i_i.

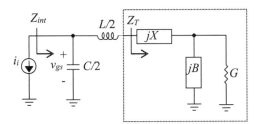

Figure 8.16 Simplified version of Figure 8.15 that explains Z_{int}. Note that Z_T consists of three elements: jX, jB, and G.

$C = 2.77$ pF, providing $\omega_c = 2.3$ GHz. Careful selection of G is required to minimize by variation of Z_{int} real over the entire bandwidth response. The improved match at ω_c (68Ω real part and zero imaginary) requires $B = 0.02076$ and $X = 10$ (for case $G = 0.018$), as shown in Figure 8.17. For the same value of $B = 0.02076$ (while G is fixed), to keep the Z_{int} imaginary null, various values of X can be chosen. However, to obtain an optimum design across the bandwidth, Z_{int} real at ω_c must be as closer to a low frequency. Therefore, by selecting an appropriate termination at any frequency, the matching can be improved. As a result, the interstage termination impedance is tapered.

The plot in Figure 8.18 shows the frequency response of the impedance Z_{int}. For comparison, the fixed termination impedance (50Ω) is included in the plot. From the plot, the real part of Z_{int} is almost constant, for example, almost 55Ω over the entire bandwidth, and the imaginary part is null. As for the conventional approach (fixed termination), strong peaking in the real part is observed as the frequency reaches close to ω_c and the imaginary part deviates to capacitance. From Figure 8.16, V_{gs} will have proportional behavior to Z_{int}.

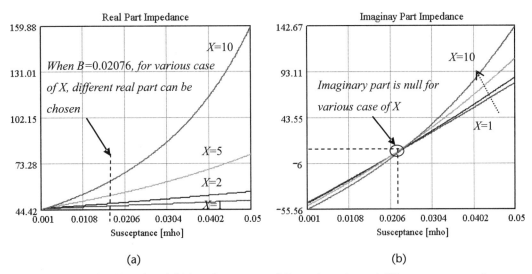

(a)   (b)

Figure 8.17 The (a) real and (b) imaginary parts of Z_{int} evaluated at cutoff frequency versus B (susceptance) for a few cases of X (for fixed G of 0.018).

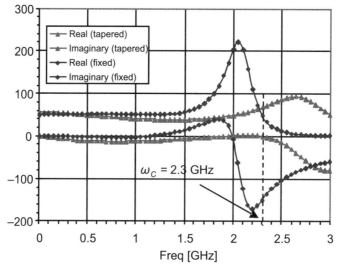

Figure 8.18 Real and imaginary parts of Z_{int} with tapered impedance (line with triangle) and fixed termination (line with diamond).

To extend the explanation of the concept for any ratio of C''_{gs}/C'_{ds}, Figure 8.19 is used. In practical application, C''_{gs} of the power transistor could be more than 2 times that of C'_{ds} of the high-f_τ transistor. The value of C''_{gs} is approximately 2.4 times that of the C'_{ds} for a gate periphery ratio of 8 times [13]. To deliver maximum power from the current source i_i, the impedance Z_{int} must have zero imaginary part and gain peaking must be minimal for stability considerations. In a conventional method (fixed termination), the imaginary part behaves at the capacitive region near ω_c, and Z_{int} seen by the current source i_i due to any ratio of C''_{gs}/C'_{ds} is given as

$$Z_{int} = \frac{\dfrac{2Z_T}{mC'_{ds}} - 2\omega^2 L Z_T + \dfrac{j3\omega L}{mC'_{ds}} - j\omega^3 L^2}{\dfrac{2}{mC'_{ds}} - \omega^2 L - \dfrac{3\omega^2 L}{m} + \omega^4 L^2 C'_{ds} + \dfrac{j2\omega Z_T}{m} + j2\omega Z_T - j2\omega^3 L C'_{ds} Z_T} \tag{8.11}$$

The benefit of the proposed Z_T network is significant when the m-ratio increases. For instance, consider $m = 2.4$. Again, a numerical example is given for illustration purposes; $L = 6.93$ nH and $C = 2.77$ pF are selected, providing $\omega_c = 2.3$ GHz.

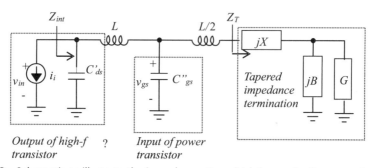

Figure 8.19 Schematic to illustrate the impedance Z_{int}, which is seen by the current source i_i.

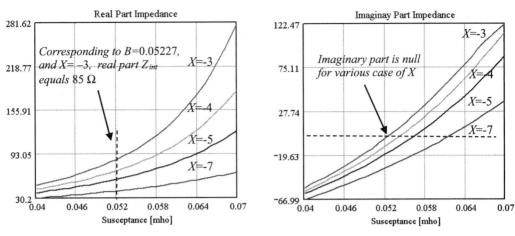

Figure 8.20 The (a) real and (b) imaginary parts of Z_{int} for the circuit shown in Figure 8.15 for m = 2.4, evaluated at ω_c versus B (susceptance) for a few cases of X (for a fixed G of 0.012).

A negative value for X and a lower value for G are needed to improve the match at ω_c. The improved match at ω_c (85Ω real part and zero imaginary) requires B = 0.05227 and $X = -3$ (for case $G = 0.012$), as shown in Figure 8.20. Careful selection of G is necessary to provide the required real part Z_{in} across the bandwidth and to ensure the real part Z_{int} with experiences minimum variation. Since G = 0.012 (which equals 83Ω), at ω_c, the real part Z_{int} is approximately 85Ω (with zero imaginary part). The X (of negative value) would be absorbed to the line inductance $L/2$. However, suitable values for G, jX, and jB depending on the design need for any ratio of m can be easily obtained through numerical analyses.

As discussed in Chapter 4, the gain in a common-source distributed amplifier cannot be increased indefinitely by adding more sections due to losses of the lines [36], and the increase in the number of sections will directly impact the cost and implementation of the size. Transconductance g_m for each section device is the most important issue for the gain. Figure 8.21 shows a small-signal equivalent circuit for the proposed circuitry to determine g_m (from Figure 8.14). The analysis will lead to determining the overall transconductance g_m due to high-f_τ and power transistors g_{m1} and g_{m2}, respectively, and the influence of Z_T, especially over the entire bandwidth frequency.

Transconductance g_m is defined as change of drain current I_D with respect to the corresponding change of gate voltage V_{GS} with drain supply voltage V_{DS} equals to a constant [44]:

$$g_m = \partial I_D / \partial V_{GS} \qquad (8.12)$$

The voltage across the input capacitor of the power transistor v_{gs2} with inclusion of interstage tapered impedance Z_T is as follows:

$$v_{gs2} = 1/j\omega C''_{gs} // (j\omega_L/2 + Z_T) \cdot i_L \qquad (8.13)$$

where $Z_T = \dfrac{1}{G + jB} + jX$ and i_L is the current flowing across L.

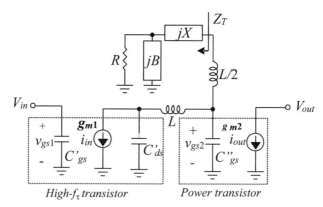

Figure 8.21   Small-signal model for gain analysis including two nonidentical transistors with interstage tapered impedance.

Substituting $i_{in} = g_{m1}v_{in}$ into (8.13), we obtain

$$i_L = \frac{1/j\omega C'_{ds}}{1/j\omega C'_{ds} + j\omega L + \left[1/j\omega C''_{gs} // \left(j\omega_L/2 + Z_T\right)\right]} \cdot g_{m1}v_{in} \qquad (8.14)$$

and thus

$$v_{gs2} = g_{m1}v_{in} \frac{1/j\omega C'_{ds} \cdot \left[1/j\omega C''_{gs} // \left(j\omega L/2 + Z_T\right)\right]}{1/j\omega C'_{ds} + j\omega L + \left[1/j\omega C''_{gs} // \left(j\omega L/2 + Z_T\right)\right]} \qquad (8.15)$$

Therefore, one can derive g_m for the circuit shown in Figure 8.21:

$$g_m = \frac{\partial i_{out}}{\partial v_{in}} = \frac{g_{m2}v_{gs2}}{v_{in}} \qquad (8.16)$$

where g_{m1} and g_{m2} are intrinsic transconductances of the respective transistors. By substituting (8.15) into (8.16), we can derive

$$g_m = g_{m1}g_{m2} \frac{1/j\omega C'_{ds} \cdot \left[1/j\omega C''_{gs} // \left(j\omega L/2 + Z_T\right)\right]}{1/j\omega C'_{ds} + j\omega L + \left[1/j\omega C''_{gs} // \left(j\omega L/2 + Z_T\right)\right]} \qquad (8.17)$$

The simplification of (8.17) by substituting $j\omega \to s,^2$ thus one can show g_m as

$$g_m = g_{m1}g_{m2} \frac{1/sC'_{ds} \cdot \left[1/sC''_{gs} // \left(sL/2 + Z_T\right)\right]}{1/sC'_{ds} + sL + /sC''_{gs} // \left(sL/2 + Z_T\right)} \qquad (8.18)$$

---

[2] The Laplace transformation $F(s)$, where $s = \sigma + j\omega$, of a continuous function $f(\omega)$ over the entire time domain($\omega = 2\pi/t$) is defined as $\int_{-\alpha}^{\alpha} f(t)e^{-st}\,dt$ [45].

Normalization of the transconductance $G(s)$ can be written as

$$\frac{g_m}{g_{m1}g_{m2}} = G(s) \tag{8.19}$$

and can be written as

$$G(s) = \frac{sL + 2Z_T}{s^4 mL^2 \left(C'_{ds}\right)^2 + s^3 mL \left(C'_{ds}\right)^2 Z_T + s^2 \left(mLC'_{ds} + 3LC'_{ds}\right) + s\left(mC'_{ds}Z_T + 2C'_{ds}Z_T\right) + 2} \tag{8.20}$$

where $m = C''_{gs}/C'_{ds}$.

Transconductance $G(s)$ is analyzed for a few cases of m (2, 2.4, and 3), when $L = 6.925$ nH; $C'_{ds} = 1.385$ pF, which was chosen to define $\omega_c = 2.2$ GHz; and Z_T is a fixed termination, for example, 50Ω. The transfer function of $G(s)$ as given in (8.20) is identified using the Scilab program. As shown in Figure 8.22(a), when $Z_T = 50\Omega$ is selected, 1 zero (only real part) and 4 poles (which are in complex conjugate location) exist, and all lie on the left half plane (LHP) and have positive real functions. The zero for any case of m does not change. The two complex conjugate poles, which are located close to the complex imaginary axis of the left s-plane, boost up the $G(s)$ at ω_c. However, as m increases, the complex conjugate poles move lower on the x-axis, reducing the bandwidth operation. This is clear evidence that with fixed termination Z_T, bandwidth operation reduces as m increases. The roots (zero and pole) are illustrated in Figure 8.22(a).

With the proposed design technique, Z_T helps to retain bandwidth operation as m increases. We can define Z_T as

$$Z_T = \frac{1 + sGL_x + s^2 BL_x}{G + sB_x} \tag{8.21}$$

where $j\omega L_x = jX$ and $j\omega B_x = jB$ from (6.4). It has been explained previously that for any m case, Z_T elements (e.g., G, B, and X) are determined numerically from (8.21).

By substituting (8.21) into (8.20), and knowing Z_T elements for any case of m, the roots of $G(s)$ are plotted in Figure 8.22(b). The complex conjugate pole at the imaginary axis is located at a higher frequency with Z_T compared to a fixed termination. It is evident from a theoretical point of view that the tapered impedance significantly improved bandwidth performance compared with fixed termination. In a similar manner, $G(s)$ from (8.20) can be analyzed in the ω domain by means of the inversion Laplace transformation.[3] Clearly, the bandwidth operation is extended significantly with this new concept.

Take note that phase synchronization is not necessary for the cascaded DPA topology [47], but the effective input capacitance C_{in} of the first high-f_τ transistor and effective output capacitance C_{opt} of the power transistor are important to form

---

[3] The inversion Laplace $f(t)$ of a continuous function $F(s)$ can be found in [45], where the complex variable $s = \sigma + j\omega$, such that $1/2\pi j \int_{\sigma_1 - j\omega}^{\sigma_1 + j\omega} F(s)e^{st} ds.$.

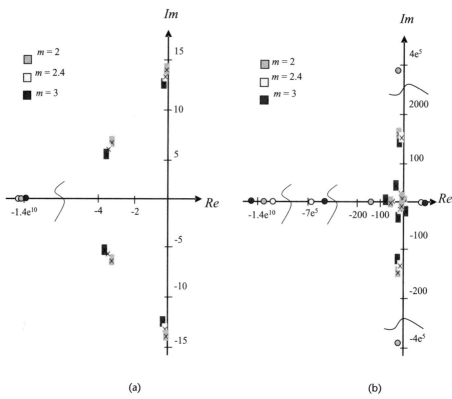

Figure 8.22 Roots of $G(s)$ for (a) fixed termination ($Z_T = 50\Omega$) and (b) tapered termination for a few cases of m.

ω_c. As shown in Table 4.1, the effective C_{in} of the first high-f_τ transistor and C_{opt} of the power transistor are lower than 3 pF, which means a ω_c of 2.2 GHz can be formed with a 50Ω load. Therefore, as shown in Chapter 4, effective gate- and drain-line elements L_i are synthesized by means of (4.58) to form the desired ω_c. Dummy drain termination is eliminated to improve the efficiency performance [16]. The elements of Z_{T1} and Z_{T2} have different values, since the termination provides a wideband matching solution between the respective interstage networks. A simplified design schematic for the DPA topology is shown in Figure 8.23.

DC bias networks, including gate and drain feeding for each transistor, are illustrated in Figure 8.23. As explained in Chapter 4, L_g–C_g and L_d–C_d networks are implemented. As the next step, integration of the dc bias network to the DPA topology is necessary while satisfying the RF-to-dc isolation over the wide bandwidth response. For the first and second high-f_τ transistors, L_d and C_d of 180 nH[4] and 33 pF,[5] respectively, are selected for bandwidth operation up to 2 GHz [14]. For the final stage, the dc feeding line is connected to a high-Q air-wound coil.[6]

---

[4] The high-Q ceramic 0603HP series chip inductors provided by Coilcraft, Inc., with a Q up to ~150 at 1.7 GHz were used.

[5] This is a broadband high-Q capacitor, size 3060, from Murata, Inc. Details can be obtained at www.murata.com.

[6] The inductance value of the air-wound coil is approximately to 43.5 nH and is from Taito Yuden, Inc.

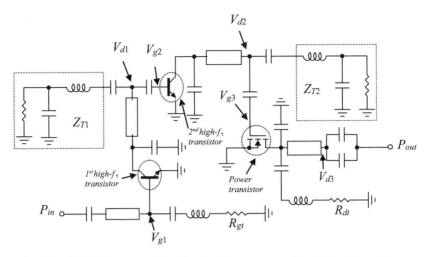

Figure 8.23   Simplified schematic of the distributed power amplifier: high-f_τ transistors cascaded to the power transistor with interstage tapered impedance. The gate and drain lines are formed with a constant-k ladder network for 50W input and output impedances.

To verify the design example, simulation with harmonic balance (HB) is carried out to understand its power performance. Nonlinear models of the devices and a passive elements model were developed. For comparison purposes, a simulation template with interstage fixed termination impedance was developed. The simulated results of drain voltage for each device and each device's power performance are shown in Figures 8.24 and 8.25, respectively. It is clear that the performance with interstage tapered termination is significantly improved compared to the case of fixed termination. The drain voltage of each transistor is kept constant over the entire bandwidth operation, as shown in Figure 8.24. For tapered termination, output power and gain of 40 dBm and 32 dB, respectively, achieved a flat response up to 2.2 GHz (refer to Figure 8.25). It is an evident that bandwidth with this new topology is extended by 500 MHz compared to the fixed termination.

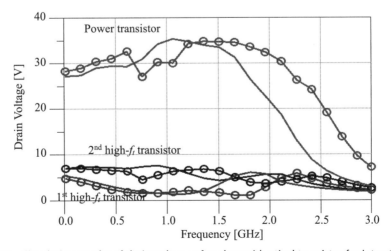

Figure 8.24   Simulation results of drain voltage of each nonidentical transistor for interstage tapered (solid line with circle) and fixed termination impedance (solid line only).

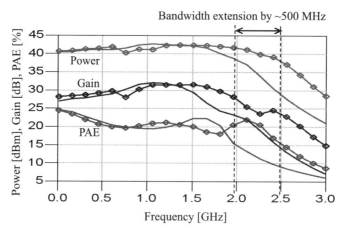

Figure 8.25 Simulation results of power performance, that is, power, gain, and PAE of a cascaded DPA with interstage tapered (solid line with circle) and fixed termination impedance (solid line only).

To pursue experimental validation of the concept technique, a prototype board for the design was developed that uses Rogers 4350B PCB material. The dc biasing terminals are bypassed to ground with multiple chip capacitors (e.g., 100 pF, 33 nF, 10 uF).[7] For the gate biasing, high-Q chip inductors (value of 220 nH, same series from Coilcraft, Inc.) are used. A series resistor of 3Ω is included in the biasing circuitry as a precaution against oscillation in the measurement level.

A four-layer, high-density Rogers PCB and the bottom layer are connected with a grounded heat sink. Component placement takes place at the top of layer 1. Layer 2 provides grounding for quasi-TEM transmission lines, for example, microstrip, and layer 4 is solid bottom ground. A via hole will connect from the top layer to the bottom layer. The prototype has an open grounding area with adequate via holes, and dc and RF routing are well isolated in the PCB for minimum spurious levels. If additional dc routing is required, then it is possible to implement that at layer 3. However, no additional dc routing is needed for this design. Modeling with a full-wave EM simulator including PCB layer stackup, via holes, indium foil, grounded heat sink, and RF connectors [11] is considered. For a GaN power transistor, a grounded heat sink is attached at the bottom layer via the indium foil[8] and multiple screws, as discussed in Chapter 4. To understand the grounding behavior of the GaN device, the contact between the screw thread and the heat sink is investigated in the EM simulator, and an important point to be noted is that the contact between the screw thread and heat sink influenced the grounding of the device, especially at higher frequencies. One way to improve the grounding is to use bigger diameter screws and more of them to trade off with the surface contact of the screws.

The layout from Cadence (ODB++ file) is imported to the CST. Discrete ports for the component pads are created, and the layout information is exported to the ADS. ADS cosimulation assisted with CST in the HB simulator (ADS environment),

---

[7] The capacitors are 600S Series ultra-low ESR, high-Q microwave capacitors, from ATC, Inc. Other capacitors the series 545-L ultra-broadband high-Q capacitors, from Murata, Inc.
[8] The indium foil is from Indium Corp., North Carolina. The thermal conductivity of the copper foil is 0.34 W/cm at 85°C. The foil part number is IN52-48SN (0.004-in. thickness).

Figure 8.26   Photograph of the DPA prototype with high-f_τ transistors cascaded to the power transistor with interstage impedance termination. The DPA size area is 38 mm × 22 mm.

where layout geometry dimensions are modified for optimum power performance. A photograph of the distributed power amplifier is shown in Figure 8.26.

Supply voltages of 5.5V and 28V are applied to the high-f_τ transistors and to the power transistor, respectively. The high-f_τ transistors are biased with I_{DQ} of 34 mA (20% I_{dss}) and 93 mA (14% I_{dss}), and the power transistor with 188 mA (4% I_{dss}), respectively. For a GaN HEMT device, one important issue is the biasing sequence. The goal while biasing the device is to stay away from areas of sensitivity to the potential instability of the device. Designers need to pay attention to how to deal with the positive gate current that will arise when a device is driven into saturation. To overcome this limitation, a resistor is used that is connected across the power supply terminals. The resistor will enable the power supply to always provide a negative current while allowing the device to source or sink current [47].

The measured results of the S-parameter are shown in Figure 8.27. The output return loss S_{22} is less than −6 dB and the input return loss S_{11} is less than −10 dB

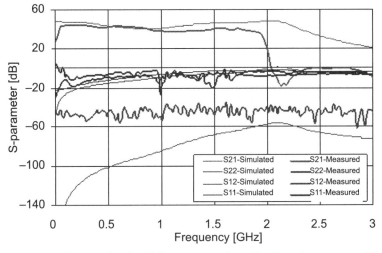

Figure 8.27   Measured versus simulated S-parameters for a distributed power amplifier having high-f_τ transistors cascaded to the power transistor with interstage tapered termination.

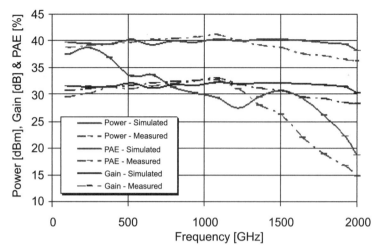

Figure 8.28 Measured versus simulated power performance for distributed power amplifier having high-f_τ transistors cascaded to the power transistor with interstage tapered termination.

across the bandwidth. Note that S_{22} is higher than -10 dB across the bandwidth, and this is because the dummy termination is eliminated to maximize power load of 50Ω. For this reason, output from the DPA will be coupled to the harmonic filter in a real application such that reflection matching can be improved. Reverse isolation S_{12} performance is better than -50 dB over the entire bandwidth operation. The measured output power, gain, and efficiency of the distributed amplifier versus frequency with the initial design (without any optimization) are shown in Figure 8.28. Output power of 10W, 32 dB gain, and a PAE of >15% across the bandwidth are recorded in the measurement level (Figure 8.28). As can be observed in Figure 8.28, output power is quite flat (10W) beyond 1.3 GHz, and slight degradation is seen beyond 1.3 GHz due to the grounding effect of the GaN device. However, good correlation between simulation and measurements is achieved through full-wave EM modeling of the complete structure.

8.3 Vectorially Combined Distributed Power Amplifier

Due to the loading effect of the bigger size transistor, dc-to-RF energy conversion is not optimum as the frequency increases toward the cutoff frequency. With these amplifiers, in fact, it is quite difficult to achieve higher output power [10–16, 20, 25, 51]. GaN HEMT grown in SiC substrate is a favorable device candidate for high output power [21, 25, 52, 53] besides the LDMOS, GaAs HBT, pHEMT, and SiGe HBT processes [12–14, 20].

The loading effect of the drain line becomes stronger the bigger the device periphery, and typically an attenuation compensation technique is used, where an active load (common-gate FET) is coupled to the common-source FET to reduce the drain-line losses dominated by the real part R_{opt} [71]. Some literature refers to this technique as a *cascode distributed amplifier* [38, 40]. However, in principle, this technique offers higher output impedance and improved reverse isolation, but additional FET device and biasing gate circuitry to the common gate are needed. Nevertheless, this is a good solution for an MMIC approach [40].

8.3.1   Overview of Vectorially Combined DPA with Load Pull Determination

To maximize power P_o from each transistor section (as shown in Figure 8.29), each current source should be loaded with an optimum load impedance R'_{opt} (for optimum excursion RF voltage V_{max}/I_{max} current swing) [55]. To utilize the V_{max} and I_{max}, which is known as power match condition, an optimum lower value of R_{load} would be selected to provide loadline match, $R'_{opt} = V_{max}/I_{max}$ [55] (Figure 8.29). It has been assumed that R_{gen} (i.e., R_{ds}) $>> R'_{opt}$ so if R_{gen} is taken into account, it would be necessary to solve the equivalent impedance ($R_{gen}//R'_{opt}$). The effective real part of the device at reference plane A (R'_{opt}) and B (R_{opt}) is different due to the fact of its shunt capacitance and package parasitic effect, as explained in Figure 4.24(b). In real applications, it is convenient to identify the real part of the device at reference plane B, which will be used to match to standard output termination (e.g., 50Ω load). The optimum impedance of the transistor $Z_{opt}(\omega)$ (which consists of a real and imaginary part) can be extracted by means of the nonlinear load pull technique [55] either by simulation or measurement, which generally exhibits frequency-dependent behavior.

As generally confirmed by experimental results on FETs, the optimum reactive part $X_{opt}(\omega)$ to be absorbed in a distributed output drain-line network is almost equivalent to a constant capacitance C_{opt} over very wide bandwidths in MMIC implementation [40], and may have nonuniform value in hybrid packaged implementation. Nevertheless, in the hybrid implementation, $X_{opt}(\omega)$, that is, intrinsic parasitic capacitances, extrinsic elements, packaging effect, and so forth, and the power optimization consist of loading the equivalent output source of the transistor with optimum power load $R_{opt}(\omega)$.

Bear in mind that to combine multicurrent sources into a single load termination, an optimum virtual impedance to each source in two directions, $Z_{u(k)}$ and $Z_{r(k)}$, must be fulfilled (as discussed in Chapter 5). The generalized design equations developed are not sufficient when device current source is loaded by $R_{opt}(\omega)$. In this section, a technique for achieving power match to each source while satisfying multicurrent sources combined into a single load termination is presented. Figure 8.30 shows a simple schematic of two current sources combined at a common node

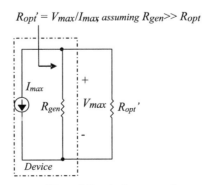

$R_{opt'} = V_{max}/I_{max}$ assuming $R_{gen} >> R_{opt}$

Figure 8.29   Optimum power condition of the device current source loaded by R_{opt}.

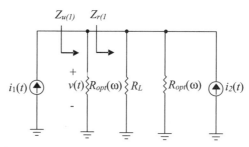

Figure 8.30 The virtual impedance seen by the current source $i_1(t)$ in both directions $Z_{u(1)}$ and $Z_{r(1)}$, respectively, when two current sources are combined to a single node.

connected to a load R_L, and each current source is loaded with optimum power load $R_{opt}(\omega)$. As shown in Figure 8.30, the virtual impedance $Z_{u(1)}$ looking into the common node with $i_2(t)$ in parallel with $R_{opt}(\omega)$ and load termination R_L is what current source vector $i_1(t)$ is loaded with. Applying the Thevenin theorem to Figure 8.30, the impedance $Z_{u(1)}$ can be derived as

$$Z_{u(1)} = v(t)/i_1(t) = \frac{\left[i_1(t) + i_2(t) \right]}{i_1(t)} \left(0.5 R_{opt} // R_L \right) \tag{8.22}$$

By substituting $i_k = I_k e^{j(wt+\theta_k)}$, where $k = 1$ and 2, we can derive the virtual impedance

$$Z_{u(1)} = 0.5 R_{opt} // R_L \left[1 + \frac{I_2}{I_1} \left(\cos\left(\theta_2 - \theta_1\right) + j\sin\left(\theta_2 - \theta_1\right) \right) \right] \tag{8.23}$$

where I_1 and I_2 represent the magnitude of the complex current source, and θ_1 and θ_2 are independent phase values, respectively.

Phase offset or in-phase combining ($\theta_2 = \theta_1$) is considered to simplify the concept, and the concept can be easily explained for $\theta_2 \neq \theta_1$. In this manner, (8.23) can be simplified as

$$Z_{u(1)} = \left(1 + \frac{I_2}{I_1} \right) \left(0.5 R_{opt} // R_L \right) \tag{8.24}$$

From (8.24), the virtual impedance $Z_{u(1)}$ seen by the current source $i_1(t)$ is directly dependent on R_L and I_2/I_1. As an example, for equal injection value $I_2/I_1 = 1$ (equal device periphery), $Z_{u(1)} \approx R_{opt}$ when R_L is set to $\geq 10 R_{opt}$. The selection of R_L reveals what characteristic impedance of transmission line $Z_{r(1)}$ needs to be designed. Note that the virtual impedance seen by current source $Z_{u(1)}$ must be close to R_{opt} while both current sources are combined at single load R_L; this meets the power match/current combining conditions. Nevertheless, $Z_{r(1)}$ will absorb the imaginary part X_{opt} to form broadband frequency operation.

In a similar manner, we can derive $Z_{u(k)}$, $k = 1, 2, \dots, n$, as shown in Figure 8.31:

$$Z_{u(k)} = \left(\frac{I_1}{I_k} + \frac{I_2}{I_k} + \frac{I_3}{I_k} + \cdots + \frac{I_n}{I_k} \right)\left(R_{\text{opt1}} /\!/ R_{\text{opt2}} /\!/ \ldots /\!/ R_{\text{opt}n} /\!/ R_L \right) \qquad (8.25)$$

Let's review the analysis for n-section, for which an identical device periphery is selected, $Z_{u(1)} = Z_{u(2)} = Z_{u(3)} \approx R_{\text{opt}} /\!/ Z_L$, and it can be approximated to R_{opt} if Z_L has reasonable termination. Figure 8.31 shows drain transmission line $Z_{r(k)}$. It must be synthesized to load each device generator by its optimum $R_{\text{opt}(k)}(\omega)$, where $k = 1,2,\ldots,n$. It is clear from Figure 8.31 that $Z_{r(1)}$ must be loaded with an optimum load resistive of the first generator R_{opt1}, and $Z_{r(n)}$ should be loaded with a Z_L value. Nevertheless, the middle section, $Z_{r(2)}$, is dependent on $Z_{u(2)}$ and $Z_{u(3)}$.

It is necessary to know the initial value of R_{opt1}, which typically can be obtained with a single device load pull determination. In a similar manner, Z_L must be performed with an initial guess of $Z_{r(1)},\ldots,Z_{r(n)}$. To obtain the initial guess, it is convenient to use a CAD simulator assisted by an optimizer (e.g., ADS), where the transistor is modeled with an ideal current source and parallel high impedance resistor. To additively combine the currents at each junction, phase synchronization between the current source and the transmission line delay $Z_{r(k)}$ is matched. From design equations of an optimum virtual impedance to each source in two directions, that is, $Z_{u(k)}$ and $Z_{r(k)}$ have behavior that adaptively reduces toward load termination. For example, $Z_{r(1)} = 50\Omega$, $Z_{r(2)} = 25\Omega$, $Z_{r(3)} = 16.7\Omega$, and $Z_{r(4)} = 12.5\Omega$ is achieved if no loading effect is taken into consideration. Therefore, we can consider this distributed power amplifier to have vectorially combined current sources with load pull determination.

Achieving high power, efficiency, and gain in the 10- to 2000-MHz operating bandwidth is a key focus, so DPA systems should be able to be coupled directly to a voltage-controlled oscillator (VCO). The VCO output is typically ~8 dBm, and the output power of 30W is the design goal over the entire bandwidth operation (40 to 2000 MHz). Thus, a gain of 37 dB is required to deliver high output power up to 30W.

A high-f_τ transistor with the lowest input parasitic capacitance C_{gs} (an ATF54143 device) is coupled to the gate line, and this may improve the loading effect of the input gate line. The achievement of high gain in the 40- to 2000-MHz operating range is adopted from Section 8.2, allowing the resulting DPA to be coupled directly to the VCO. Each section has two nonidentical high-f_τ transistors (ATF54143 and ATF511P8), respectively, from Avago, Inc., which are cascaded to the power

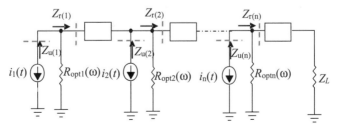

Figure 8.31 Synthesize drain transmission line to load each device generator with its optimum $R_{\text{opt}(k)}(\omega)$.

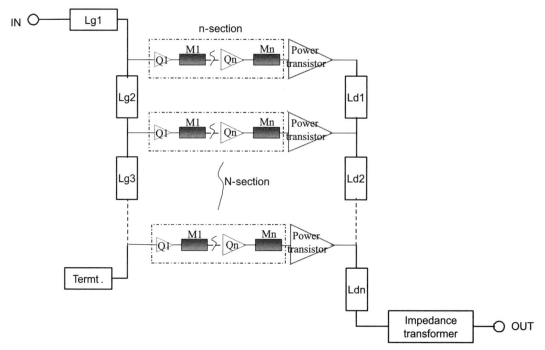

Figure 8.32 General diagram of high-power DPA circuit proposal. The terms *n* and *N* refer to the number of cascaded stages and sections, respectively. An impedance transformer may be included when the termination is less than 50Ω.

transistor (GaN device from CREE, Inc.) with interstage tapered impedance. As discussed earlier, the optimum number of *n*-sections to maximize power at any frequency is 3...4 [11]; therefore, three sections are used in this design to deliver 30W output power (assuming each GaN device contributes ~10W). The general diagram for this DPA topology is shown in Figure 8.32. The terms *n* and *N* refer to the number of cascaded stages and sections, respectively.

The design methodology discussed in Section 4.5 is applied in this section. A GaN device (CGH40010F) is used in this work, and the V_{bk} of the device is ~73V. The drain loading effect of the device is very significant. Effective input and output real and imaginary parts of the packaged devices $X_{in}(\omega)$ and $X_{opt}(\omega)$ (of high-f_τ and power transistors), respectively, are extracted as explained in Section 4.5, and the values are tabulated in Table 4.1. For instance, $X_{opt}(\omega)$ of the second output high-f_τ transistor will be matched to the $X_{in}(\omega)$ of the input power transistor (GaN device) by means of the theoretical approach presented in Section 8.2.

The effective input capacitance of the first high-f_τ transistor C_{in} and effective output capacitance of the power transistor C_{out} are important to form ω_c. As shown in Table 4.1, the effective input capacitance of the first high-f_τ transistor and output capacitance of the power transistor are lower than 3 pF, which means ω_c of 2.2 GHz can be formed with a 50Ω load. Therefore, effective gate and drain-line elements L_i are synthesized by means of (4.48) to form the desired ω_c. Dummy drain termination is eliminated to improve the efficiency performance [20, 21]. To achieve 30W of output power, a PAE of >35%, and a gain of more than 35 dB (since the input drive is ~8 dBm), and to achieve bandwidth up to 2.2 GHz, $N = 3$ and $n = 3$ are

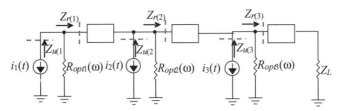

Figure 8.33 Synthesized drain transmission line to load each device generator by its optimum $R_{opt(k)}(\omega)$ for $n = 3$. Note that Z_L is identified for optimum power performance in a CAD simulator.

proposed. It is important to note that the DPA gain-bandwidth response is dominated by the gate line and cascaded stages, whereas the power-efficiency response is mainly contributed by the drain line. Therefore, design work focused on synthesizing the gate-line network and interstage tapered impedance, followed by a drain-line network to achieve power performance for the broadband range up to 2 GHz. An approach demonstrated in [13] was adopted for the gate line, whereby phase synchronization is achieved with a nonuniform gate-line design and the gate-line impedance is adaptively reduced. A power match and current combining technique is implemented at the drain line.

DC bias networks include a gate and drain feeding design for each high-f_τ transistor from Section 4.5, where the L_g–C_g and L_d–C_d networks from Figure 4.20 are implemented. Integration of the dc bias network to the DPA topology is necessary while satisfying the RF-to-dc isolation over the wide bandwidth response. It is beneficial to measure RF-to-dc isolation over a wide frequency response of the DPA to ensure the design of broadband bias networks. For the first and second high-f_τ transistors, L_d and C_d of 180 nH[9] and 33 pF,[10] respectively, are selected for bandwidth operation up to 2 GHz [14]. The dc gate biasing terminals are bypassed to ground with multiple chip capacitors (e.g., 100 pF, 33 nF, 10 uF, etc)[11] for each transistor. The dc feeding lines for the final stage are connected to a high-Q air-wound coil (from Coilcraft, Inc.),[12] and high-Q chip inductors (value of 220 nH)[13] for the first and second stages from Coilcraft, Inc.

To begin the drain-line synthesis, it is important to identify R_{opt} of the power transistor for optimum dc-to-RF energy conversion. For single-ended dc-to-RF energy conversion of the power transistor (CGH40010F), $R_{opt}(\omega) \approx 40\Omega$ (average value across bandwidth) is obtained from the load pull simulation technique. Bear in mind that the optimum load impedance has frequency-dependent behavior. Hence, $Z_{r(1)} \gg 40\Omega$, and $Z_{r(2)}$ and $Z_{r(3)} \gg Z_L$ are optimized in a CAD simulator for optimum power performance up to 2 GHz; refer to Figure 8.33 for the drain-line synthesis. A simplified design schematic for the new DPA topology is shown in Figure 8.34.

---

[9] The high-Q ceramic 0603HP series chip inductors were provided by Coilcraft, Inc. They have a Q up to ~150 at 1.7 GHz.
[10] This is a broadband high-Q capacitor, size 3060, from Murata, Inc. Details can be obtained at www.murata.com.
[11] The capacitors are 600S Series ultra-low ESR, high-Q microwave capacitors, from ATC, Inc. Other capacitors the series 545-L ultra-broadband high-Q capacitors, from Murata, Inc.
[12] The electrical specification of the broadband choke: $L = 1.3$ uH $\pm$ 10%, DC resistance (DCR) = 12.6 mΩ, $I_{rms} = 4$A.
[13] The series of the choke is 0603CS_XNL (1008HS), from Coilcraft, Inc.

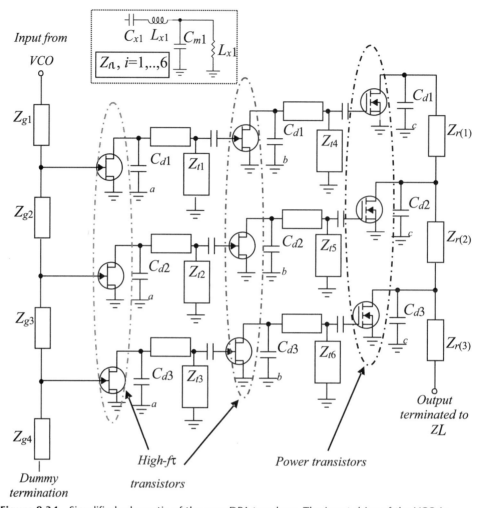

Figure 8.34 Simplified schematic of the new DPA topology. The input drive of the VCO is ~8 dBm, and a nonuniform gate line is adopted from [41]. Two nonidentical high-f_τ transistors (ATF54143 and ATF511P8) are cascaded to the power transistor (GaN device). Output impedance is terminated to Z_L, which will be coupled to a transformer/filter.

To verify the design example, simulation with HB is carried out to understand the design's power performance. Nonlinear models of the devices and a passive elements model were developed. Supply voltages of 5.5V and 28V were applied to the high-f_τ transistors and to the power transistor, respectively. The high-f_τ transistors are biased with I_{DQ} of 34 mA (20% I_{dss}) and 93 mA (14% I_{dss}), and the power transistor with 188 mA (4% I_{dss}), respectively. The input drive throughout the measurement level is fixed to a low level (~8 dBm).

The simulated results of the S-parameter and power performance (i.e., power, PAE, and gain) are shown in Figures 8.35 and 8.36, respectively. The small-signal gain S_{21} is quite flat over the frequency range (about 40 dB) and small peaking occurred outside the band. Good reverse isolation S_{12} of the distributed power amplifier is more than –60 dB, and the input and output return loss is better than

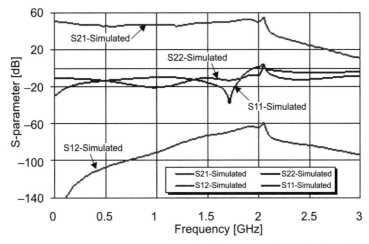

Figure 8.35 Simulation analysis of S-parameters for the new topology distributed power amplifier across 0.1 to 2 GHz. A Z_L of ~14Ω is required for optimum performance.

−10 dB, respectively, over the frequency. The large-signal performance (i.e., output power) is achieved at a constant 45 dBm (30W), with a flat gain of 37 dB up to 2 GHz. The PAE is more than 40%, but in general an average PAE of ~52% was achieved at the simulation level.

To validate the concept technique experimentally, a prototype board for the design was developed that uses Rogers 4350PCB material. Because additional dc routing is needed to connect each gate of high-f_τ and power transistors, layer 3 is used. An open grounding area with adequate via holes, and dc and RF routing are well isolated in the PCB for minimum spurious levels. As part of EMC requirements, it is necessary to have the dc layer below the RF grounding layer [56, 57]. Hence,

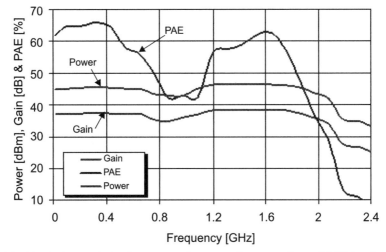

Figure 8.36 Simulation analysis of power performance (power, gain, and PAE) for the new DPA topology across 0.01 to 2 GHz. A Z_L of ~14Ω is required for optimum performance.

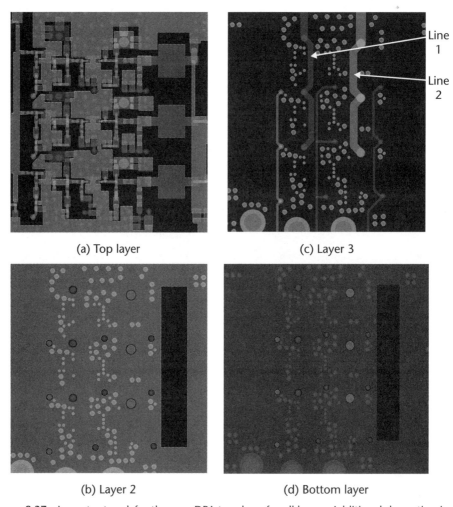

(a) Top layer   (c) Layer 3

(b) Layer 2   (d) Bottom layer

Figure 8.37 Layout artwork for the new DPA topology for all layers. Additional dc routing is done via layer 3, where layer 2 is RF grounding. (a) Top layer, (b) layer 2, (c) layer 3, and (d) bottom layer.

layer 2 and layer 4 have solid grounding planes. The thickness of the dc routing is computed to carry adequate dc current, and the width thickness is shown in Figure 8.37. For an example, in layer 3, thickest routing (line 2) is used for the drain supply of the second high-f_τ transistor, and the routing shown in line 1 is for the drain supply of the first high-f_τ transistor. Layout artwork for the new DPA topology (for all layers from Cadence) is shown in Figure 8.37.

Two diameter screws, which are 5 and 1.4 mm, are mounted to hold the grounded heat sink chassis. The modeling of the screws and the foil are reused (from Section 8.2), and the RF connector (from Section 5.5) is reused. The inductance of the gate and drain lines is realized in a lumped transmission line with the aid of CST. The layout structure (from Figure 8.37) was imported from Cadence, and the line properties (i.e., length and width) are optimized to obtain an optimum value for the inductance and Q-factor over a wide frequency range. Because the length is fixed

Figure 8.38   Photograph of the new DPA topology prototype. The DPA size is 38 mm × 32 mm.

(because of power transistor length dimension to 9.5 mm), only the width will be optimized. From the CST simulation, self-resonance of the line is kept higher than 4 GHz for the uniform result of inductance and a reasonably high Q-factor.[14]

For the high-f_τ transistors and power transistors, supply voltages of 5.5V and 28V are applied, respectively. A bias current IDQ of 34 mA (20% *Idss*) and 93 mA (14% *Idss*) are biased for the high-f_τ transistors and the power transistor with 188 mA (4% *Idss*), respectively.

First, the measurement work of the prototype board was started by identifying the optimum load impedance to be terminated at point *a'* in Figure 8.38. A test fixture for line calibration[15] was built to de-embed the transmission line length [58, 59]. Figure 8.39 shows the example measured results of the load pull impedance contour for 100 MHz and 1 GHz, respectively. The highest power is recorded at an impedance of $25 + j39\Omega$ and $25 - j18\Omega$ for 100 MHz (40W) and 1.9 GHz (25W), respectively. The imaginary impedance data indicated that at a low frequency the line has higher capacitance and the inductance varies as the frequency increases. Figure 8.40 shows the comparison between measured versus simulated results for the highest output power with the load pull technique across the bandwidth. However, the optimum performance (output power, gain, and PAE) is achieved at the ~12Ω region at the measurement level across the entire bandwidth (Figure 8.41). The output of 25W, and PAE results of 45% to 56% are recorded at the measurement level, while a flat response for the operating gain is obtained. Although some imaginary part exists (the variation ±$j4\Omega$), the real part of 12Ω is sufficient to be terminated at the output of the distributed power amplifier. The following section explains the design of a broadband transformer (12Ω to 50Ω) with a minimum insertion loss (~1 dB) within the operating bandwidth frequency.

---

[14] The Q-factor is defined by center frequency over bandwidth operation. A Q-factor of 500 is required.
[15] Thru, reflect, and line (TRL) calibration is the standard type of calibration for a load pull determination system.

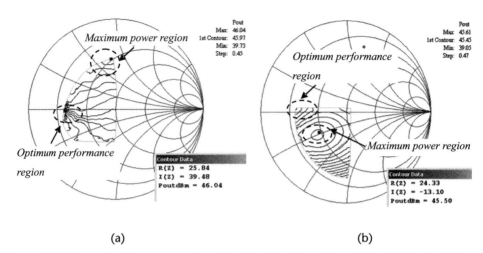

(a)   (b)

Figure 8.39 Example of measurement results of load pull impedance determination for 100 MHz and 1 GHz, respectively. The maximum power occurred at $\Re\{Z_l\}$ of 25Ω.

With the same prototype board, measurement with 50Ω termination was carried out to understand its performance. Measured versus simulated results of S-parameters and power performance of the topology terminated to 50Ω are presented in Figures 8.42 and 8.43, respectively. The output return loss S_{22} is less than −8 dB and the input return loss S_{11} is less than −12 dB across the bandwidth. Bear in mind that S_{22} is higher than −10 dB across the bandwidth, and this is because the dummy termination is eliminated to push all the current from each device to a single load termination. As a result, output reflection matching is not well optimized. The measured isolation S_{12} is about 60 dB, and the small-signal gain S_{21} is more than

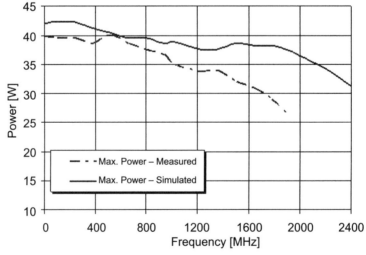

Figure 8.40 Measurement results for the highest output power with load pull impedance determination across the entire bandwidth. The maximum power occurred at $\Re\{Z_l\}$ of ~25Ω at the measurement level.

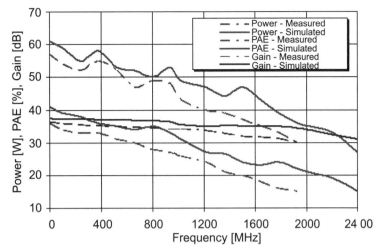

Figure 8.41 Measurement results of optimum power performance (power, gain, and PAE) with load pull impedance determination across the entire bandwidth. The maximum power occurred at $\Re\{Z_l\}$ of ~12Ω at the measurement level.

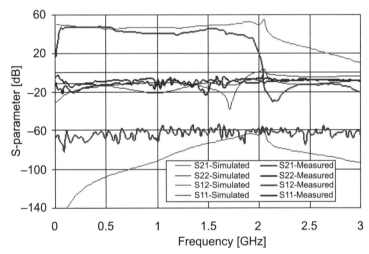

Figure 8.42 Measured versus simulated S-parameters for the DPA topology terminated to 50Ω load.

40 dB across the bandwidth. No small-signal peaking is recorded at the measurement level (near ω_c), although a small value is observed at the simulation level. An output power of ~15W and a PAE of 15% to 42% are recorded at the measurement level with 50Ω termination. At 50Ω termination, the highest output power is 18W at low frequency, and at 2 GHz, the power is degraded to 10W.

Since the measurements were done with low input drive to meet SDR applications, power characterization (power and PAE) at 1 GHz by sweeping the input drive was carried out, as shown in Figure 8.44. The results indicated that the output power can hit ~18W with a PAE of 34% with 50Ω termination. The optimum power happened at an input drive of ~10 dBm (slightly higher than output level from VCO

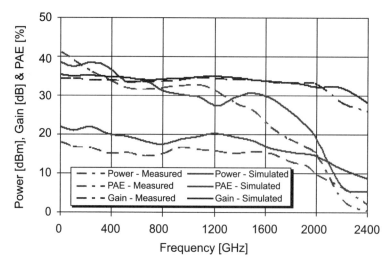

Figure 8.43 Measured versus simulated power performance for the DPA topology terminated to 50Ω load.

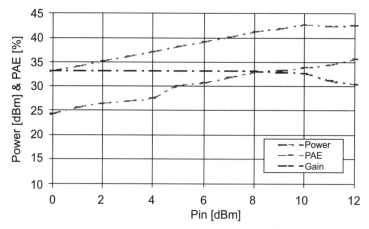

Figure 8.44 Measured results of power performance (power and PAE) with sweeping of the P_{in} drive (at 1 GHz) of the new DPA topology.

capability). Keep in mind that the input drive required for a DPA may change with frequency behavior. As the frequency increases, higher drive is needed. Therefore, a RF input drive adjustment circuitry to provide adequate RF signal strength to the distributed power amplifier under constant envelope modulation [60] can be implemented. Implementation of capacitors at the drain line that is, 22 μF (tantalum capacitor)[16] and 10 pF, 33 pF, 470 pF, and 22 nF (ceramic capacitors) takes place as a precaution against low-frequency parasitic oscillation. No oscillation is reported and the DPA operation is very stable with 50Ω termination. Nevertheless, further work on a stability check with a 4:1 VSWR will be carried out.

---

[16] The 22-uF tantalum capacitor is from Vishay, Inc. (part number 595D226X0050R2T). Its operating voltage is 60 V.

8.3.2   Impedance Transformer Design via Real-Frequency Technique

This section explains the use of impedance transformation to fulfill the vectorially combined distributed power amplifier from Section 8.3.1. In designing wideband communication systems, the use of impedance transformers and filters is inevitable. Usually, filters are designed between resistive terminations, say, R_1 and R_2, to restrict the frequency band of operations. As shown in Figure 8.45, R_1 designates the idealized internal resistance of the Thevenin driving source E_G, the load which dissipates transferred signal power P_L over the prescribed frequency band of operation [61–65]. In practice, the source side may represent the output of the distributed power amplifier, which may be a low resistance such as $R_1 = 12\Omega$, and the load may be a standard termination like $R_2 = 50\Omega$.

The classical filter literature [61–65] has well established that passband filters are constructed based on the lowpass prototype using lowpass-to-bandpass transformations that in turn double the number of elements of the original lowpass prototype. For many RF applications, it is customary to design a lossless two-port, which transforms a resistive termination R_1 to R_2 to provide maximum power transfer over a prescribed band of operation as shown in Figure 8.46. This configuration is neither an ideal transformer nor an ideal filter. It is the combination of both. Hence, we call it a transformer/filter.

For transforming resistance R_1 to R_2, an ideal transformer/filter must have a flat transducer power gain $T_0 = 1$ over the passband $B = f_2 - f_1$; or equivalently over the angular frequency band $B(\omega) = \omega_2 - \omega_1$ as depicted in Figure 8.47. In these descriptions, f_2 and f_1 are the upper and the lower cutoff frequencies of the transformer/filter, and the angular cutoff frequencies are specified by $\omega_2 = 2\pi f_1$ and $\omega_2 = 2\pi f_2$.

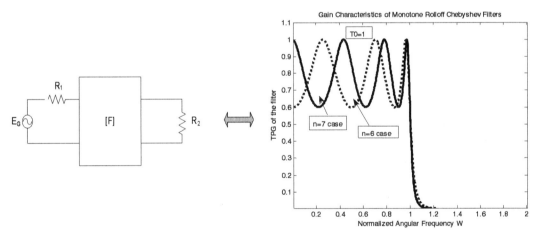

Figure 8.45   Classical filter problem for transformation from R_1 to R_2, to transform the lowpass monotone roll-off Chebyshev filter response.

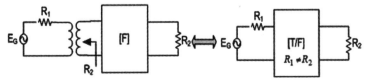

Figure 8.46   An ideal transformer with a filter, which constitutes a *transformer/filter*.

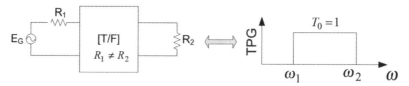

Figure 8.47 An ideal transformer/filter with transducer power gain (TPG) characteristics.

To the researcher's knowledge, there is no known analytic form of a lowpass prototype transfer function that approximates the idealized characteristic of a transformer/filter as concluded above. In practice, however, a transformer/filter can be designed using readily available CAD tools such as Spice, AWR, ADS, and so forth. In the design process, the user first selects a proper circuit topology with unknown element values, then initializes the element values. Eventually, using a nonlinear optimization algorithm, unknown element values are determined to approximate the ideal transducer gain characteristic. When dealing with a few elements in the circuit topology, say, up to three elements, this ordinary approach may be sufficient to construct a narrow bandwidth transformer/filter. Unfortunately, if the bandwidth becomes wide enough, the optimization process becomes highly nonlinear in terms of the element values. In this case, one needs to employ state-of-the-art approaches such as real frequency techniques [61–65].

In this chapter, we would like to design a transformer/filter that can transform a resistance $R_1 = 12\Omega$ to $R_1 = 50\Omega$ over a 100-MHz to 2.2-GHz bandwidth. In fact, in this problem, $R_1 = 12\Omega$ represents the output resistance of a distributed amplifier that is supposed to deliver its maximum power to a standard 50Ω termination. Hence, we face a typical design problem for using a transformer/filter over a wide frequency band. In the course of the design process, we try to use readily available CAD tools by selecting proper circuit topology. Due to the highly nonlinear nature of the problem, an optimization scheme for the design was not successful. Eventually, we employed the Real Frequency Direct Computational Technique (RF-DCT), which yields an excellent solution for the transformer/filter problem under consideration. RF-DCT may be considered a semianalytical procedure for constructing lossless two-ports for a preassigned gain performance [66–69].

Referring to Figure 8.48, in RF-DCT, the lossless transformer/filter is fully described in terms of its Darlington's driving point impedance. Darlington proved that any positive real impedance can be realized as a lossless two-port in resistive termination [66].

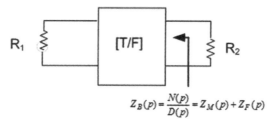

$$Z_B(p) = \frac{N(p)}{D(p)} = Z_M(p) + Z_F(p)$$

Figure 8.48 Darlington's description of a lossless two-port transformer/filter, where $p = \sigma + j\omega$ (complex variable) [66].

In the present case, $Z_B(p)$ must be determined to approximate the idealized transformer/filter characteristics in such a way that, when it is synthesized in the Darlington sense, the lossless transformer/filter is obtained in desired termination R_1. Positive real (PR) impedance such as $Z_B(p)$ from Figure 8.48 can be expressed as the summation of a minimum reactance $Z_M(p)$ and a Foster $Z_F(p)$ function, $Z_M(p)$ + $Z_F(p)$. The minimum reactance impedance function $Z_M(j\omega)$ is expressed as

$$Z_M(j\omega) = R_M\left(\omega^2\right) + jX_M(\omega) \tag{8.26}$$

where the real part $R_M(\omega^2)$ is a nonnegative even function in the angular frequency ω, and by definition, a minimum reactance function $Z_M(j\omega)$ is free of the right half plane (RHP) and $j\omega$ poles [66].

From the design point of view, the rational form of $R_M(\omega^2)$ specifies a lumped-element network topology for the lossless transformer/filter when it is terminated in a resistance. The general form of $R_M(\omega^2)$ is given by

$$R_M\left(\omega^2\right) = \frac{A_1\omega^{2n} + A_2\omega^{2(n-1)} + \cdots + A_n\omega^2 + A_{n+1}}{B_1\omega^{2n} + B_2\omega^{2(n-1)} + \cdots + B_n\omega^2 + 1} \geq 0; \forall \omega \tag{8.27}$$

This form corresponds to highly complicated circuit topologies depending on the values of the numerator coefficients A_i; $i = 0$. In this case, there is no way to control the termination resistance at the far end of the synthesis. On the other hand, a simpler form is given by

$$R_M\left(\omega^2\right) = \frac{R_1}{B_1\omega^{2n} + B_2\omega^{2(n-1)} + \cdots + B_n\omega^2 + 1} \geq 0; \forall \omega \tag{8.28}$$

which yields an n-element LC lowpass ladder circuit topology terminated in $Z_M(0)$ = $R_M(0) = R_1$ as desired.

At this point, we should note that once $R_M(\omega^2)$ is specified as in (8.28), eventually the closed-form of $Z_M(p)$ of (8.29) yields the full degree rational PR function in complex variable p, as follows:

$$Z_M(p) = \frac{a_1p^{n-1} + a_2p^{n-2} + \cdots + a_{n-2}p + a_n}{b_1p^n + b_2p^{n-1} + \cdots + b_np + b_{n+1}} \tag{8.29}$$

Synthesis of (8.28) yields a lowpass LC ladder as shown in Figure 8.49. It is noted that the LC ladder starts with a shunt capacitor C_1 since $Z_M(p)$ is a minimum

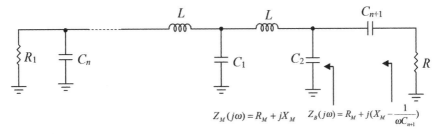

Figure 8.49   Circuit topology dictated by the real part of the positive real impedance Z_B.

reactance function. Depending on the value of integer n, the last component is either a capacitor C_n (n = odd case) or an inductor L_n (n = even case). For the problem under consideration, a simpler and meaningful form of the Foster part of the driving point impedance $Z_B(p)$ is given as

$$Z_F(j\omega) = jX_F(\omega) = -\frac{1}{\omega C_{n+1}}$$

(8.30)

Then, one can deduce the PR impedance of Darlington's driving point as

$$Z_B(j\omega) = R_M + j\left(X_M - \frac{1}{\omega C_{n+1}}\right)$$

(8.31)

where (8.29) and (8.30) fully describe a bandpass lossless two-port structure for a transformer/filter terminated in specified resistance R_1.

Referring to Figure 8.47, the transducer power gain of the transformer/filter is given by

$$T(\omega) = \frac{4R_M R_2}{\left(R_M + R_2\right)2 + \left(X_M + X_F\right)^2}$$

(8.32)

Over a specified angular frequency band $B(\omega) = \omega_2 - \omega_1$, ideally, $T(\omega) = 1$. Thus, within the passband, let's define an error function $\varepsilon(\omega)$ such that

$$\varepsilon(\omega) = \left(R_M + R_2\right)^2 + \left(X_M + X_F\right)^2 - 4R_M R_2 \qquad \omega_1 \leq \omega \leq \omega_2$$

(8.33)

or

$$\varepsilon(\omega) = \left(R_M - R_2\right)^2 + \left(X_M + X_F\right)^2 \geq 0, \quad \omega_1 \leq \omega \leq \omega_2$$

(8.34)

The crux of RF-DCT is to determine $R_M(\omega)$ and $X_F(\omega)$ such that the error function $\varepsilon(\omega)$ is minimized over the band of interest [69]. Obviously, this is a nonlinear optimization problem. The success of the nonlinear optimizations depends on the degree of nonlinearity of the error function. If the error function is quadratic in terms of the unknowns, it is possible to hit the global minimum; otherwise, the solution may be complicated [69]. Now, let us investigate the degree of nonlinearity of the optimization by starting with the selected topology. In this case, let x_i designate the unknown element values of the selected topology. Then, the driving point impedance $Z_B(p)$ is expressed as

$$Z_B(p) = \frac{x_{n+1}}{p} + \cfrac{1}{x_1 p + \cfrac{1}{x_2 p + \cfrac{1}{x_3 p + \cfrac{1}{\cdots + \cfrac{1}{R_1}}}}}$$

(8.35)

When we generate the real part $R_M(\omega)$ from (8.29), it can be shown that the leading term B_1 of (8.28) is $B_1 = ((x)x_2...x_n)^2$, which describes a $2n$ degree of nonlinearity as opposed to direct generation of B_1. Similarly, the other coefficients B_2, B_1, ..., B_n exhibit a descending degree of nonlinearity $\{(2n - i); i = 2, 3, ..., n\}$ in terms of the unknown elements values. As far as RF-DCT is concerned, in (8.31), $R_M(\omega)$ and $X_F(\omega)$ are the unknowns of the optimization problem. In other words, coefficients $B_i; i = 1, 2, ..., n$ of (8.28) are determined to minimize $\varepsilon(\omega)$ over the passband. Once B_i is initialized, $X_M(\omega)$ is evaluated by means of (8.26). In this case, the Foster part X_F is selected in such a way that it practically cancels X_M; that is, $X_M + X_F \approx 0$.

8.3.2.1 Numerical Evaluation of Hilbert Transformation and Real Frequency Line Segment Technique

Once $R_M(\omega)$ has been initialized, $X_M(\omega)$ can be generated via the Hilbert transformation integral. In RF-LST, the real part is approximated by means of straight lines as shown in Figure 8.50. Let $\{R_j, \omega_j; j = 1,2,...N\}$ be the selected sampling points of $R_M(\omega)$. In this case, $R_M(\omega)$ is expressed by

$$R_M(\omega) = \left\{ \begin{array}{ll} a_j\omega + b_j & \omega_j \leq \omega \leq \omega_{j+1}; j = 1,2,(N-1) \\ 0 & \omega \geq \omega_N \end{array} \right\} \tag{8.36}$$

where $a_j = \dfrac{R_j - R_{j+1}}{\omega_j + \omega_{j+1}} = \dfrac{\Delta Rj}{\omega_{j+1} - \omega_j}$, with $b_j = \dfrac{(R_{j+1})\omega_j - (R_j)\omega_{j+1}}{\omega_j - \omega_{j+1}}$ and $\Delta R_j = R_{j+1} - R_j$.

Using (8.36), the Hilbert transformation integral reveals the imaginary part $X_M(\omega)$:

$$X_M(\omega) = \sum_{j=1}^{N-1} B_j(\omega)\Delta R_j \tag{8.37}$$

such that

$$B_j(\omega) = \frac{1}{\pi\left(\omega_1 - \omega_{j+1}\right)}\left[F_{j+1}(\omega) - F_j(\omega)\right] \tag{8.38}$$

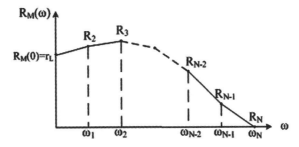

Figure 8.50 Piecewise linearization of $R_M(\omega)$.

with

$$F_j(\omega) = \left(\omega + \omega_j\right)\ln\left(\left|\omega + \omega_j\right|\right) + \left(\omega - \omega_j\right)\ln\left(\left|\omega - \omega_j\right|\right) \tag{8.39}$$

It is interesting to note that, in fixing r_L as specified by the design problem of the transforming filter, transducer power gain optimization can be carried out over the break points $\{R_j; j = 2,...R_{N-1}\}$ for the prefixed break frequencies $\{\omega_j; j = 1,2,... \omega_N\}$. In this case, the error function $\varepsilon(\omega) = (R_M - R_2)^2 + (X_M + X_F)^2$ is quadratic in terms of the unknown break points $\{R_i; i = 1,2,3,...,(n-1)\}$. Obviously, under the current optimization scheme, residues of the Foster function $X_F(\omega)$ of (8.30) are included among the unknowns. Thus, RF-LST results in idealized data points for $Z_M(j\omega) = R_M(\omega) + jX_M(\omega)$ and the analytic form of the Foster function X_F (i.e., $X_F = -k_0/\omega$). Once data points for $R_M(\omega)$ have been generated, they can be modeled as a nonnegative even rational function as in (8.28) by means of any regression algorithm, which in turn leads to an analytical form for $Z_M(p)$ of (8.29); at this point, practical generation of minimum reactance functions becomes crucial. Therefore, instead of using the integral equation from [67], it may be preferable to implement the parametric approach to generate $Z_M(p)$ from $R_M(\omega)$ as outlined in the following subsection [68].

8.3.2.2 Approach for Generating a Minimum Reactance Function from Its Real Part

In this method, $Z_M(p)$ is expressed in terms of its poles, which are all located in the LHP as follows:

$$Z_M(p) = \frac{a_1 p^{n-1} + a_2 p^{n-2} + \cdots + a_{n-2} p + a_n}{b_1 p^n + b_2 p^{n-1} + \cdots + b_n p + b_{n+1}} = R_\infty + \sum_{i=1}^{N_p} \frac{k_i}{p - p_i} \tag{8.40}$$

Its even part is given as

$$R_M\left(-p^2\right) = \frac{1}{2}\left[Z_M(p) + Z_M(-p)\right] \text{ or } R_M\left(-p^2\right) = \frac{1}{2}\left\{2R_\infty + \sum_{i=1}^{n}\left[\frac{k_i}{p - p_i} + \frac{k_i}{-p - p_i}\right]\right\} \text{ or}$$

$$R_M\left(-p^2\right) = R_\infty + \sum_{i=1}^{n}\left[\frac{k_i p_i}{\left(p - p_i\right)^2}\right]$$

$$\tag{8.41}$$

Thus, the residues can directly be computed from (8.41) as

$$k_i = \left.\frac{\left(p^2 - p_i^2\right)\left(R_M\left(-p^2\right)\right)}{pi}\right|_{p=pi} \tag{8.42}$$

with

$$R_\infty = R_M(-p^2) \tag{8.43}$$

In this approach, $R_M(-p^2)$ is initialized (i.e., known) and it is specified as in (8.41) replacing ω^2 with $-p^2$. Once all of the residues have been computed, by straightforward algebraic manipulations, $Z_M(p)$ is generated in its rational form for synthesis purposes.

8.3.2.3 Actual Design Principle of the Impedance Transformer

From Section 8.3.1, the distributed power amplifier has a measured output imped-ance of ~12Ω, which is supposed to drive a $R_L = 50\Omega$ load over a finite frequency band of 100 MHz to 2.2 GHz. As it stands, this problem describes a typical design for an impedance transformer/filter. In this case, the output of the distributed ampli-fier is considered to be the Thevenin generator E_G with internal resistance $R_G = 12\Omega$. Referring to (8.40), we can choose the impedance normalization number as $R_o = 50\Omega$, which makes the normalized load $r_L = R_L/R_o = 1$. Similarly, the normalized generator resistance r_G becomes $r_G = R_G/R_o = 12/59 = 0.24$.

Actual frequencies may be normalized with respect to the upper edge of the frequency passband. Hence, f_o is selected as $f_o = 2.2$ GHz. In this case, normalized lower and upper edge angular frequencies become $\omega_1 = 0.0455$ and $\omega_2 = 1$, respec-tively. For the construction of the lossless transformer/filter, we use a main Matlab program called "Transfilter.m." "Transfilter.m" takes all user-defined inputs to mini-mize the objective function. Initials for the coefficients can be determined by means of the real frequency line segment technique. This is an extra computational step that may not be desirable. On the other hand, it is verified that an ad hoc choice on the initial values such as $\{x_i = +1 \text{ or } -1; i = 1, 2, ..., n, n + 1\}$ is sufficient to obtain a successful optimization. In fact, this is how we initiated the optimization for the problem under consideration. "Isqnonlin.m" returns to the main program with opti-mized polynomial coefficients $\{x_1, x_2, ..., x_n\} = \{C_i; i = 1, 2, ..., n\}$ and x_{n+1}, which in turn yields the normalized value of the series capacitor $C_{n+1} = 1/x_{n+1}^2 > 0$. After the optimization, Matlab polynomials $a(p) = [a_1\ a_2\ a_3\ a_n\ a_{n+1}]$ and $b(p) = [b_1\ b_2\ b_3\ b_n\ b_{n+1}]$ are determined. Then, $Z_M(p) = a(p)/b(p)$ is synthesized and, eventually, the resulting transducer power gain in decibels and the transformer/filter circuit with optimized element values are printed as a lossless LC ladder in unit termination. Let's review the main program "Transfilter.m" with the following inputs.

Inputs:

- Normalized value of $r_G = 0.24$.
- Initial values of the polynomial coefficients. Note that here we used five poly-nomial coefficients, which will result in a five-element LC ladder network when $Z_M(p)$ is synthesized.
- Initial value for the series capacitor $C_{n+1} = 1/x_{n+1}^2 = 10$, which corresponds to $x_{n+1} = 0.31623$.
- Lower edge of the angular frequency band: $\omega_1 = 0.045$ which corresponds to 99 MHz to bias the optimization in favor of $f_1 = 100$MHz (or normalized frequency = 100MHz/2.2GHz = 0.045455).

- Upper edge of the angular frequency band: $\omega_2 = 1$.
- Flat gain level of transducer power gain $T_o = 0.99$. (Actually, ideal value of $T_o = 1$. However, we prefer to work with $T_o = 0.99$ to reduce the gain fluctuations within the passband.)

Results of Optimization:

- Optimized unknown vector: $x = [2.2069\ -0.41107\ -7.7351\ 0.871\ 3.8151\ 0.085061]$, which reveals optimized coefficients of the auxiliary polynomial $C(\omega)$ such that

$$C = [2.2069\ 0.41107\ 7.7351\ 0.871\ 3.8151\ 0.085061]$$

and $x_{n+1} = 0.085061$, meaning that the series capacitor $C_{n+1} = 1/(0.085061)^2 = 138.21$.

- Coefficients of polynomial of $P_n(\omega^2)$:

$$C = [-4.87\ -33.9\ -75.95\ -59.084\ -16.29]$$
$$P_n(\omega^2) = (1/2)[C^2(\omega) + C^2(-\omega)] = B_1\omega^{2n} + B_2\omega^{2(n-1)} + \cdots + B_n\omega^2 + 1$$

- Analytic-rational form of minimum reactance function $Z_M(p) = a(p)/b(p)$, where

$a(p)$ as a Matlab polynomial vector:

$$a = [0\ \ 1.65\ \ 1.0653\ \ 2.2253\ \ 0.84655\ \ 0.45312]$$

and

$b(p)$ as a Matlab polynomial vector:

$$b = [1\ \ 0.\ \ 0.64563\ \ 3.696\ \ 2.0286\ \ 2.2769\ \ 0.45312].$$

Note that the leading coefficient of $a(p)$ is zero, which means that the degree of $a(p)$ is one degree lower than that of $b(p)$—as it should be. Furthermore, $a_6 = b_6 = 0.45312$, which yields $R_1 = 1$ as desired.

- Synthesis of $Y_M(p) = 1/Z_M(p) = b(p)/a(p)$ by long division yields the following normalized element values:

$$Y_M(p) = C_1 p + \cfrac{1}{L_2 p + \cfrac{1}{C_3 p + \cfrac{1}{L_4 p + \cfrac{1}{C_5 p + 1}}}} \tag{8.44}$$

with

$$C_1 = 0.6061;\ L_2 = 0.7029;\ C_3 = 2.870;\ L_4 = 1.165\text{e};\ C_5 = 1.549\text{e};\ r_L = 1.00$$

The resulting circuit diagram is shown in Figure 8.51. Actual capacitors are given by $C_{iA} = C_{iN}/2\pi f_o R_o$; similarly, actual inductors are given by $L_{iA} = L_{iN}R_o/2\pi f_o$ where C_{iN} and L_{iN} represent the normalized values of capacitors and inductors,

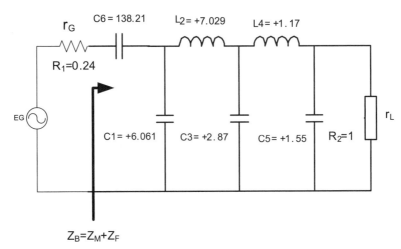

Figure 8.51   Circuit topology of transformer/filter with optimized element values.

respectively. Normalization numbers are specified as $R_o = 50\Omega$ and $f_o = 2.2$ GHz. Hence, we have $C_{1A} \cong 0.88$ pF, $L_{2A} \cong 2.55$ nH, $C_{3A} \cong 4.15$ pF, $L_{2A} \cong 4.22$ nH, $C_{5A} \cong 2.24$ pF, and $C_{6A} \cong 200$ pH. Finally, actual termination resistance is $r_L = 50\Omega$. The gain performance of the impedance transforming filter is -0.0432 and -2.1454 dB at 2.076 GHz and 100 MHz, respectively. The average gain is given by $T_{\text{average}} = -1.094 \pm 1.0511$ dB.

The PCB used was Rogers 4350 with $\varepsilon_r = 3.66$ and a thickness of $h = 0.762$ mm. The inductors were realized using high characteristic impedance transmission lines printed on the board like microstrip lines. For accurate design, inductors were modeled employing the CST, in which PCB properties were included. The Q-factor and inductance values of the transmission line were optimized in CST to end up with a flat gain response over the band of interest. A top metal thickness of 2 oz with gold plating is applied in the transmission line for high-power handling requirements beyond 30W.

Modeling for the capacitors was performed. Careful layout design with practical considerations were taken into account. Over high-frequency operations in particular, the diameter size of the via holes and spacing between them require special attention. In the present case, via-hole diameter was taken as 0.15 mm, connecting the component's grounding from top layer to the bottom layer, and the separation between holes was set to 0.3 mm. Adequate numbers of via holes were placed to provide good electrical grounding for the shunt capacitors. Using equivalent models for inductors and capacitors, the gain performance of the transformer/filter was simulated on ADS Momentum. The circuit layout was imported from Cadence. Figure 8.52 depicts measured and simulated gain performance for the transformer/filter under consideration. As designed, port 1 is terminated in 12Ω. Port 2 is connected to 50Ω. In the course of measurements, first, the scattering parameters of the transformer/filter were measured between 50Ω terminations using SMA RF connectors. To end up with accurate measurement results, each SMA connector is regarded as a separate two-port.

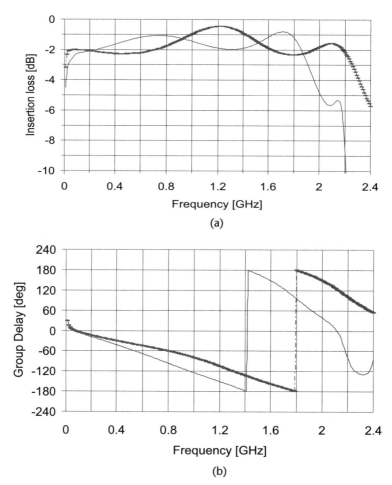

Figure 8.52 Simulated (thin line) versus measured (thicker line) performance of transformer/filter: (a) insertion loss and (b) group delay within operating bandwidth.

However, bear in mind that the original transformer/filter is driven with R_G = 12Ω instead of 50Ω. Therefore, to end up with the actual performance of the transformer/filter as designed, the measured scattering parameters of port 1 must be extracted from the measured S-parameters of the complete system. Then, the actual scattering parameters are generated with respect to the original input port normalization number R_G = 12Ω, which in turn yields the desired electrical performance of the system as it is driven by the distributed amplifier. In fact, this is what we have done. First, we measured the 50Ω based scattering parameters of the SMA connectors used in both port 1 and port 2. At this point, measurement results were compared with those of the model provided with Molex, Inc. It has been observed that measured data show excellent correlation with the SMA connector model given by Molex (as discussed in Section 5.5). Then, the measured S-parameters of the input SMA connector were extracted from the measured scattering parameters of the two-port, resulting in R_o = 12Ω based S-parameters.

Note that all S-parameter measurements were carried out using an HP Network Analyzer (HP6778). All of the above computations were automatically completed on

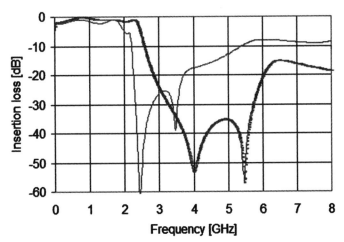

Figure 8.53   Simulation performance harmonic filtering up beyond passband. The thick line reflects the use of discrete high-Q inductors and the thin line is for a lumped transmission line.

the ADA CST/Momentum platform, which is interfaced with the network analyzer. Results are plotted in Figure 8.52. The thin line corresponds to the gain performance of the transformer/filter, which is simulated with the equivalent models of transmission line inductors and discrete high-Q capacitors.[17] Finally, the thicker line is the measured transducer power gain response in dB of the actual transformer/filter implemented with transmission line inductors and discrete capacitor. It is seen that the measured gain in the pass band is about -1.1 dB with about ±0.9 dB fluctuations as expected. Hence, measurements reveal excellent agreement with simulations.

Furthermore, Figure 8.53 also reveals that use of discrete lump inductors (thick line) does not change the gain performance much within the passband; however, their use does suppress the harmonics in the stopband as they should do. The thick line shows the performance of the simulation obtained with high-Q discrete inductors from Coilcraft, Inc.[18] For comparison, a lumped transmission line simulation is illustrated by the thin line.

8.4   Drain-Line High-Power Device Loading Compensation

Signal losses dominated by the positive FET resistances (i.e., R_{gs} and R_{ds}) increase proportionally with a bigger device periphery. If these loss mechanisms are somehow mitigated with attenuation compensation, the gate and drain lines can accommodate additional gate periphery and more sections and, thus, output power can be increased. The cascode topology is a well-known one for attenuation compensation, as shown in Figure 8.54. Several works have been reported on high output power and efficiency with a cascode topology in a distributed amplifier [21, 40, 70]. The motivation of this section is to provide a feasible study to achieve high output

---

[17] The capacitors are 600S Series ultra-low ESR, high-Q microwave capacitors, from ATC, Inc.
[18] The inductor series is 0603CS_XNL (1008HS) from Coilcraft, Inc.

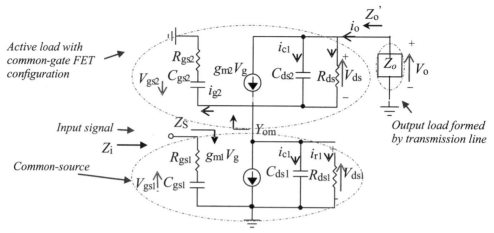

Figure 8.54 Simplified cascode topology, in which a common-source FET is coupled to the active load with a common-gate configuration.

power by employing a cascode, with a combination of vectorially combined current sources while keeping the load termination close to 50Ω. Clearly, the proposed topology offers an alternative to the DPA solution of Section 8.3 for achieving high output power.

An explanation is given for how the loading effect of the common-source power transistor is resolved when the transistor (e.g., GaN) is loaded into the output transmission line network while preserving current combining to a single load termination (~50Ω). As a result, impedance transformation can be eliminated.

The resulting input and output impedance, Z_i and Z_o, respectively, of the cascode topology shown in Figure 8.54 is given as

$$Z_i = R_{gs1} + j\omega C_{gs1} \tag{8.45}$$

$$Z_o = \frac{R_{ds2}}{1 + j\omega C_{ds2}R_{ds2}}\left[1 - \frac{g_{m2}Z_S}{1 + j\omega C_{gs2}\left(R_{gs2} + Z_S\right)}\right] + \frac{Z_S\left[1 + j\omega C_{gs2}R_{gs2}\right]}{1 + j\omega C_{gs2}\left(R_{gs2} + Z_S\right)} \tag{8.46}$$

where Z_S is the output termination impedance of a common-source FET. Note that C_{gd} is neglected for reasons of simplicity.

Input impedance Z_i requires a similar match as conventional common-source distributed amplifier. Expanding (8.46), the real part of Z_o can be written as

$$\text{Re}\{Z_o\} = \frac{R_{ds2}}{1 - \omega^2/\omega_d^2} - \frac{g_{m2}R_{ds2}Z_S}{\left(1 - \omega^2/\omega_d^2\right)\left(1 - \omega^2/\omega_g^2\left(\frac{R_{gs2} + Z_S}{R_{gs2}}\right)^2\right)}$$

$$+ \frac{Z_S\left(\frac{R_{gs2} + Z_S}{R_{gs2}}\right)^2 + Z_S\left(\omega^2/\omega_g^2\right)}{\left(1 - \omega^2/\omega_g^2\right)} \tag{8.47}$$

where $\omega_d = 1/R_{ds2}C_{ds2}$ and $\omega_g = 1/R_{gs2}C_{gs2}$. Note that (8.47) as derived above is similar to that given in [71]. The term that provides negative resistance from (8.47) is

$$Z_{nr} = -\frac{g_{m2}R_{ds2}Z_S}{\left(1 - \omega^2/\omega_d^2\right)\left(1 - \omega^2/\omega_g^2\left(\dfrac{R_{gs2} + Z_S}{R_{gs2}}\right)^2\right)} \qquad (8.48)$$

and other terms are passive in nature. The term Z_{nr} is directly proportional to g_{m2} to provide attenuation compensation to the positive real part of Z_o. For $(\omega/\omega_d)^2 > 1$, we can expect the Z_{nr} of (8.48) to become positive, which has no benefit in attenuation compensation. Similarly, the passive terms of (8.47) can be negatively resistant for $(\omega/\omega_d)^2 > 1$ and $(\omega/\omega_g)^2 > 1$, causing oscillation. The imaginary part is as follows:

$$\mathrm{Im}\{Z_o\} = \frac{g_{m2}R_{ds2}Z_S\omega/\omega_d}{\left(1 - \omega^2/\omega_d^2\right)\left(1 - \omega^2/\omega_g^2\left(\dfrac{R_{gs2} + Z_S}{R_{gs2}}\right)^2\right)} + \frac{\omega/\omega_g\left(\left(\dfrac{R_{gs2} + Z_S}{R_{gs2}}\right)^2 - 1\right)}{\left(1 - \omega^2/\omega_g^2\right)}$$

$$(8.49)$$

With further simplification of Figure 8.54 for low frequencies, analysis leads to the topology given in Figure 8.55. In general, $Z_{od} = \sqrt{\left(L_d/C_{ds2}\right)\left(1 - \omega^2/\omega_c^2\right)^{-1}}$ will absorb the effective capacitance (C_{ds1} and C_{ds2}) and output line inductance L_d, which synthesized the artificial transmission line. Thus, Z_{od} is approximated to 25Ω at a low frequency since both arms are terminated with 50Ω.

At low frequency (let $\omega = 0$), output impedance $Z'_o = R_{ds2} + Z_s(1 - g_{m2}R_{ds2})$ can be estimated from (8.47). The output impedance ratio between a cascode and common-source configuration is given by Z_o/Z_S, which is factored by $(R_{ds2} + Z_s(1 - g_{m2}R_{ds2}))/(g_{m1}R_{ds1})$.

Impedance Z_1 seen by current generator g_{m1} is $Z_1//Z_{om} = R_{ds1}//(R_{ds2} + Z_{od})/(1 - g_{m2}R_{ds2})$. The value of Z_{om} is typically lower than R_{ds1}, and thus most current $g_{m1}V_{gs1}$ flows in the source of the upper FET and total transconductance $G_m \gg g_{m1}$. Power delivered by the lower generator is estimated as $(g_{m1}V_{gs1})^2Z_1$. The current injection of current generator g_{m2} depends on the voltage drop across the first generator. According to the power conversion law, power delivered by current generators is equal to power absorbed by the resistive elements as shown here:

$$\left(g_{m1}V_{gs1}\right)^2Z_1 + \left(g_{m2}V_{gs2}\right)^2Z_2 = V_{ds1}^2/R_{ds1} + V_{ds2}^2/R_{ds2} + V_o^2/Z_{od} \qquad (8.50)$$

where $A_V \approx V_o/V_{gs1}$ can be deduced from (8.50).

Let's consider two FETs that are identical and have $R_{dsi} = 100\Omega$, $g_{mi} = 100$ mS, and $V_{gsi} = 1$V, where i is 1, 2. From the analysis, Z_1 is approximately 11.1Ω since Z_{om} is 12.5Ω; output current $i_o \approx g_mV_{gs1}$. Hence, current gain $A_i \approx 1$ and voltage gain

$A_V \approx 22$. The output impedance ratio of Z_d/Z_S is 2.3 at a low frequency. However, the output impedance at a higher frequency range can be maintained as close to 2.3 (at low frequency) with a properly intermatched design between the common-source and common-gate FET. As an example, an inductor L_p (Figure 8.58) is introduced for this reason.

From Figure 8.55, Z_{od} will absorb the effective capacitance of the cascode topology, from (8.49), and output line inductance L_d, which is synthesized from the artificial transmission line to form the desired ω_c. It has been shown analytically that the output impedance of the cascode is ~2 times higher than for the common-source configuration for identical device selection. When the loading effect is resolved, it is beneficial that current combining to a common load will be achieved without an additional transformation network. Figure 8.56 illustrates the multicurrent sources (cascode configuration) that are loaded into the output drain transmission line. It is confirmed experimentally that the real part of the output impedance of the die model is very close to the R_{ds}. Current gain $A_i \approx 1$ indicates that the current contribution is similar to a common-source configuration; hence, the biasing condition for the common-gate is crucial to deliver similar current injection.

Validation of the concept above for equal magnitude and in-phase combining for $n = 3$ is selected, leading to $\Re\{Z_{u(1)}\} = \Re\{Z_{u(2)}\} = \Re\{Z_{u(3)}\} = 150\Omega$, and $\Re\{Z_{r(1)}\} = 150\Omega$, $\Re\{Z_{r(2)}\} = 75\Omega$, and $\Re\{Z_{r(3)}\} = 50\Omega$, and an imaginary part does not exist. For simplicity, a hypothetical drain terminal of a transistor has been modeled as an ideal current source with parallel resistance (effective real part of the cascode) [70], as shown in Figure 8.57. The imaginary part of Z_o (typically capacitance) will be absorbed in $Z_{r(k)}$. An example of output impedance Z_o for a cascode GaN die

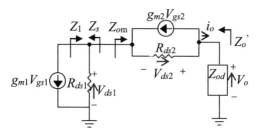

Figure 8.55 Simplified cascode topology analysis at low frequency.

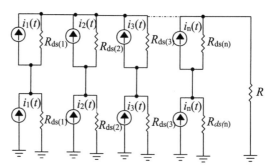

Figure 8.56 Current combining to a common node R, with cascode configuration, for n sections.

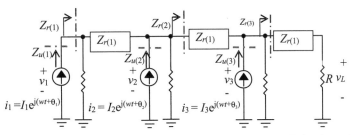

Figure 8.57 Circuit showing multicurrent sources to combine at a single load termination, where $R = 50\Omega$.

model showed that the real part is increased by a factor of 2, and the imaginary part is reduced by factor of 2. To additively combine the currents at each junction, phase synchronization between the current source and the transmission line is crucial. Since transmission lines delays $\theta_{d(k)}$ vary linearly with frequency, making the current source delays also vary with frequency would guarantee delay matching between the sources and the transmission lines. The cascode delivers two times higher power than the common-source topology across the frequency, where all the sources are presented with exactly the required hypothetical optimum impedance of 150Ω at all frequencies.

First, a dc biasing condition is established to satisfy the analytical presentation in the previous section. The main dc source voltage V_{drain} is fed to the cascode, and the biasing gate voltage V_{g2} of the common-gate transistor determines the effective V_{d2} to allow drain current to flow across the cascode configuration, as illustrated in Figure 8.58. Applying Kirchoff's law (voltage loop) to Figure 8.58, dc biasing can be deduced as

$$V_{g2} = V_{gs2} + V_{d2} \tag{8.51}$$

where V_{gs2} refers to the gate source of the common-gate transistor. Therefore, the total voltage of V_{d1} and V_{d2} will determine the V_{drain} value. For example, with the GaN die model, a V_{drain} value of 32V is needed to generate a similar drain current in the cascode configuration, wherelse 28V for the common-source configuration. As explained in [13], in the absence of L_p, circuit bandwidth is primarily limited by the pole associated with the internal node of the cascode cells. Therefore, the output

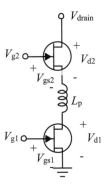

Figure 8.58 DC biasing condition for cascode topology.

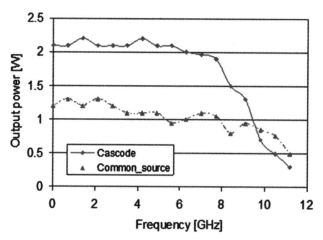

Figure 8.59 Power performance between cascode and common-source topology.

impedance of the cascode topology, especially at higher frequency operation, can be improved with inclusion of L_p.

Performance results with $n = 3$ sections indicated a power performance of a flat 2W from dc-7 GHz with cascode topology, and the power is improved by factor of ~1.8 over the common-source topology (Figure 8.59). A simplified design schematic of the cascode with the vectorially combined current sources is presented in Figure 8.60. Since C_{opt} (the effective output of common gate (CG) FET) is smaller than

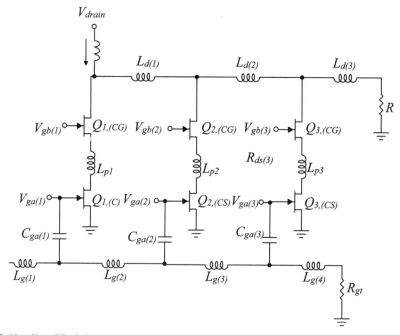

Figure 8.60 Simplified design schematic of a cascode with a current combining approach to a single load termination.

C_{in} (the input capacitance of a common source (CS) FET), phase synchronization between the gate and drain lines is achieved by using a capacitively coupled technique [49] and a nonuniform gate-line design as shown in [16].

References

[1]   B. Kim, H. Q. Tserng, and H. D. Shih, "High Power Distributed Amplifier Using MBE Synthesized Material," *IEEE Microwave and Millimeter-Wave Monolithic Integrated Circuits Symp.*, pp. 35–37, 1985.

[2]   P. H. Ladbrooke, "Large Signal Criteria for the Design of GaAs FET Distributed Power Amplifier," *IEEE Trans. Electron Devices*, Vol. ED-32, No. 9, pp. 1745–1748, Sep. 1985.

[3]   B. Kim and H. Q. Tserng, "0.5W 2–21 GHz Monolithic GaAs Distributed Amplifier," *Electronics Lett.*, Vol. 20, No. 7, pp. 288–289, Mar. 1984.

[4]   M. J. Schindler et al., "A 15 to 45 GHz Distributed Amplifier Using 3 FETs of Varying Periphery," *IEEE GaAs IC Symp. Technical Digest*, pp. 67–70, 1986.

[5]   J. B. Beyer et al., "MESFET Distributed Amplifier Design Guidelines," *IEEE Trans. Microwave Theory and Techniques*, Vol. MTT-32, No. 3, pp. 268–275, Mar. 1984.

[6]   S. N. Prasad, J. B. Beyer, and I. S. Chang, "Power-Bandwidth Considerations in the Design of MESFET Distributed Amplifiers," *IEEE Trans. Microwave Theory Techniques*, Vol. 36, pp. 1117–1123, July 1988.

[7]   Y. Ayasli et al., "2-to-20 GHz GaAs Traveling Wave Power Amplifier," *IEEE Trans. Microwave Theory Techniques*, Vol. 32, No. 3, pp. 290–295, Mar. 1984.

[8]   K. B. Niclas et al., On Theory Performance of Solid-State Microwave Distributed Amplifiers, *IEEE Trans. Microwave Theory Techniques*, Vol. MTT-31, No. 6, pp. 447–456, June 1983.

[9]   T. McKay, J. Eisenberg, and R. E. Williams, "A High Performance 2–18.5 GHz Distributed Amplifier Theory and Experiment," *IEEE Trans. Microwave Theory and Techniques*, Vol. 12, No. 34, pp. 1559–1568, Dec. 986.

[10] K. Narendra et al., "Cascaded Distributed Power Amplifier with Non-Identical Transistors and Interstage Tapered Impedance," *40th European Microwave Conf.*, Sep. 2010, pp. 549–522.

[11] K. Narendra et al., "Design Methodology of High Power Distributed Amplifier Employing Broadband Impedance Transformer," *IEEE Int. Conf. of Antenna, Systems and Propagation*, Sep. 2009.

[12] K. Narendra et al., "pHEMT Distributed Power Amplifier Employing Broadband Impedance Transformer," *Microwave J.*, Vol. 51, pp. 76–83, June 2013.

[13] K. Narendra et al., "High Efficiency Applying Drain Impedance Tapering for 600mW pHEMT Distributed Power Amplifier," *IEEE Int. Conf. on Microwave and Millimeter Wave Technology*, pp. 1769–1772, Apr. 2008.

[14] L. Zhao et al., "A 6 Watt LDMOS Broadband High Efficiency Distributed Power Amplifier Fabricated Using LTCC Technology," *IEEE MTT-S Int. Microwave Symp. Dig.*, pp. 897–900, June 2002.

[15] K. Narendra et al., "Vectorially Combined Distributed Power Amplifier with Load Pull Determination," *Electronics Lett.*, Vol. 46, Issue. 16, pp. 1137–1138, Aug. 2010.

[16] K. Narendra et al., "Vectorially Combined Distributed Power Amplifiers for Software-Defined Radio Applications," *IEEE Trans. Microwave Theory Technique*, Vol. 60, No. 10, pp. 3189–3200, Oct. 2012.

[17] K. B. Niclas, R. R. Pereira, and A. P. Chang, "On Power Distribution in Additive Amplifiers, "*IEEE Trans. Microwave Theory and Techniques*, Vol. 38, No. 3, pp. 1692–1699, Nov. 1990.

[18]   C. Duperrier et al., "New Design Method of Uniform and Nonuniform Distributed Power Amplifiers," *IEEE Microwave Theory Techniques*, Vol. 49, No. 12, pp. 2494–2499, Dec. 2001.

[19]   M. Campovecchio et al., "Optimum Design of Distributed Power FET Amplifiers: Application to 2–18 GHz MMIC Module Exhibiting Improved Power Performances," *IEEE MTT-S Int. Microw. Symp. Dig.*, San Diego, pp. 125–128, June 1994.

[20]   L. Zhao, A. Pavio, and W. Thompson, "A 1 Watt, 3.2 VDC, High Efficiency Distributed Power PHEMT Amplifier Fabricated Using LTCC Technology," *IEEE MTT-S Int. Microwave Symp. Dig.*, pp. 2201–2204, June 2003.

[21]   B. M. Green et al., "High-Power Broadband AlGaN/GaN HEMT MMICs on SiC Substrates," *IEEE Trans. Microw. Theory and Techniques*, Vol. 49, No. 12, pp. 2486–2493, Dec. 2001.

[22]   A. Martin et al., "Balanced AlGaN/GaN Cascode Cells: Design Method for Wideband Distributed Amplifiers," *Electronics Lett.*, Vol. 44, No. 2, pp. 116–117, Jan. 2008.

[23]   K. Narendra, E. Limiti, and C. Paoloni, "Dual Fed Distributed Power Amplifier with Controlled Termination Adjustment," *Progress in Electromagnetic Research*, Vol. 139, pp. 761–777, May 2013.

[24]   X. Jiang and A. Mortazawi, "A Broadband Power Amplifier Design Based on the Extended Resonance Power Combining Technique," *IEEE MTT-S Int. Microwave Symp. Dig.*, pp. 835–838, June 2005.

[25]   J. Gassmann et al., "Wideband High Efficiency GaN Power Amplifiers Utilizing a Nonuniform Distributed Topology," *IEEE MTT-S Int. Microwave Symp. Dig.*, pp. 615–618, June 2007.

[26]   S. N. Prasad and A. S. Ibrahim, "Design Guidelines for a Novel Tapered Drain Line Distributed Power Amplifier," *Proc. 36th European Microwave Conf.*, pp. 1274–1277, Sep. 2006.

[27]   C. S. Aitchison et al., "The Dual-Fed Distributed Amplifier," *IEEE MTT-S Int. Microwave Symp. Dig.*, Vol. 2, pp. 911–9914, June 1988.

[28]   C. S. Aitchison, N. Bukhari, and O. S. Tang, "The Enhanced Performance of the Dual-Fed Distributed Amplifier," *19th European Microwave Conf.*, pp. 439–444, Sep. 1989.

[29]   J. Y. Liang and C. S. Aitchison, " Gain Performance of Cascade of Single Stage Distributed Amplifiers," *Electronics Lett.*, Vol. 31, No. 15, pp. 1260–1261, July 1995.

[30]   S. D. Agostino and C. Paoloni, "Design of High Performance Power Distributed Amplifier Using Lange Couplers," *IEEE Microwave Theory Techniques*, Vol. 42, No. 12, pp. 2525–2530, Dec. 1994.

[31]   S. D. Agostino and C. Paoloni, "Innovative Power Distributed Amplifier Using the Wilkinson Combiner," *IEE Proc. Microwave Antennas and Propagation*, Vol. 142, No. 2, 1995.

[32]   S. D. Agostino and C. Paoloni, "Power Distributed Amplifier Based on Interdigital Combiners," *Electron Lett.*, Vol. 29, No. 23, pp. 2050–2051, July 1993.

[33]   K. W. Eccleston, "Compact Dual-Fed Distributed Power Amplifiers," *IEEE Trans. Microwave Theory Techniques*, Vol. 53, No. 3, pp. 825–891, Mar. 2005.

[34]   K. W. Eccleston, "Output Power Performance of Dual-Fed and Single-Fed Distributed Amplifiers," *Microwave Opt. Technology Lett.*, Vol. 27, No. 4, pp. 281–284, Nov. 2000.

[35]   F. Noriega and P. J. Gonzalez, "Designing *LC* Wilkinson power splitters," *RF Design*, Aug. 2002.

[36]   I. Bahl, *Lumped Elements for RF and Microwave Circuits*, Norwood, MA: Artech House, 2003.

[37]   D. Kajfez, Z. Paunovic, and S. Pavlin, "Simplified Design of Lange Coupler," *IEEE Trans. Microwave Theory Techniques*, Vol. MTT-26, No. 10, pp. 806–808, Oct. 1978.

[38]   R. C. Liu et al., "Design and Analysis of DC-14-GHz and 22-GHz CMOS Cascode Distributed Amplifiers," *IEEE J. Solid State Circuits*, Vol. 39, No. 8, pp. 1370–1374, Aug. 2004.

[39] X. Guan and C. Nguyen, "Low Power Consumption and High Gain CMOS Distributed Amplifiers Using Cascade of Inductively Coupled Common-Source Gain Cells for UWB Systems," *IEEE Trans. Microwave Theory and Techniques*, Vol. 54, No. 8, pp. 3278–3283, Aug. 2006.

[40] J. P. Fraysse et al., "A 2W, High Efficiency, 2–8GHz, Cascode HBT MMIC Power Distributed Amplifier," *IEEE MTT-S Int. Microwave Symp. Dig.*, pp. 529–532, June 2000.

[41] J. Shohat, I. D. Robertson, and S. J. Nightingale, "10Gb/s Driver Amplifier Using a Tapered Gate Line for Improved Input Matching," *IEEE Microwave. Theory Techniques*, Vol. 53, No. 10, pp. 3115–3120, Oct. 2005.

[42] S. D. Agostino and C. Paoloni, "Design of a Matrix Amplifier Using FET Gate-Width Tapering," *Microwave Opt. Technol. Lett.*, Vol. 8, pp. 118–121, Feb. 1995.

[43] P. Dueme, G. Aperce, and S. Lazar, "Advanced Design for Wideband MMIC Power Amplifiers," *Proc. IEEE GaAs Integrated Circuits Systems*, Oct. 7–10, 1990, pp.121–124.

[44] K. Narendra et al., "Adaptive LDMOS Power Amplifier with Constant Efficiency," *IEEE MTT-S Int. Microwave and Optoelectronics Conf.*, pp. 861–864, Sep. 2007.

[45] M. E. V. Valkenburg, *Network Analysis*, 3rd ed., Upper Saddle River, NJ: Prentice Hall, 1974.

[46] J. Y. Liang and C. S. Aitchison, "Gain Performance of Cascade of Single Stage Distributed Amplifiers," *Electronics Lett.*, Vol. 31, No. 15, pp. 1260–1261, July 1995.

[47] Technical Report of GaN, Fujitsu.

[48] B. Thompson and R. E. Stengel, "System and Method for Providing an Input to a Distributed Power Amplifier System," U.S. Patent 7,233,207, June 2007.

[49] Y. Ayasli et al., "Capacitively Coupled Traveling Wave Power Amplifier," *IEEE Microwave Theory Techniques*, Vol. MTT-32, No. 12, pp. 1704–1709, Dec. 1984.

[50] C. Xie and A. Pavio, "Development of GaN HEMT Based High Power High Efficiency Distributed Power Amplifier for Military Applications," *IEE MILCOM 2007*, pp. 1–4, Oct. 2007.

[51] R. W. Chick, "Non-Uniformly Distributed Power Amplifier," U.S. Patent, 5,485,118, Jan. 1996.

[52] S. Lee et al., "Demonstration of a High Efficiency Nonuniform Monolithic GaN Distributed Amplifier," *IEEE MTT-S Int. Microwave Symp. Dig.*, pp. 549–522, June 2000.

[53] S. Lin, M. Eron, and A. E. Fathy, "Development of Ultra Wideband, High Efficiency, Distributed Power Amplifiers Using Discrete GaN HEMTs," *IET Circuits Devices Systems*, Vol. 3, No. 3, pp. 135–142, 2009.

[54] K. Narendra et al., "High Performance 1.5W pHEMT Distributed Power Amplifier with Adjustable Inter-Stage Cascaded Network," *Microwave Techniques, COMITE 2008*, pp. 1–4, Apr. 2008.

[55] S. C. Cripps, *Power Amplifier for Wireless Communications*, 2nd ed., Norwood, MA: Artech House, 2006.

[56] M. I. Montrose, *EMC and the Printed Circuit Board—Design, Theory, and Layout Made Simple*, New York: Wiley-Interscience, 1996.

[57] M. R. A. Gaffoor et al., "Simple and Efficient Full-Wave Modeling of Electromagnetic Coupling in Realistic RF Multilayers PCB Layouts," *IEEE Trans. Microwave Theory Techniques*, Vol. 50, No. 6, pp. 1445–1457, June 2002.

[58] A. Rumiantsev, R. Doerner, and S. Thies, "Calibration Verification Procedure Using the Calibration Comparison Technique," *36th European Microwave Conf.*, pp. 489–491, Sep. 2006.

[59] A. Aldoumani et al., "Enhanced Vector Calibration of Load-Pull Measurement Systems," *83rd ARFTG Microwave Measurements Conf.*, pp. 1–4, 2014.

[60] N. K. Aridas, K. Macwien, and L. Joshua, "Power Control Circuit and Method," U.S. Patent 7,342,445, Mar. 2008.

[61]   H. J. Carlin, "Gain-Bandwidth Limitation on Equalizers and Matching Networks," *Proc. IRE*, pp. 1676–686, Nov. 1954.

[62]   A. I. Zverev, *Handbook of Filter Synthesis*, New York: John Wiley, 1967/2005.

[63]   Matthaei, G., E. M. T. Jones, and L. Young, *Microwave Filters, Impedance-Matching Networks, and Coupling Structures*, Dedham, MA: Artech House, 1980.

[64]   A. Williams and F. Taylor, *Electronic Filter Design Handbook*, 4th ed., New York: McGraw-Hill, 1988.

[65]   R. Schaumann and M. E. Valkenburg, *Design of Analog Filters*, New York: Oxford University Press, 2001.

[66]   B. S. Yarman, *Design of Ultra Wideband Power Transfer Networks*, New York: John Wiley, 2010.

[67]   S. Yarman, "Broadband Matching a Complex Generator to a Complex Load," Ph.D. thesis, Cornell University, 1982.

[68]   H. J. Carlin and B. S. Yarman, "The Double Matching Problem: Analytic and Real Frequency Solutions," *IEEE Trans. Circuits and Systems*, Vol. 30, pp. 15–28, Jan. 1983.

[69]   H. J. Carlin, "A New Approach to Gain-Bandwidth Problems," *IEEE Trans. Circuits and Systems*, Vol. 23, pp. 170–175, Apr. 1977.

[70]   P. Heydari, "Design and Analysis of a Performance-Optimized CMOS UWB Distributed LNA," *IEEE J. Solid-State Circuits*, Vol. 42, No. 9, pp. 1892–1905, Sep. 2007.

[71]   S. Diebele and J. B. Beyer, "Attenuation Compensation in Distributed Amplifier Design," *IEEE Trans. Microwave Theory and Techniques*, Vol. 37, No. 9, pp.1425–1433, Sep. 1989.

About the Authors

Narendra Kumar is a senior member of IEEE and a Fellow of IET. He received his doctorate degree in electrical engineering from RWTH Technical University Aachen, Aachen, Germany. He has been with Motorola Solutions since early 1999. He holds several U.S. patents, all of which were assigned to Motorola Solutions. Since early 2011, he has been actively involved in giving RF and microwave power amplifier courses to many organizations in Asia and Europe. He serves as a reviewer for several technical journals and has participated as a keynote speaker at several IEEE and PIER conferences. He is the author or coauthor of more than 60 papers and two books related to RF and microwave power amplifiers. Currently, he is an associate professor at the University of Malaya and leads microwave and terahertz research activities.

Andrei Grebennikov is a senior member of IEEE. He received his Dipl. Eng. in radio electronics from the Moscow Institute of Physics and Technology and Ph.D. in radio engineering from the Moscow Technical University of Communications and Informatics in 1980 and 1991, respectively. He gained academic and industrial experience working with Moscow Technical University of Communications and Informatics (Russia), Institute of Microelectronics (Singapore), M/A-COM (Ireland), Infineon Technologies (Germany/Austria), and Bell Labs, Alcatel-Lucent (Ireland) as an engineer, researcher, lecturer, and educator. He was guest professor at the University of Linz (Austria) and presented short courses and tutorials as an invited speaker at the International Microwave Symposia, European and Asia-Pacific Microwave Conferences, Institute of Microelectronics, Singapore, Motorola Design Centre, Malaysia, Tomsk State University of Control Systems and Radioelectronics, Russia, and Aachen Technical University, Germany. He is an author and coauthor of more than 100 papers, 25 European and U.S. patents, and seven books dedicated to RF and microwave circuit design.

Index

Recent Titles in the Artech House Microwave Library

Microwave Network Design Using the Scattering Matrix,
Janusz A. Dobrowolski

Microwave Radio Transmission Design Guide, Second Edition, Trevor Manning

Microwave Transmission Line Circuits, William T. Joines, W. Devereux Palmer, and
Jennifer T. Bernhard

Microwaves and Wireless Simplified, Third Edition, Thomas S. Laverghetta

Modern Microwave Circuits, Noyan Kinayman and M. I. Aksun

Modern Microwave Measurements and Techniques, Second Edition,
Thomas S. Laverghetta

Neural Networks for RF and Microwave Design, Q. J. Zhang and K. C. Gupta

Noise in Linear and Nonlinear Circuits, Stephen A. Maas

Nonlinear Microwave and RF Circuits, Second Edition, Stephen A. Maas

Q Factor Measurements Using MATLAB®, Darko Kajfez

QMATCH: Lumped-Element Impedance Matching, Software and User's Guide,
Pieter L. D. Abrie

Passive RF Component Technology: Materials, Techniques, and Applications,
Guoan Wang and Bo Pan, editors

Practical Analog and Digital Filter Design, Les Thede

Practical Microstrip Design and Applications, Günter Kompa

Practical Microwave Circuits, Stephen Maas

*Practical RF Circuit Design for Modern Wireless Systems, Volume I: Passive Circuits
and Systems,* Les Besser and Rowan Gilmore

*Practical RF Circuit Design for Modern Wireless Systems, Volume II: Active Circuits
and Systems,* Rowan Gilmore and Les Besser

*Production Testing of RF and System-on-a-Chip Devices for Wireless
Communications,* Keith B. Schaub and Joe Kelly

Radio Frequency Integrated Circuit Design, Second Edition, John W. M. Rogers and
Calvin Plett

RF Bulk Acoustic Wave Filters for Communications, Ken-ya Hashimoto

RF Design Guide: Systems, Circuits, and Equations, Peter Vizmuller

RF Linear Accelerators for Medical and Industrial Applications, Samy Hanna

RF Measurements of Die and Packages, Scott A. Wartenberg

The RF and Microwave Circuit Design Handbook, Stephen A. Maas

RF and Microwave Coupled-Line Circuits, Rajesh Mongia, Inder Bahl, and Prakash Bhartia

RF and Microwave Oscillator Design, Michal Odyniec, editor

RF Power Amplifiers for Wireless Communications, Second Edition, Steve C. Cripps

RF Systems, Components, and Circuits Handbook, Ferril A. Losee

The Six-Port Technique with Microwave and Wireless Applications, Fadhel M. Ghannouchi and Abbas Mohammadi

Solid-State Microwave High-Power Amplifiers, Franco Sechi and Marina Bujatti

Stability Analysis of Nonlinear Microwave Circuits, Almudena Suárez and Raymond Quéré

Substrate Noise Coupling in Analog/RF Circuits, Stephane Bronckers, Geert Van der Plas, Gerd Vandersteen, and Yves Rolain

System-in-Package RF Design and Applications, Michael P. Gaynor

Terahertz Metrology, Mira Naftaly, editor

TRAVIS 2.0: Transmission Line Visualization Software and User's Guide, Version 2.0, Robert G. Kaires and Barton T. Hickman

Understanding Microwave Heating Cavities, Tse V. Chow Ting Chan and Howard C. Reader

Understanding Quartz Crystals and Oscillators, Ramón M. Cerda

For further information on these and other Artech House titles, including previously considered out-of-print books now available through our In-Print-Forever® (IPF®) program, contact:

Artech House Publishers
685 Canton Street
Norwood, MA 02062
Phone: 781-769-9750
Fax: 781-769-6334
e-mail: artech@artechhouse.com

Artech House Books
16 Sussex Street
London SW1V 4RW UK
Phone: +44 (0)20 7596 8750
Fax: +44 (0)20 7630 0166
e-mail: artech-uk@artechhouse.com

Find us on the World Wide Web at: www.artechhouse.com